The Animal in Ottoman Egypt

The Animal in Ottoman Egypt

ALAN MIKHAIL

OXFORD
UNIVERSITY PRESS

OXFORD
UNIVERSITY PRESS

Oxford University Press is a department of the University of Oxford.
It furthers the University's objective of excellence in research, scholarship,
and education by publishing worldwide.

Oxford New York
Auckland Cape Town Dar es Salaam Hong Kong Karachi
Kuala Lumpur Madrid Melbourne Mexico City Nairobi
New Delhi Shanghai Taipei Toronto

With offices in
Argentina Austria Brazil Chile Czech Republic France Greece
Guatemala Hungary Italy Japan Poland Portugal Singapore
South Korea Switzerland Thailand Turkey Ukraine Vietnam

Published in the United States of America by
Oxford University Press
198 Madison Avenue, New York, NY 10016

Library of Congress Cataloging-in-Publication Data
Mikhail, Alan
The animal in Ottoman Egypt / Alan Mikhail.
pages cm
Includes bibliographical references.
ISBN 978-0-19-931527-7 (hardback)
1. Animals and civilization—Egypt. 2. Human-animal relationships—Egypt.
3. Egypt—History—1517-1882. I. Title.
QL85.M55 2013
304.2'709620903—dc23
2013036490

1 3 5 7 9 8 6 4 2
Printed in the United States of America
on acid-free paper

Man is by nature a political animal.

—Aristotle

CONTENTS

ACKNOWLEDGMENTS

An unnamed donkey from 1697 changed my life. I saw him in a court record one day in 2005. He looked at me. I looked at him. I wrote this book. Just-so stories are of course never just-so. Between an unnamed donkey and this book stand many humans and some nonhumans too.

Among the latter, numerous institutions have generously helped fund the research and writing of this book. I am honored to thank Yale University, my intellectual home; the Andrew W. Mellon Fellowship of Scholars in the Humanities at Stanford University; the Department of History at the University of California, Berkeley; the Institute of Turkish Studies; and the American Research Center in Egypt.

For about a decade, the research for this book happily took me to many archives and libraries. For their assistance and patience, I thank the staffs of the Archivio di Stato in Venice; the Bancroft Library at the University of California, Berkeley; the Başbakanlık Osmanlı Arşivi in Istanbul; the Beinecke Rare Book and Manuscript Library at Yale University; Dār al-Kutub al-Miṣriyya in Cairo; Dār al-Wathā'iq al-Qawmiyya in Cairo; Ma'had al-Makhṭūṭāt al-'Arabiyya in Cairo; the Malta Study Center of the Hill Museum and Manuscript Library at Saint John's University; the National Archives of the United Kingdom in Kew; the Süleymaniye Kütüphanesi in Istanbul; the Topkapı Sarayı Müzesi Arşivi in Istanbul; the Topkapı Sarayı Müzesi Kütüphanesi in Istanbul; and the Yale Center for British Art. For permission to reproduce images from their collections, I thank the Beinecke Library; the Museum of Fine Arts, Boston; the National Archives of the United Kingdom; the Süleymaniye Kütüphanesi; the Topkapı Sarayı Müzesi Kütüphanesi; and the Worcester Art Museum.

Countless colleagues and friends read and commented on parts of this book; invited me to present some of it; and generally offered advice, support, and assistance. Among the many, I want to single out Emad Abou-Ghazi, Seven Ağır, Abbas Amanat, Antonis Anastasopoulos, Ali Anooshahr, Karl Appuhn,

Ali Asani, On Barak, Beth Baron, Ned Blackhawk, Maurits van den Boogert, Antoine Borrut, Richard W. Bulliet, Indrani Chatterjee, George Chauncy, Deborah Coen, Merve Çakır, Murat Dağlı, J. P. Daughton, Diana K. Davis, Corrie Decker, Beshara Doumani, Fabian Drixler, Laura Engelstein, Tolga Esmer, Khaled Fahmy, Suraiya Faroqhi, Carter V. Findley, Paul Freedman, Erica Fudge, Christiane Gruber, Magdi Guirguis, Valerie Hansen, Robert Harms, Lynette Hart, Marcia Inhorn, Ayfer Karakaya-Stump, Arash Khazeni, Sara Kippur, Jennifer Klein, Naomi R. Lamoreaux, Thomas W. Laqueur, Mary Lui, Ussama Makdisi, Joseph G. Manning, John R. McNeill, Joanne Meyerowitz, Mary Miller, Timothy Mitchell, Hande Özkan, Leslie Peirce, Peter C. Perdue, Christine M. Philliou, Kenneth Pomeranz, Sara Pursley, Paul Sabin, Robert A. Schneider, Omnia El Shakry, Jonathan Sheehan, Edith Sheffer, Amy Singer, K. Sivaramakrishnan, Kahraman Şakul, Robert L. Tignor, Francesca Trivellato, Judith Tucker, Kate Wahl, Nicole Watts, Sam White, and Ali Yaycıoğlu. The dogs of Wooster Square (and sometimes their human owners as well) made many long summer afternoons in New Haven both happier and illuminating of some of the contemporary human-animal world.

Kathleen Kete, Dina Rizk Khoury, and Harriet Ritvo each read a full version of this manuscript and improved it immeasurably.

My parents put up with an academic son writing a book about animals. I thank them for that and everything else.

As she always does, Susan Ferber at Oxford University Press masterfully sailed this ship into port. Her careful reading and editing of the entire text helped turn a manuscript into a book. I cannot thank her enough. Molly Morrison and her team at Newgen magically made pages out of megabytes and saved me from many costly errors. I am very grateful.

I am also grateful for the permission to revise portions of the following two articles for use in this book: "Animals as Property in Early Modern Ottoman Egypt," *Journal of the Economic and Social History of the Orient* 53 (2010): 621–52; "Unleashing the Beast: Animals, Energy, and the Economy of Labor in Ottoman Egypt," *American Historical Review* 118 (2013): 317–48.

PREFACE

Three Species

I stepped in Teddy's shit at 9:17 p.m. I had just thrown out some watermelon rinds and was walking back from the kitchen to catch the end of a commercial break in the Egyptian melodrama that was all the rage—something about a long-lost son who had returned home after twenty years of working on ships around the world. Teddy was the most important creature in my family's apartment, and he wasn't happy that I was in town for two weeks. He barked and howled, tore apart some of my socks, and generally did all he could to make my life difficult. He was cute and curly haired. I admit it; I loved him. How couldn't you?

A cocker spaniel, Teddy's life was better than that of most other living things in Egypt. While some lucky humans ate leftovers, Teddy got rotisserie chicken delivered warm from down the street. My family never let Teddy leave their immaculately kept eighth-story apartment, save for trips to the countryside, because—it was said—Cairo's streets were too dirty for him. He relieved himself—except, evidently, when I was visiting—in a corner of my aunt and uncle's bathroom covered with newspaper and dedicated to sparing Teddy the chore and disgust of going outside a few times a day. Teddy was the master of the house, and I was intruding on his territory. My aunt and uncle's attention was temporarily diverted to me, and Teddy wanted to send a message that he was still in charge.

Statements asserting his deep Egyptian nationalist sentiments notwithstanding, Teddy was bred and born in Russia. Like so many Italians, Maltese, Turks, French, Dutch, British, and others before them, Russians were the latest group in the early 2000s to be given the respect and cultural cachet that derived from being foreign in Egypt. They had money and liked Egypt's warm weather. Some Egyptian women lamented the fact that even many of Egypt's belly dancers were now Russian. A brand on the inside of Teddy's left ear evidenced his own

Russian lineage. His status was thus secure as part of a new foreign elite in Egypt. He was bourgeois and knew it.

The reading room in the Egyptian National Archives was, like the rest of Egypt in 2006, a dictatorship. Authority was clearly embodied in a single person, power was arbitrarily enforced, and ostensible "rules" controlled both everything and nothing. Access to archival documents and historians' careers hinged on personalities, whims, and moods that changed daily. Historians were to come to the archives in neatly pressed shirts, sit at one of twenty-four desks (four rows of six), and work carefully over their documents using only pencil and with their mobile phones on silent.

One spring day, the glasshouse of the reading room's order was rattled by a pleasant chirp emanating from behind a row of imposing though dilapidated brown shelves lining the front wall. Had Tito given these shelves to Nasir in the 1960s as part of some non-aligned gift exchange? Had they been assembled domestically from shoddy materials left over from a synthetic materials factory in the Egyptian Delta that produced office supplies for the government? My many moments of contemplative stares over the years hadn't figured out their provenance. Whatever the case, the archive's employees—who knew this furniture best—immediately took notice of the noise coming from behind the shelves. Immensely bored and criminally underpaid, the reading room staff quickly and understandably ignored the centuries-old materials they had ordered from the storage depot and us researchers reading those materials. Discussions ensued. These were, however, soon interrupted by the dramatic sight of the chirping culprit scurrying from behind one of the shelves to the ceiling—a small, brown, truly unimpressive urban gecko.

The matriarch of the reading room shuddered upon seeing the creature but quickly recomposed herself. The reptilian challenger to the order she oversaw soon circled back down from his upward climb to relative safety in the dark behind the shelves. More conversations. Voices got louder, phone calls were made, tensions mounted. This unexpected action in the reading room raised us researchers from our uncomfortable chairs and brought us to the front of the room. We were yelled at to keep our distance.

A plan was hatched. Two burly men from the storage depot were called up to the reading room to pry out the shelf behind which the gecko was thought to be hiding. They were to kill or capture—whichever was easiest. The men took their positions. We held our collective breath as they began to inch the shelf away from the wall. Once sufficiently out of place, one of the men stuck his head in the revealed space only to reemerge with the disappointing news that there was nothing there. More yelling. The men were told to take the shelf out completely to look more thoroughly. They did. Still nothing. A quick peek behind the shelves on either side also revealed nothing.

With this excitement subsiding, the head of the reading room, in a clear need to reassert her power on the heels of this defeat, yelled at all of us to return to our chairs since there was clearly nothing to see. As I walked back to my seat, I felt necessarily quiet satisfaction that this gecko had done what so many researchers could only imagine in moments of daydreaming. It had snubbed the head of the reading room and gotten away with it. I arrived at my chair, sat back down, grabbed my pencil, heard the wood of my chair creak, and returned to the animals of early modernity.

I killed a deer. I didn't mean to, but I did. I was visiting my parents in The Woodlands, Texas—an upper middle-class suburb north of Houston where people drive very big cars and where Sarah Palin's *America by Heart: Reflections on Family, Faith, and Flag* had sold well in supermarket checkout lines. I had borrowed my father's relatively small—by local standards—new car to have dinner with a friend and was driving back around midnight on an empty two-lane road when an unassuming and elegant doe bounced out in front of me. I couldn't brake quickly enough, and in the bright of the headlights I saw the deer ricochet off the front of the car. I was shaken. She was dead.

Emerging from my initial shock, I realized that I had kept the car idling in the middle of the road for several minutes. Thankfully, as far as I can remember, no other cars came by. I pulled over to the side of the road, got out of the car, and walked over to the deer who was now lying in a drainage ditch. She was bloody and motionless. I saw her tail many yards away from the rest of her body. It was messy and awful. I didn't mean to kill this creature, but I did.

I was visiting my parents for a week before heading to Cairo for more research. Were the animals trying to stop me? Was I being told that I wasn't cut out for this kind of work because of the blood on my hands? How could I write critically about human-animal relations when I was so clearly on the side of the humans? I was thoroughly discombobulated.

I had grown up in this suburb, the product of white flight, underpaid Mexican labor, widespread development of forestland, and a massive influx of cash in the 1970s. Humans had pushed into deer's habitats, poured concrete over where they used to eat, shot them, and then had the gall to call the place "The Woodlands." The idea was that we had somehow conquered nature, whatever that trite phrase means, and successfully separated ourselves from the insecurity and random violence of the animal world. I knew better of course. Still, it took the shock of an accident and an unnecessary death to make real to me (yet again) that we are, have been, and always will be deeply intertwined with the other animals around us.

Even more so than I had already been throughout my life, I was party to the killing of an innocent and threatened creature. What business did I have trying to write about relationships between humans and animals?

The Animal in Ottoman Egypt

Introduction

Cephalopods in the Nile

Origin of man now proved. Metaphysics must flourish. He who understands baboon would do more towards metaphysics than Locke.

—Charles Darwin

The smartest of all invertebrates, cephalopods are masters of camouflage.[1] Squid, cuttlefish, and octopus can all change their shape, size, color, skin pattern, texture, brightness, and luminance in the flash of an eye to fit into any space or match any surrounding so as to be nearly invisible. Such adaptations have served these sea creatures well in mating, hunting, and hiding. To the unassuming predator or unaware scuba diver, a world of teeming life might otherwise appear a barren landscape.

Like those other animals who swim right past the oceans' cephalopods, historians have all too often mistakenly assumed a vastly complex world of humans and animals to be a more desolate and mundane terrain of exclusively human action, agency, and history.[2] The fact is that the overwhelming majority of humans have lived in very close proximity to animals. Humans have eaten animals, loved them, worshiped them, represented them, dressed like them, cared for them, rode them, and built cosmographies around them. Yet these animals rarely figure into the histories we tell of past societies. The human-animal relationship is one of the most historically consequential of all past phenomena. Understanding that relationship and how it changed is crucial to the study of any place and any time.

This book examines the multiple changing relationships between humans and animals in a place that affords perhaps the longest documentary record of the human-animal relationship. Egypt was home to the world's first zoo (circa 2500 BCE), one of the oldest religions to incorporate animal forms, and perhaps the first domesticated dogs.[3] During the crucial centuries of Ottoman rule in Egypt between 1517 and 1882, the changing relationships between humans and animals were central to Egypt's transformation from an early modern society fully ensconced in an Ottoman imperial system to a nineteenth-century centralizing

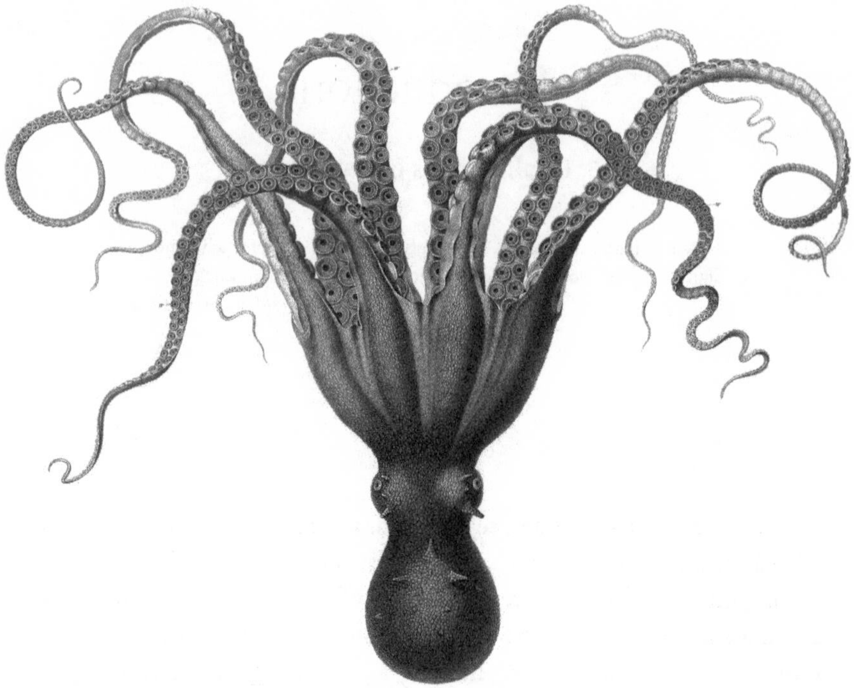

Figure I.1 Egyptian Octopus. Commission des sciences et arts d'Egypte, *Histoire naturelle*, vol. 1, pt. 3 of *Description de l'Égypte, ou, recueil de observations et des recherches qui ont été faites en Égypte pendant l'éxpédition de l'armée française, publié par les ordres de Sa Majesté l'empereur Napoléon le Grand* (Paris: Imprimerie impériale, 1809–1828), Zoologie, Céphalopodes, pl. 1. Beinecke Rare Book and Manuscript Library, Yale University.

state.[4] Egypt in this period transitioned from a society characterized primarily by intense human-animal interaction to one in which this relationship was no longer the basis of commercial and social life.[5] The result of this transition was a fundamental reordering of political, economic, social, and ecological power. This book thus explains one of the most important historical transformations of the last five hundred years through the story of changes to some of the most historically significant human relationships—those with other animals.

The central question is why and how Egypt moved from being the most lucrative province of the Ottoman Empire in the early modern period to one of the most important centralizing bureaucracies around the Mediterranean in the nineteenth century. Nearly all early modern agrarian societies between roughly 1750 and 1850 experienced some form of intense transformation involving commercialization and modernization of their economies, urbanization, and integration into global commodity and trade networks. While Egypt's experience was clearly unique, analyzing its particular history illuminates how changes to human-animal relations both shaped and reflected the global transition of early modern societies to more modern forms of governance, economy, and society at the turn of the nineteenth century.

Historians of Egypt and the Ottoman Empire have focused on the period from 1770 to 1830 to understand and explain the many significant social, economic, and political changes that characterized these decades. It was at the end of the eighteenth century that the massive accumulation of land by a small group of rural leaders began to transform Egypt and other parts of the Ottoman Empire.[6] In the years that followed, the most centralized government in centuries emerged in Cairo. It undertook various wars of expansion; eventually founded schools, ministries, hospitals, military colleges, and other institutions of state; and built roads, bridges, canals, and other infrastructural works.[7] The population of Egypt's two major cities—Cairo and Alexandria—also rapidly increased between 1770 and 1830.[8] Why and how all of this happened has proven one of the most vexing and pressing questions facing historians of Ottoman Egypt, as well as scholars of other early modern agrarian economies across the globe that experienced similar changes.[9]

Three general explanations have been given for this transition. Some historians have pegged these massive social, political, and economic transformations to the

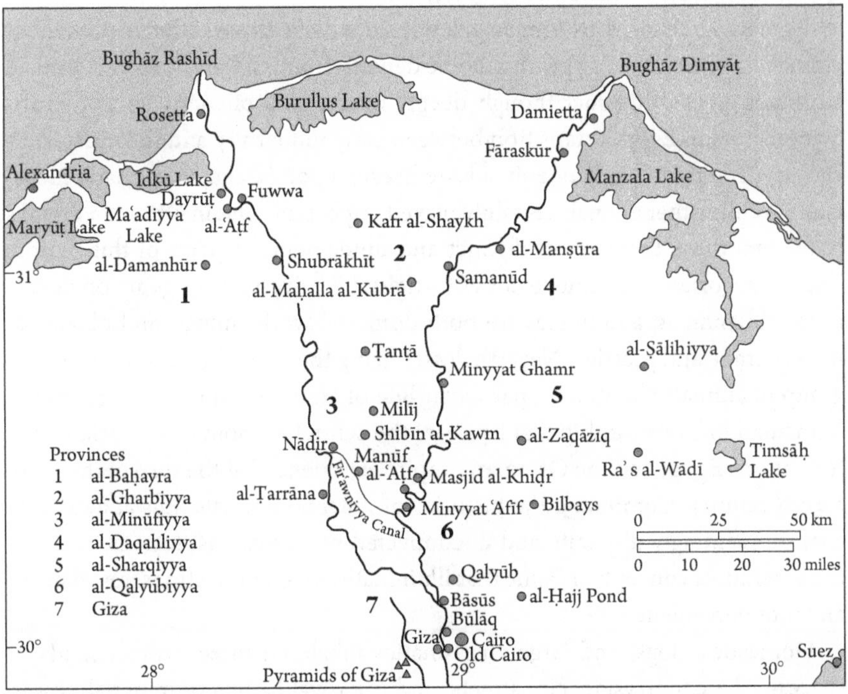

Nile Delta

French invasion of Egypt from 1798 to 1801.[10] With Napoleon's armies came a lesson in European success—the entrenchment of what Marshall Hodgson calls "the apparatus of the Enlightenment on Egyptian soil."[11] More discipline, more order, and more training meant greater military, economic, and political power. Thus the French expedition, this story goes, precipitated a process of emulation that Egypt is in many ways still trying to work out.[12] A second narrative, one identified with a nationalist school of Egyptian historiography, credits Mehmet ʿAli—the Ottoman provincial governor from 1805 to 1847 and considered by some "the founder of modern Egypt"—with these changes.[13] His persistence in forging a modern state eventually "pulled" Egypt—whether the populace liked it or not—into modernity. A third explanation points to longer-term structural processes in Egyptian society, commerce, and politics. In this account, Egyptian merchants developed capitalist modes of production and commercial relations before the French ever arrived in Egypt; Egyptian intellectuals independently advanced sophisticated arguments about logic, science, and government; and Egyptian religious figures proposed secular notions about society before ever encountering the Enlightenment.[14] While all three of these attempts to explain the enormous and critically important changes that occurred at the turn of the nineteenth century have proven fruitful for advancing our understanding of Ottoman and Egyptian history, they have also distracted us from seeing and analyzing other crucial phenomena, forces, and events.

By tracing changes to human relationships with three different classes of animals in Ottoman Egypt, this book departs from earlier work and instead examines a set of distinct though deeply interrelated phenomena as the primary factors in Egypt's transition between early modernity and the nineteenth century. The three kinds of animals are livestock, dogs, and charismatic megafauna. While other animals certainly played important roles in Ottoman Egypt's transition, these three reveal distinct and fundamental aspects of this period. These categories are of course not watertight. Like livestock, dogs are obviously domestic animals, and horses are both domestic work animals and objects of intense trade and prestige. Nevertheless, tracing the individual histories of each group of animals allows for a particular line of argument about changes to the human-animal relationship that lays bare the central economic and social transformations Egypt and the Ottoman Empire experienced at the turn of the nineteenth century. Combining the histories of different animals thus allows us to analyze seemingly disparate and unconnected processes, institutions, and histories within a conceptual frame that illuminates their connections, resonances, and interdependence.

Domestics, dogs, and large charismatics inhabited three distinct kinds of spaces—the countryside, city streets, and the royal haunts of imperial power, respectively—and thereby highlight three distinct geographies: the village, the province, and the early modern empire. Domestic animals explain changes in the nature of rural labor; dogs elucidate changes in understandings of urban sanitation, health, and the human and animal body; and charismatic megafauna index changes in global trade and economic modes of exchange. Each part of this book is dedicated to one of these classes of animals and addresses the same transitional period at the turn of the nineteenth century, but each part also covers a different chronological scope. Part I zooms in on the decades around the turn of the nineteenth century; Part II widens the lens to span the early years of Islam in the seventh century to the middle of the nineteenth century; and Part III covers the entire period of Ottoman rule in Egypt from the early sixteenth century to the late nineteenth.

Part I focuses on domesticated livestock like oxen, water buffalo, and donkeys to show how the character of rural labor fundamentally changed at the turn of the nineteenth century. In the early modern period, humans and animals worked side by side on relatively small-scale repair and infrastructure projects throughout the countryside.[15] The caloric energy afforded by both human and animal muscle was wholly sufficient to make the countryside function.[16] A few decades later, labor pools in Egypt were almost completely human, with corvée (forced labor) eventually emerging as the preferred means of labor in nineteenth-century rural Egypt. This change in labor regime resulted from reductions in domestic animal populations; the advent of ever more complex public works projects; and new conceptions of efficiency, order, and reproducibility as principles of rural labor. As domestic animals were increasingly deemed less useful for their labor, they were progressively cut out of definitions of the productive social and economic realm and were recast as either useless dumb animals or, at best, sources of meat. These changes in the nature of rural labor in Egypt degraded the lives of both beast and man.

Contrary to the conventional thinking that dogs were always considered ritually impure in Islam and Muslim societies, Part II shows that the actual historical record of Muslims' writings about and interactions with dogs is on the whole much more positive. Certain thinkers praised dogs for their loyalty and trustworthiness. They were valued for their productive social roles as protectors of flocks, combatants during war, and—above all—consumers of garbage in Egypt's cities. Ottoman officials actively encouraged the increase and maintenance of dog populations in Cairo by putting out food and water troughs and by punishing those who committed violence against them. This human-canid cooperation drastically changed in the early nineteenth century. With increasing immigration from the Egyptian countryside to Cairo and Alexandria and the crystallization of an understanding of disease as emanating from the miasmas of standing water, rotting flesh, and garbage, dogs became targets of eradication campaigns. For the first time since the founding of Cairo in the seventh century, dogs were actively removed from the city. What resulted was an increase in the use of violence against dogs, an imperative to come up with some other solution for Cairo's garbage, a reworking of urban space, and, I argue, an evolutionary divergence among dog species in Egypt.

The final group of animals this book addresses are charismatic megafauna. Elephants, lions, and tigers—and lesser numbers of giraffes, hippopotami, and certain kinds of birds—had been coming to Egypt for millennia. In the early modern period, these animals represented an important facet of the Indian Ocean trade among Mughal South Asia, Safavid Iran, and Ottoman Egypt. Sovereigns exchanged these animals as diplomatic gifts in an economy of reciprocity, meant both to cement alliances and to impress upon rivals one's abilities to master, control, and distribute nature's wealth. In the nineteenth century, the charismatic animal economy moved from gift exchange to market-driven capitalist commercial relations and from the Indian Ocean to the Mediterranean. Because of its geographic location and centrality to earlier fauna circulation networks, Egypt emerged as one of the chief hinges of the global nineteenth-century trade in exotic animals to European—and later, North American—zoos. Egypt was not only an exporter or way station for these animals, but also became an active consumer of them as it founded multiple animal institutions that served to mediate the human experience of these large creatures. The school of veterinarian medicine, the silk industry, hunting preserves, and particularly the zoo became the cornerstones of a new Egyptian relationship with charismatic megafauna—one characterized by separation, capitalist exchange, and cages.

A New Empire

The Ottoman Empire is changing. Thanks to several decades of imaginative, rigorous, and conceptually rich scholarship, many long-cherished narratives of imperial rise and fall, sixteenth-century efflorescence, early modern decline, and nineteenth-century reform have fallen by the wayside in favor of profoundly new understandings of both the empire's chronology and geography. In this new historiographical topography, the turn of the nineteenth century emerges as one of the most crucial periods. In the decades between 1770 and 1830, the empire supposedly moved from early modernity to the modern—from empire to nation-state, subject to citizen, Ottoman to Turkish, old to new.[17] While clearly something very significant happened at the turn of the nineteenth century—1850 indeed looked very different from 1750 nearly everywhere in the empire—historians have yet to fully grasp the significance of these years.[18] Because much Ottoman historiography, including that about Egypt, remains within traditional temporal frameworks like "early modern" or "the nineteenth century," few scholars take the century-straddling period from 1770 to 1830 as their explicit and specific temporal unit of analysis.[19] These decades are usually either the end of a story about the early modern period or the beginning of a story about the modern period—the end of Ottoman rule in Egypt (and Greece, Bulgaria, Syria, and so on) or the beginning of modern Egypt (and its proxies). To understand early modernity or the nineteenth century, however—never mind the relationship between the two—we must obviously fully grasp this middle period of Ottoman history on its own terms and not simply as one or another bookend.

The term *early modern* in this book refers to the period from roughly the early sixteenth century to the beginning of the nineteenth century. This was an era characterized by the presence of large imperial powers whose rule was based on reciprocal relationships between sovereigns and subjects most often mediated through taxation and military service.[20] In the Ottoman case, these relationships were articulated through the idea of the circle of justice, which made clear that the state upheld order so that its subjects could cultivate resources in peace and then return cash and manpower to the state to reinforce its continuity and well-being.[21] This feedback loop of imperial power meant that the early modern Ottoman Empire was a political power connected to millions of localities throughout its realms through specific and often temporary, tendentious, tensorial, and taxing relationships of push and pull that proved productive for both subject and state and durable over centuries. Linguistic, religious, cultural, or even political differences were unproblematic from the perspective of the state as long as order and revenue were maintained.

The early modern world was also one of intense global interconnection.[22] In historian John Richards's definition, early modernity entailed the rise of globalized movement and trade across the seas, a world economy, a doubling of the earth's population, and large stabilized states that used land more intensely and employed new technologies—gunpowder,[23] printing presses,[24] and New World crops[25]—to make all of this possible.[26] Early modernity also involved, for the very first time, a consciousness of the world as a unitary whole. Only in the sixteenth century does the writing of world history emerge as an identifiable genre of historiographical practice.[27] Within these general political, economic, and intellectual parameters, the more specific characteristics of this period included subsistence agriculture,[28] non-exploitative and non-irreversible relationships between humans and their environments,[29] and modes of economic exchange based on reciprocity and personal trust but also self-serving economic interests and imperial rivalry.[30] The early modern state was clearly essential to all of this. Although it may seem anachronistic to call the Ottoman Empire an early modern *state*, the term highlights the fact that the political, legal, economic, and natural resource management structures that made up the polity are the precise sites where differences and changes between early modernity and modernity are most visible.[31] In the early modern period, the state was generally more of a framework of social and economic relations rather than their central and final arbitrator. In other words, the state was not simply the sultan's imperial divan. Rather, it included the entire bureaucratic apparatus of the imperial administration, from merchants who engaged in commercial relations and peasants who grew the empire's foodstuffs, to soldiers and animals that helped sustain an Ottoman system of economics and governance.

By contrast, modern states—in Ottoman Egypt and elsewhere—are characterized by the centralization and attempted monopolization of bureaucratic, economic, political, and military power. The modern state, Max Weber famously said, is "a human community that (successfully) claims the *monopoly of the legitimate use of physical force* within a given territory."[32] Modern polities have built large armies, educational institutions, physical infrastructure, economic tools meant to accumulate wealth, and standardized legal strictures and structures. Modern states have the ability to enter into personal, familial, and small-scale social relationships like never before and to gather the minutest information about populations and individuals.[33] While the power of modern societies and political formations allows for the creation of all sorts of political and educational institutions, social forms, legal frameworks, economic arrangements, and military capabilities and possibilities, the detrimental and nefarious aspects of modern states are also many: disciplinary power; carceral politics; mass violence; the medicalization of society; environmental degradation; policing; and sexism, racism, and other discriminatory practices.[34] In other words, to borrow from Michel Foucault, "the Enlightenment did not make us mature adults."[35]

The supposed rupture of the year 1800 is therefore clearly a historiographical rather than a historical phenomenon.[36] Standing in the face of this purported separateness, the sources, histories, actors, processes, and ideas on either side of 1800 plainly show this period to contain within it a great degree of continuity and connection across time. The nineteenth century, simply put, came from the eighteenth century. As banal and obvious as this sentence is, it is worth stating directly since so much of the historiography of the Ottoman Empire suggests otherwise. Focusing on both local specificities and wider imperial processes over the entire period from 1770 to 1830 thus helps to overcome many inherited historiographical inventions—"the Ottoman period in the Arab lands" or the "modern Middle East," for example—that mask much more than they illuminate.

Eco-Empire

The Ottoman Empire's early modern history was not a pristine moment of unvariegated state and society relations waiting patiently for the forceful ideas, actors, and wars of the nineteenth century; rather, it was a dynamic and conflicted period in its own right.[37] Too often, the early modern period is taken as an undifferentiated and unspoiled historical forest ready and waiting for the chainsaws of modernity, whether these machines are defined as progress or loss.[38] Early modernity is likewise frequently portrayed as an epoch of fluidity, even democratization, free from the rigid identitarian politics and violence of the nineteenth century.[39] Not only do such descriptions ignore the important and obvious power relations at play in the early modern period, but they—perhaps more significantly—serve to remove early modernity from history.

With an eye toward the ways environmental historians have shown how nature makes humans and humans make nature, Ottoman historical realities suggest it is useful to conceive of the empire as an ecosystem, a complex nexus of relationships among resources, peoples, ideas, and places in which each element of the community is dependent on every other. A change or perturbation in any one part of the environment affects each component of the whole.[40] The idea of the Ottoman Empire as an ecosystem foregrounds how the smallest and largest of imperial actors were connected through means of trade, administration, and mutual reliance. The example of irrigation in Ottoman Egypt, for instance, clearly shows how farmers in often very remote parts of the empire were in constant dialogue with the palace in Istanbul and how the two worked together to make the countryside productive.[41] Peasants used the empire, and the empire used peasants. Similarly, the empire was enmeshed in all sorts of quotidian and mundane matters—marriage, housing construction, personal correspondence, sex, and inheritance. These binds and bonds of empire clearly involved conflict; taxation and conscription were the most common sites of these struggles in the early modern period.[42]

Whether the focus is productivity, tension, contestation, or violence, viewing the Ottoman Empire and other polities as ecosystems allows for nuanced and capacious understandings of politics, economy, and society.[43] From the perspective of empire-as-ecosystem, most everything within and around an imperial realm comes to be connected, which then allows for analyses of how these connections formed and functioned and then were challenged, reconstituted, and ultimately defined. This kind of analysis could operate at many levels. A study of Ottoman ecologies could focus on pushes and pulls within the relatively small confines of the palace and its internal dynamics, or could work to elucidate how Ottoman merchants in the Mughal Empire affected economic networks in Ottoman Salonica.

Viewing the Ottoman Empire as an ecosystem means that Egyptian (or Turkish or Arab or Balkan) history cannot be understood as somehow separate from Ottoman history, or vice versa. Ottoman history made Egyptian history, and Egyptian history made Ottoman history. An ecological perspective encompasses the broad Ottoman world that set the rough parameters for almost everything that occurred in the eastern Mediterranean, Anatolia, and the Balkans for many centuries. Thus, an ecological approach to empire reveals how the empire's variegated geographies, overlapping chronologies, and connected histories functioned across space and time, and how small changes in one part of the empire affected places, ideas, and peoples across the imperium and beyond.[44] Alongside new conceptualizations of the Ottoman Empire's chronology, the idea of the empire as an ecosystem therefore opens up new geographies for Ottoman history.

Such an ecological approach to empire also integrates multiple kinds of imperial actors—many of whom have remained in the shadows of Ottoman history. For example, large households and other corporate bodies in various parts of the empire have recently been shown to be crucial to the constitution of wealth and the creation of economic, familial, and social networks across vast swaths of the empire's territory and beyond.[45] As an important subset of this work, historians have shown how families in the Ottoman world were central and constitutive points of contact between individuals and imperial institutions like courts and the military.[46] This scholarship has shown how various strands of the Ottoman world were intertwined with each other, mutually reliant, and tied together in integral, lasting, and often unseen ways. These new histories, moreover, allow for a subtler treatment of the turn of the nineteenth century as a period of connection and change rather than one of simple and clean rupture.

Inspired and influenced by recent work in Ottoman historiography and environmental history, this book shows that animals, like numerous other imperial actors, were integral to the functioning and transformation of the Ottoman world at the turn of the nineteenth century. Studying how and why their multiple relationships with humans metamorphosed crystallizes both the continuities and changes that occurred throughout this period. Egypt is the perfect case study for such a history. From its initial incorporation into the empire in the first half of the sixteenth century until its final takeover by British forces in 1882, Egypt was the most significant province of the empire, the foundation of the entire Ottoman edifice of governance and economics.[47] It was the most lucrative province of the empire[48] and its largest single supplier of foodstuffs.[49] Egypt was crucial to Ottoman rule in the Mediterranean,[50] Hijaz,[51] Red Sea, and Indian Ocean.[52] In the late eighteenth and early nineteenth centuries, some of the largest military threats to the empire's stability came from Egypt.[53] Many Ottoman functionaries in Istanbul maintained large economic interests in Egypt,[54] as did

Venetian merchants.[55] And the province was crucial for Ottoman expansionist efforts in Africa in the nineteenth century.[56] Because Egypt was so vital to the entirety of the Ottoman imperial system (and beyond), it consequently had greater potential than any other province to threaten, weaken, or otherwise impact the empire as a whole.

The Animal Historian

This book focuses on the productive social and economic functions of animals—how they enabled and shaped certain economic relations, made possible tasks that otherwise could not have been undertaken, and were enmeshed in all sorts of social relationships. In short, this book is a materialist analysis of what animals did in Ottoman Egypt rather than a study of what they represented or symbolized.[57] In the same way that one could not adequately understand, for instance, the history and development of the twentieth-century American city without addressing the material, economic, and social history of the automobile, rural Egypt cannot be understood without giving animals their due as central shapers of the early modern world.[58]

One of the foremost challenges facing the animal historian is whether it is possible to enter the world of, say, a dog to understand her historical experience. What conceptual, biological, empirical, and intellectual leaps are required to enter the dog world?[59] This question in many ways points to the basic task of any historian—understanding how historical subjects experienced the past. As the following pages demonstrate, the historical experiences of donkeys, horses, dogs, tigers, and elephants in 1850 were entirely different from what they had been in 1650. Understanding how, when, and why these historical actors' experiences changed therefore illuminates a great deal about the social, economic, political, and interspecies histories of this period. It also pushes the limits of the historical craft itself by attempting to grasp how the past was variously experienced and shaped by species other than humans.

I am under no illusion, however, that it is possible for me to fully understand (let alone explain) what it was like for a cow, dog, or elephant to live in Ottoman Egypt or to experience the turn of the nineteenth century, life on an Egyptian farm, or the treacheries of a sea voyage from Goa to Suez.[60] What I do hope is that this book contributes to making possible an understanding of animals' subjective experiences of the past.[61] By fleshing out a picture of the economic and social worlds Ottoman-Egyptian animals inhabited and by explaining the ways the human-animal relationship changed at the turn of the nineteenth century, I hope to lay some of the groundwork for future animal histories that will no doubt be able to draw on advances in the fields of ethology and animal behavior and to benefit from the ever-increasing availability of sources about the rural Ottoman world.

Animal history derives from the same impulse as those histories that push beyond the elite human realm to ask questions of human subalterns—peasants, workers, slaves, or minorities. History is not limited to the human realm, and self-reflexive intentionality is not a prerequisite to historical agency. Historians of climate change have clearly shown how the nonhuman forces of weather have impacted and directed both human and nonhuman histories.[62] Animals and a multitude of other historical agents have similarly affected history without necessarily intending to do so.

If we accept the historical agency of nonhuman animals, certain methodological issues immediately arise.[63] Dogs do not write. Water buffalo do not keep diaries. How then do we write their histories? Following the lead of historians of subaltern humans, animal historians have asked this question in multiple ways.[64] Can the mosquito speak?[65] Can the subaltern bark?[66] Animal histories, like so much else about the past, must thus be narrated largely through the source materials left by sovereign classes, which in this case have the complication of being across not a class, status, educational, racial, or gender divide, but a species divide. Although the vast majority of historical work has been concerned with evoking the worlds of humans, there is little reason to think that humans somehow have a better entry point into the worldview of humans living hundreds or thousands of years ago in unseen places speaking unknown languages than into the worlds of animals living both today and in the past. Why is attempting to understand human history somehow easier or more accessible than understanding the experiences and worldviews of animals?[67] Are not both historical projects similar exercises in historical imagination and the use and evaluation of source material? The tools we historians use to imagine the experiences of other living and dead humans—information gleaned from contemporary sources, historical context, and even imagination—can likewise be marshaled to imagine—and in turn understand—the experiences and histories of other living and dead animals.

Paw Prints

These methodological and conceptual challenges demand an imaginative and capacious use of available source materials. This book therefore utilizes many very different kinds of sources. Most of them are quite familiar to Ottoman and Egyptian historians. Indeed, one does not have to go very far to find sources about animals. To take one example, Evliya Çelebi, whose travel account is perhaps the single most important Ottoman Turkish chronicle for the study of the Ottoman Empire in the seventeenth century, devotes page after page in his final section

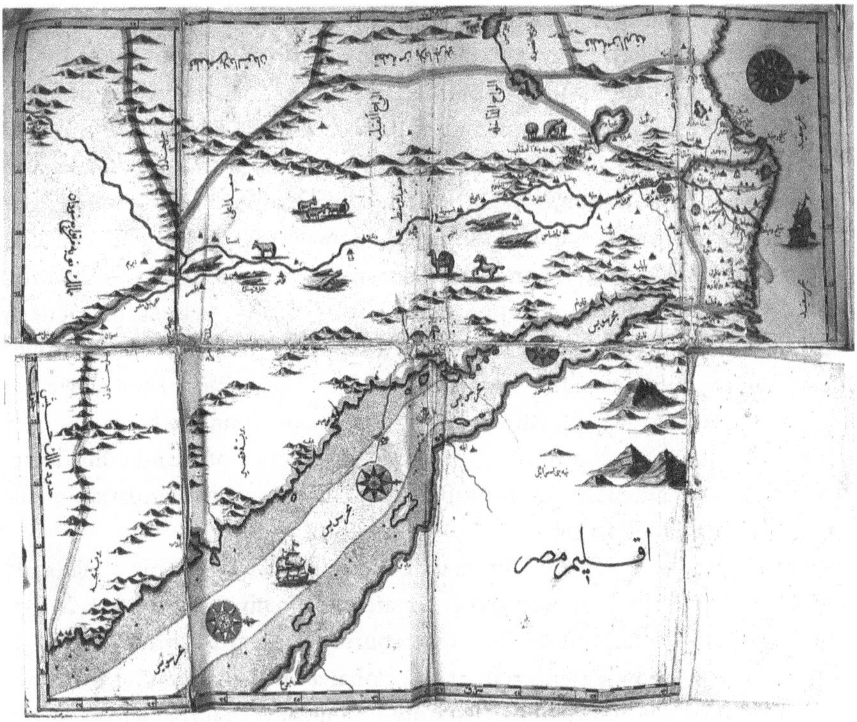

Figure I.2 Ottoman Egypt and Its Animals. Ahmed ibn Hemdem Süheyli, *Tarih-i Mısır ül-Cedid*, Süleymaniye Kütüphanesi, Hüsrev Paşa 353/2, 121. Used by permission.

on Egypt to—among other more mundane animal matters—the intricacies of crocodile sex (including interspecies sex between crocodiles and humans)[68] and the proper incubation of chicken eggs.[69] Likewise, even a foundational text like the Ottoman Kanunname of Egypt (Kanunname-i Mısır), the imperial legal code formalizing and outlining Ottoman rule in the province, makes clear the need to manage the role of domestic animals as one of the primary bases of Egypt's vast wealth.[70] Archival materials, unpublished manuscripts, chronicler accounts, pictorial representations, and literary sources all offer a surfeit of evidence for the human-animal relationship in the Ottoman Empire. Animals are indeed crawling, galloping, creeping, lurking, and slithering all over the sources Ottoman and Egyptian historians have been using for generations.[71]

The empirical heart of this book relies on archival sources from the central repository of the records of the Ottoman Empire; legal cases from rural Islamic courts spread throughout early modern Egypt; administrative documents from both the central Ottoman state and the Egyptian imperial bureaucracy in Cairo; unpublished manuscript sources from over half a dozen countries; and some additional British, Maltese, Venetian, and Chinese archival materials and other sources. The Ottoman Turkish sources for this study, mostly firmans and other kinds of imperial orders and correspondence, outline the early modern trade in charismatic megafauna from India and Iran to Egypt and Istanbul, outbreaks of various epizootics, animal thefts, hunting, the use of animals in warfare, and various other issues about the roles of animals in the Ottoman Empire.[72] The mostly Arabic archival materials from Cairo offer different insights into the social and economic histories of animals in Ottoman Egypt.[73] Estate inventories document the vast wealth peasants acquired through their ownership of large domesticated animals. Records of material transport, construction projects, and agricultural production show the extreme importance of animal labor to the early modern economy. Legal cases involving theft, sale, rent, and other uses of animals evidence the complexities of the economic, social, and legal roles of animals in the Egyptian countryside. And administrative documents from the beginning and middle of the nineteenth century paint an intricate picture of the founding logic, function, and maintenance of institutions like the school of veterinarian medicine, the zoo, and silk farms.

These varied and copious Ottoman Turkish and Arabic archival materials are supplemented by European archival materials, chiefly from London, Malta, and Venice.[74] Records produced by various bureaucratic structures of the British Empire help to show Egypt's role in the global animal economy of the nineteenth century and evidence its centrality to animal quarantine efforts in the Mediterranean and Red Seas. Maltese and Venetian sources help to illustrate Egypt's pivotal political and economic role in the early modern Mediterranean in the realms of, for example, piracy, the slave economy, and the trade and distribution of Yemeni coffee.[75] Because neither Ottoman nor Egyptian history can be adequately grasped without accounting for the larger global context of which the empire was such an integral part, any history of Ottoman Egypt must draw on both Ottoman and Egyptian archives, as well as many other sources. Account books, imperial orders, philosophical treatises, estate inventories, colonial records, religious texts, bureaucratic documents, chronicler histories, equestrian manuals, court cases, consular reports, and miniature paintings all have something to say about animals in Ottoman Egypt. Because animals were everywhere in Ottoman Egypt, they are everywhere in its sources.

Until approximately the second half of the nineteenth century, the social world of almost all people—rural and urban, rich and poor, Egyptian and not—was characterized by their immediate, near-constant, and necessary engagement with various kinds of animals. The history of the nineteenth-century transition away from these intense early modern social and economic human-animal relations is partly the history of new definitions of society and the economy as realms without animals. Remember that Weber, perhaps not purposefully but nevertheless not irrelevantly, identified the modern state as a *human* community.

Defining the contours and character of the productive social world of Ottoman Egypt was one of the most important outcomes of the period from 1770 to 1830. This book examines a fundamental aspect of the processes that created this nineteenth-century social, economic, and political world—the large-scale, though by no means total, removal and exclusion of animals from society. This "disappearance of animals from daily life"—to use art critic John Berger's words—while never complete, would continue to be central to the subsequent history of Egypt in the twentieth and early twenty-first centuries.[76] The labor of animals was deemed obsolete at the turn of the nineteenth century; their bodies became threatening sources of disease and annoyance; and their economic worth came to be defined by capitalist market relations instead of reciprocal exchange and an economy of wonder. The modernizing state forever cleaved the human-animal relationship.

This history also shows how what happened to animals in Ottoman Egypt would eventually happen to certain kinds of humans. Just as livestock, dogs, and elephants were stripped of their constructive social and economic functions in the early nineteenth century, so too were Egyptian peasants, the uneducated, the disabled, the poor, the sick, the criminal, and the itinerant cut out of the productive social and economic realms of Egypt later in the century. As their animal counterparts were confined in veterinary holding pens and the zoo, these humans would be subjected to similar nineteenth-century projects of enclosure—the prison, asylum, conscription camp, hospital, and school. As the social became more strictly, vigorously, and narrowly defined, fewer and fewer human and nonhuman beings were given access to it and its borders came to be more intensely policed and defended through violence, coercion, and discipline. Ottoman Egypt's transition to modernity was thus a wrenching and painful experience for most animals and humans alike. The story of what happened to the animal in Ottoman Egypt is therefore also the story of what happened to the human—a story of the animal in all of us.

PART ONE

BURDENED AND BEASTLY

1

Early Modern Human and Animal

As a beast of toil an ox is fixed capital.
—Karl Marx

On May 16, 1792, al-Ḥājj Muṣṭafā ibn al-Marḥūm al-Ḥājj Darwīsh ʿIzzat ibn ʿUbayd from the city of Damanhūr in the northwestern Egyptian subprovince of al-Baḥayra came to that city's court to file a claim against al-Shābb ʿAlī ibn Sīdī Aḥmad al-Muzayyin from the nearby village of Surunbāy.[1] Five days earlier, al-Ḥājj Muṣṭafā had purchased an ox (*thaur*) from al-Shābb ʿAlī for the price of thirty-five riyāls, a significant amount of money for a peasant in this period. Upon taking this ox home, al-Ḥājj Muṣṭafā immediately put the powerful animal to work on his waterwheel. He secured the ox into the waterwheel's harness and watched it move the enormous weight of the wheel a few turns. Without such an animal, moving water from a well or basin into irrigation ditches was nearly impossible. Confident in his new ox's understanding of its laboring role, al-Ḥājj Muṣṭafā retired for the evening with the hope that his rice fields would be adequately watered when he awoke the next morning.

Instead, al-Ḥājj Muṣṭafā arose the next day to find his newly purchased ox dead. According to al-Ḥājj Muṣṭafā's testimony in the court's records, the animal had fallen during the night into the small pond around the waterwheel. The ox had died from the impact of the fall, as a result of drowning, or from a combination of the two. Seeking redress, al-Ḥājj Muṣṭafā appeared in his local Islamic law court to request reimbursement of the thirty-five riyāls he had paid to the defendant al-Shābb ʿAlī. The text of this case does not cite any specific reason why al-Ḥājj Muṣṭafā made this rather weak request in the court. He perhaps meant to imply that there was some undisclosed defect with the animal at the time of purchase. Or perhaps he felt that since he had only purchased the animal a few hours before it died, he could somehow link the animal's death to its seller.[2] Not surprisingly, al-Shābb ʿAlī denied any responsibility for the fate of the animal after its sale and demanded that the plaintiff bring forth proof (*thubūt*) that he was somehow implicated in the

Figure 1.1 Irrigating Egypt. Commission des sciences et arts d'Egypte, *État moderne,* vol. 2, pt. 2 of *Description de l'Égypte,* Arts et métiers, pl. 5. Beinecke Rare Book and Manuscript Library, Yale University.

animal's death. When al-Ḥājj Muṣṭafā was unable to make any logical connection between the animal's seller and its death, the judge ruled against al-Ḥājj Muṣṭafā and ordered him not to make any further claims on al-Shābb ʿAlī.

Although this case is legally quite straightforward—a plaintiff made a claim, the defendant denied the claim and asked the plaintiff to prove it, the plaintiff could not prove the claim, and the judge sided with the defendant—it encapsulates many significant aspects of the social and economic history of animals in the early modern Ottoman Empire. This case comes at the end of a period in Egypt in which animals were mainstays of the rural economy and the primary modes of labor in the Egyptian countryside. In the absence of steam engines, trucks, or other mechanized forms of technology, animals served as the heavy-lifters, stores of energy, and long-distance transporters needed to power the early modern economy.[3] These agricultural actors addressed fundamental needs in the Egyptian countryside.[4] They moved water from canals and streams to fields; plowed the soil; and transported grains, foodstuffs, information, and people from villages to towns and markets.[5] Roads, doorways, gates, and various other structures were constructed to accommodate the passage of animals. Waterwheels were designed and built to fit these large mammals' bodies. And volumes of food shipments were limited by the amount animals could carry on their backs.

In short, domesticated animals were essential to all aspects of life in early modern Ottoman Egypt and to the economic fortunes of both Egyptian peasants and the wider Ottoman state. Consider, for example, that the earliest administrative divisions of agricultural land in the empire were based on the amount of land a family could farm with a yoked pair of oxen.[6] Social and economic historians of the Ottoman Empire have long worked to elucidate the immense wealth accrued by imperial subjects and merchants through the ownership of and trade in various commodities. Paper, silk, coffee, and myriad other material goods have all rightly been considered key components in the economic history of the empire.[7] Animals deserve no less.[8] While livestock—a term that in and of itself points to the economic importance of animals—were similar to these other commodities, they were also very different in several respects.[9] As living, moving, sentient, and procreating assets, animals indeed challenged numerous aspects of property law and ownership rights in Ottoman Egypt.[10]

Food production lay at the heart of why animals were so important in Ottoman Egypt, and why Ottoman Egypt—the largest grain-producing region in the empire—was vital to the entire imperial system. The utilization of nonhuman forms of labor in rural Egypt was thus an integral part of the effort to sustain human life, since the need to grow and transport food demanded the efforts and caloric output of both animals and humans.[11] Getting food into the stomachs of people across the expanse of the Ottoman Empire was a driving force behind imperial rule in Egypt and behind this early modern empire's harnessing of animal power and technology—plows, waterwheels, harnesses, and the like—to maximize its overall levels of agricultural production.[12] At the same time, animals were also status symbols in rural Ottoman Egypt and represented significant investments of capital.[13] Individuals in the Ottoman Empire did not own land as property; the state owned all agricultural land, while rural cultivators, tax farmers, and others held the usufruct rights to cultivate parcels of that land. Because peasants—the overwhelming majority of the population of Egypt and the rest of the empire—could not purchase land and seldom owned slaves or other large forms of property, animals represented some of the most sizeable forms of agricultural capital.[14]

This chapter explicates the intense economic and social stakes of animal wealth and labor in rural Ottoman Egypt. Egypt's countryside was in this regard quite similar to other rural societies around the world in which animals and humans lived in close proximity and collectively and collaboratively engaged in productive labor.[15] Given their centrality to the agricultural economy, animals were subject to various kinds of legal disputes in Egypt's rural Ottoman courts involving financial transactions, estate inventories, thefts, marriages, and public works projects. Moreover, because they could move, die, lactate, and procreate, disputes involving domesticated ungulates—the vast majority of animals employed in rural Ottoman Egypt—represent some of the most challenging and unique cases about property in the archival record.[16] In this way they help

Figure 1.2 Water Buffalo Plow. Frederik Ludvig Norden, *Voyage d'Égypte et de Nubie, par Frederic Louis Norden, ouvrage enrichi de cartes & de figures dessinées sur les lieux, par l'auteur même*, 2 vols. (Copenhagen: Imprimerie de la Maison Royale, 1755), 2: pl. 56. Beinecke Rare Book and Manuscript Library, Yale University.

delineate notions of wealth, property, and nonhuman nature in the empire. Ottoman Egypt's nexus of intense human-animal interactions changed radically at the end of the eighteenth century, around the time al-Ḥājj Muṣṭafā's ox died.

The Animal Economy

Livestock constituted the largest single source of wealth in the early modern Egyptian countryside.[17] Their ubiquity in the general agricultural economy of Ottoman Egypt structured the lives and livelihoods of all social strata; indeed, nearly all classes of rural society participated in a wide variety of often quite sophisticated animal ownership arrangements. Rural elites—amirs, members of military households, shaykhs, rich *multazim*s (tax farmers), and imperial officials—often possessed large numbers of animals that they put to good use as laborers. Likewise, Egypt's peasant farmers—the vast majority of the rural population—often owned a few or a dozen animals of their own.[18] Individuals could buy shares (or even futures) in animals, and one could sell an animal and still hold the ownership rights to that animal's offspring.[19] Livestock were literally live "stocks" for Egyptians.[20]

At the top of the economic hierarchy of these quadrupeds was the enormous Egyptian *jāmūsa* or buffalo cow.[21] Buffalo cows were by far the most expensive and valuable animal of frequent use in the Egyptian countryside.[22] In an eighteenth-century estate inventory from the court of al-Baḥayra, both a buffalo

Table 1.1 **Animals in the Estate of al-Ḥājj Ḥasan ibn al-Marḥūm al-Sayyid ʿAlī al-Sharīf**

Animal	*Quantity*	*Total Price (niṣf fiḍḍa)*	*Unit Price (niṣf fiḍḍa)*
Ox	2	131	65.5
Buffalo cow	2	100	50
Cow	1	16	16
Camel	1	15	15
Ewe	8	32	4

cow and a camel were priced about six times higher than an ox.[23] Indeed, the buffalo cow and the camel were among the most expensive single items in this entire estate belonging to the deceased al-Shaykh Ibrāhīm ibn al-Shaykh Ramaḍān al-Shayūnī al-Khaḍrāwī.[24] In a case of inheritance from roughly the same period, a buffalo cow was valued at two and a half times the price of a cow and over three times the price of an ox.[25]

When al-Ḥājj Ḥasan ibn al-Marḥūm al-Sayyid ʿAlī al-Sharīf from the village of Bīwīṭ in al-Baḥayra died at the end of the eighteenth century, nearly his entire inheritable estate (assembled over the last decade of his life) consisted of animals.[26] Of the 298 niṣf fiḍḍa that made up the total value of this estate, animals accounted for 294 niṣf fiḍḍa; the other 4 niṣf fiḍḍa consisted of a quantity of rice.[27] Table 1.1 shows that oxen and buffalo cows were the most valued animals in this particular estate.[28] The value of the other animals was markedly lower.

On the rare occasion that horses appeared in the estates of deceased peasants or residents of subprovincial cities, they were sometimes—but not always—judged to be more valuable than buffalo cows.[29] Horses were, however, quite rare in the Egyptian countryside and mostly possessed by the rich and military elite.[30] They were often considered lavish gifts.[31] For example, horses were customarily given as part of the gift package new governors to Egypt received upon their arrival in the province.[32] This reward partly compensated imperial governors for the animals they were not allowed to bring by ship to their new post.[33] Horses were also the preferred animal for riding between various regions of rural Egypt.[34]

In the case of a 1739 inventory of the estate of Muḥammad ibn ʿAbd al-Raḥman ibn Muḥammad Sulīkar, the combined value of his buffalo cow and her calf was greater than all his other possessions, save his stocks of rice, wood, and salt.[35] Indeed, only those items possessed in large bulk quantities proved to be worth more than his one buffalo cow and calf. To purchase an animal as expensive as a buffalo cow, peasants often had to pool their resources. In these cases, peasants bought shares in an animal allowing them partial rights to its labor, milk, and

Figure 1.3 Horses and Other Animals in the Residence of the Fabulously Wealthy ʿUthmān Bey. Commission des sciences et arts d'Egypte, *État moderne*, vol. 1, pt. 2 of *Description de l'Égypte*, Le Kaire, pl. 50. Beinecke Rare Book and Manuscript Library, Yale University.

other productive capacities. In an inheritance case, a particularly wealthy man bequeathed to each of three heirs a third of one buffalo cow, and to each of two other heirs half of another buffalo cow. Tellingly, each of these shares was more expensive than the full price of the man's one "red cow" (*baqara ḥamrā'*) and her calf.[36] In another case, two men came to court each to buy half of a buffalo cow from another man.[37] Splitting shares in an animal was of course not limited to buffalo cows and commonly occurred with other animals as well.[38]

In the economic hierarchy of animals in the Egyptian countryside, donkeys, cows, and camels came beneath she-camels (*nāqa*s) and buffalo cows.[39] Although cows could produce milk—as could she-camels and buffalo cows—they were not extremely useful as work animals.[40] They were big, slow, not easily ridden, and required a great deal of food and space to maintain.[41] The milking potential of cows was thus the main reason Egyptian peasants considered them of value and generally more expensive than male camels and donkeys, both of which were very easily acquired in Egypt.[42] For example, an estate inventory from the southern city of Manfalūṭ lists a cow at eight times the price of a donkey, two times the price of a calf, and sixteen times the price of a sheep.[43] In sum then, based on the relative prices of these animals in Ottoman-Egyptian estate inventories, the hierarchy of animals in the Egyptian countryside was crowned

Table 1.2 **Cow Ownership of ʿAlī Muṣṭafā ʿĪsāwī**

Percentage of Cow Owned	*Price (niṣf fiḍḍa)*
Half	175
Half (plus the right to the cow's offspring)	310
Fourth	82
Half	225
Half	300
Third	150
Fourth	82
Half	310
Fourth	125
Fourth	50
Half	320
Half	230
Half	200
Third	150
5 and 2/3 Cows	**2709**

by the buffalo cow, the ox, and the she-camel (and occasionally the horse). Next came cows and then donkeys, male camels, and calves.[44]

Tables 1.2 through 1.5 summarize the rather estate of ʿAlī Muṣṭafā ʿĪsāwī, who died in Isnā in 1759.[45] His estate's considerable animal holdings, representing over 95 percent of his total wealth, give further details about the relative values of animals of moderate expense in the middle of the eighteenth century.

As these tables show, cows were both the most common and the most expensive animals in ʿAlī Muṣṭafā ʿĪsāwī's estate and were subject to similar sorts of sharing agreements as other animals.[46] Cows and other quadrupeds were clearly big business in the early modern world. They were the vital caloric motors

Table 1.3 **Donkey Ownership of ʿAlī Muṣṭafā ʿĪsāwī**

Percentage of Donkey Owned	*Price (niṣf fiḍḍa)*
Whole	60
Third	60
Half	75
1 and 5/6 Donkeys	**195**

Table 1.4 **Calf Ownership of ʿAlī Muṣṭafā ʿĪsāwī**

Percentage of Calf Owned	*Price (niṣf fiḍḍa)*
Fourth	50
Half	30
Fourth	50
1 Calf	**130**

powering much of rural Ottoman Egypt. Without the energy their muscles provided, the countryside clearly could not have functioned. The early modern Egyptian countryside was therefore a world characterized by cooperative, constructive, and deeply intertwined relationships between humans and animals mediated mostly through their shared labor in the agricultural realm and the human ownership of animals. Simply put, animals were valuable because they provided the energy humans needed to undertake their daily labor and production needs.

Examination of other estate inventories offers further insight into the costs of animals relative to other nonliving things, and even to humans, and thus gives a fuller sense of the economic worth of animals in Ottoman Egypt. Both cows and buffalo cows were generally much more valuable than most household items. For example, the combined value of a cow and a young calf owned by one Muṣṭafā al-Faqīr ibn Muḥammad ʿAqaṣ was 4,430 niṣf fiḍḍa.[47] His copper kettle was valued at only 90 niṣf fiḍḍa, an antique *ṭarbūsh* (a kind of hat) was priced at 120 niṣf fiḍḍa, a red shawl cost 20 niṣf fiḍḍa, and a *zaʿabūṭ* (a common woolen peasant garment) was priced at 60 niṣf fiḍḍa. Another estate from 1749 consisted of a 20-percent share in a donkey, which was priced much higher than either the deceased's set of wooden spoons or his large number of eggs.[48]

The rare instances of probate cases in Egypt's rural courts dealing with estates that included humans provide a glimpse as to how animals, humans, and inanimate objects were relatively situated in an early modern economic and social hierarchy of value. An example comes from the court of the Mediterranean port city Rosetta in 1741.[49] The wealthy deceased Aḥmad ibn Aḥmad, who was also

Table 1.5 **Average Value per Animal**

Animal	*Average Price (niṣf fiḍḍa)*
Cow	478
Donkey	106
Calf	130

known as al-Sayyid Yūnī, owned a great deal of rice, two cows, a black concubine, a sizeable number of dates and figs, and some bricks and lime.[50] Of these, the most expensive item was his large quantity of rice, followed by his black concubine, who cost 2,000 niṣf fiḍḍa. The individual prices of the rice and concubine were dwarfed by the costs of the preparation of the deceased's body and its burial (*tajhīz wa takfīn*). These postmortem expenses, which occasionally appear as part of probate cases, ran to 5,000 niṣf fiḍḍa. The total cost of his two cows was 1,000 niṣf fiḍḍa, and together the figs and dates cost 500 niṣf fiḍḍa. The brick and lime were also listed together as costing 500 niṣf fiḍḍa.

This case lists the human concubine—four times as costly as a cow—alongside all the other items in the estate without any distinguishing marker to set her apart from the nonhuman entries.[51] Similar to that of a shawl or a donkey, this concubine's color was even identified. Thus, within an economic arrangement in which human owners controlled both human and nonhuman property, no sort of divide between the human and the nonhuman informed the legal treatment of owned slaves.[52] In other words, concubines, slaves, physical objects, and animals alike were all entered into the realm of the commodity to be evaluated, priced, bought, and sold.[53] Indeed, the language used to describe slaves in Ottoman Egypt, like elsewhere, often employed the use of animalized vocabulary. For example, the Ottoman writer Muṣṭafā ʿĀlī, who visited Egypt in 1599, described black slaves in the province as "herds of studhorses."[54] He continued on to say that "when they wander along like a herd of animals, naked from head to foot, they would pass by a fountain spring and would, like bears, fill their palms and cupped hands with water, and drink."[55]

Slaves and concubines were conceived of in literally the same legal and descriptive terms as animals and inanimate physical objects. This shows that economic and social function determined the treatment and worth of living beings. Slaves and livestock were on the same spectrum as other commodities because they undertook productive and useful labor in the countryside. Work thus forged social and political relationships and contributed to the shared existence of possessed humans and nonhumans in an economically stratified society.

Consuming Animals

Animals that produced milk and could be utilized as efficient laborers in agricultural cultivation, such as buffalo cows and she-camels, were the most valuable beasts of burden in rural life.[56] Animals that produced milk but were less useful as agricultural workers followed buffalo cows and she-camels in terms of economic value in the Egyptian countryside.[57] Like human farmers, animals proved their value through their contributions to food production. Unlike Egyptian peasants, however, some animals possessed the added economic potential of producing milk that could be made into various consumable goods. Indeed, she-camels' value derived in large measure from the fact that they served as sources of milk for communities of rural cultivators.[58] For example, the most expensive item in the estate of ʿAlī ʿAbd al-Qādir al-Ḥamdanī (d. 1759) was a she-camel worth 1,133 niṣf fiḍḍa.[59] With its large amounts of potassium, iron, and other vitamins, camel milk was generally considered more nutritious than cow's milk and was often part of the Egyptian peasant diet.[60] Camels were, moreover, historically among the most prized living possessions in all of Egypt, especially among the Bedouin.[61]

Animals were also killed for their meat, though this was a rather rare occurrence in the early modern Egyptian countryside, since peasants seldom, if ever, killed their own work animals for meat.[62] Eating meat was a luxury few could afford, and thus consuming animal flesh came to be associated with the wealthy.[63] Animals were, for example, always slaughtered as part of the celebration around the arrival of a new Ottoman provincial governor in Cairo.[64] And once settled in their new post, these governors and their retinues received 180 *okke* (509.1 lbs.) of veal every day from the city's slaughterhouse (*silihāne*).[65]

When a rich man in Isnā endowed a *waqf* (pious endowment) on the island of Aṣfūn before his death, he instructed that horses, camels, cows, buffalo cows, and sheep be kept on hand to be killed and served as food to those guests who came to the *waqf*'s diwan on special occasions.[66] The endowment of this rather grand complex and the provisions the endower made for the feeding of guests was a sure means of preserving prestige and status in his community after death.[67]

Not only were land animals consumed in the countryside, but so too were fish and birds. There was a successful fish market in Rosetta[68] and another in Old Cairo.[69] A very lucrative *muqāṭaʿa* (tax farm) known as Baḥayrat al-Samak was charged with regulating fishing and hunting rights on a large lake near Damietta.[70] A similar *muqāṭaʿa* was established by Ḥasan Paşa in 1786–87 on the Lake of Maṭariyya near Cairo.[71] In return for this tax farm, the area's *multazim* had to pay the state 200,000 paras annually.[72] The lake's circumference was a distance of several days' journey, and the people who lived around it earned their income by selling both different kinds of fish from the lake and the many sorts of sea birds who frequented the area.[73] Fishing was also regulated on the shores of Būlāq and Old Cairo, in canals in and around Cairo, at the mouth of the Nile's branches in Rosetta and Damietta, in lakes formed by the Nile's overflow in Cairo, and in Lake Maʿdiyya in Alexandria.[74] The *multazims* of these areas usually collected two fish from each fisherman as payment and were entitled to 15 to 20 percent of the revenues raised from the sale of fish in their area. There was also a corporation of fishmongers in the markets of Cairo.[75] Despite this evidence, the centrality of the Nile to peasant life, and the high amounts of protein

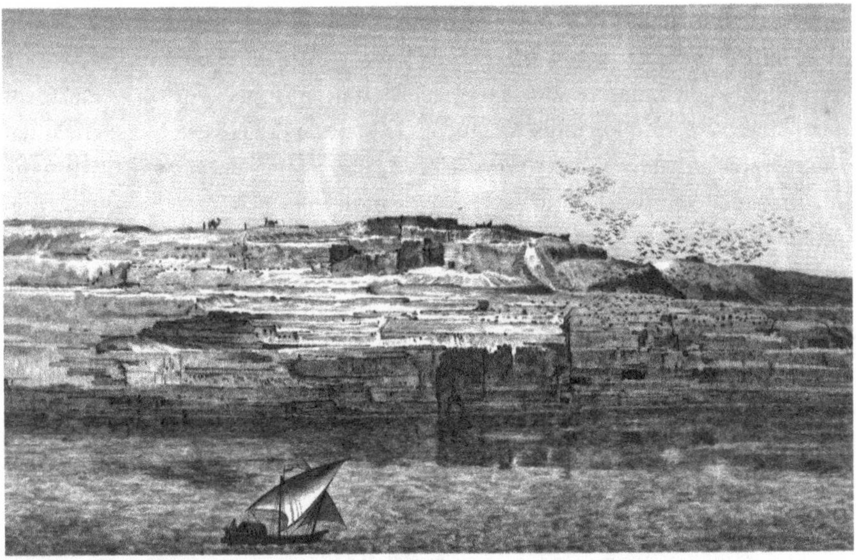

Figure 1.4 Migratory Birds in Middle Egypt. Commission des sciences et arts d'Egypte, *État moderne*, vol. 1, pt. 2 of *Description de l'Égypte*, Égypte moyenne, pl. 7. Beinecke Rare Book and Manuscript Library, Yale University.

and other nutrients in fish, aquatic animal life is strangely underdiscussed as a source of food in the early modern period.[76]

Many peasants also kept pigeons for the market and their own personal consumption.[77] Eating them was not the only way to utilize pigeons as sources of food. Their nutrient-rich droppings were thought to be extremely useful for the cultivation of melons, for example.[78] Overall, fowl were cheaper and more readily available in Egypt than elsewhere in the Ottoman Empire. Visiting Egypt in 1599, Muṣṭafā ʿĀlī observed that "geese, chickens, young pigeons, are extremely inexpensive. Young pigeons are sold for one para apiece, well-fed chickens for two paras apiece which according to the current prices in the Turkish provinces (*vilāyet-i Rūm*) would cost two and three aspers each."[79] Nevertheless, in a rural society in which most food was grown in the soil, the luxury of eating meat—whether fish, fowl, or fodder-farmed—was generally one enjoyed by the rich or reserved for special occasions such as feasts or weddings.

Animals at Work

Above and beyond their abilities to produce food, animals' useful labor was the main reason they were so coveted by Egyptian farmers and the Ottoman administration of Egypt.[80] For example, the exceptionally rich Bedouin shaykh and amir Humām ibn Yūsif ibn Aḥmad ibn Muḥammad ibn Humām ibn Ṣubayḥ ibn Sībīh al-Hawwārī of Upper Egypt had over twelve thousand head of cattle to work in the cultivation of sugarcane on his estate.[81] He also employed cattle for threshing grain, running mills, and turning waterwheels. His estate also included many water buffalo and dairy cows. Here again, animals proved essential to the maintenance and function of Egypt's irrigation infrastructure, the aquatic lifeline of the Ottoman Empire's breadbasket.

In the realm of irrigation, animals were also often used for digging and dredging canals and reinforcing canal embankments. In August 1774, camels, donkeys, and horses were employed in the re-digging and re-dredging of a series of canals in the province of al-Daqahliyya that ran to the villages of Minyyat Ṭalkhā and Minyyat Ḥaḍr from the large canal known as al-Baḥr al-Ṣaghīr.[82] Camels were also used to clear debris and mud that hindered the proper function of wells throughout the countryside.[83] For their part, buffalo cows were often utilized in the maintenance of canal embankments, as in a village near Sandūb in the subprovince of al-Daqahliyya in 1653.[84] The expenses incurred in this repair were to be paid by the villagers overseeing this work, who were then to be reimbursed by the imperial bureaucracy. The largest expense these villagers reported was renting buffalo cows to do the heavy labor.

In another case concerning the repair of two waterwheels in Damanhūr in the subprovince al-Baḥayra, the Ottoman administration of Egypt again conceptualized both animals and humans as representing the same unit of expense—laborer.[85] During the four years required to repair the waterwheels, from September 1790 to the summer of 1794, much of the over thirty-seven niṣf fiḍḍa spent on this work went toward the costs of bricks, wood, and lime. Human and animal workers were also listed on this inventory of expenses. Workers and builders were needed to move and arrange materials, and donkeys were used to haul mud to and from the worksite.[86] Similar to the conceptual propinquity of concubines and animals in the estate inventories discussed above, human and animal laborers in this case were listed in almost the same breath as belonging to the same analytical category of laborer. Undertaken in 1794, this case (like the next one) was in many ways part of the transitional period of human and animal labor at the end of the eighteenth century.

The entire water supply of the village of Shūbar in the subprovince of al-Gharbiyya came from what the village's dam could capture and store during the Nile's annual flood.[87] A repair of this dam at the end of the eighteenth century called for both human and animal labor: one hundred men and sixty-two oxen. These laborers were to devote themselves to hauling and positioning straw and dirt in the dam. This work required bags (*kīs*) to carry the straw and dirt, and rakes (*jarārīf*) to move and collect the dirt. Provisions of beans (*fūl*) sustained these human and animal laborers during their work. The casting of these laborers on the pages of this work register testifies to the fact that food was not differentiated as going to peasants or oxen. The collapsing of humans and animals into a single unit of living laborer again suggests that in the realm of work on irrigation features in the Egyptian countryside, the Ottoman bureaucracy did not differentiate between human and animal workers. Both were, in the end, simply different kinds of laboring units. Human and nonhuman laborers were thus abstracted by the Ottoman administration as possessions of the state to be enumerated, moved around, and configured as needed for Egypt's irrigation network.[88]

Not only did nonhuman animals perform the arduous tasks of transporting building materials to and hauling unwanted dirt from worksites, but they were also necessary for the actual operation of many irrigation works, especially waterwheels. The muscles of large domesticated agricultural animals were the vital sources of power that turned waterwheels to keep rural Egypt irrigated. As evidenced by numerous cases from rural Egyptian courts, including that of al-Ḥājj Muṣṭafā and his ox, animals such as cows and buffalo cows were expensive and coveted partly for their ability to move waterwheels.[89] Camels were rarely used to turn waterwheels and seem to have been reserved for transporting items; they, for example, often hauled lumber through the desert between Cairo and Suez for the construction of boats in the Red Sea port city.[90] These animals were also employed on irrigation worksites—and in other construction projects—to remove excess dirt and materials.[91]

Ottoman Egypt ran on animals.

Animals in Question: Fraud and Theft

Thus far this chapter has shown how livestock were in widespread use throughout rural early modern Ottoman Egypt and how the countryside was literally built and run on their backs. Animals were not only used in this agricultural economy to produce food and to carry out useful work, however. They also represented some of the largest forms of capital investment for all classes of rural society. As visible, significant, and coveted markers of economic and social status, livestock were not surprisingly subject to property disputes, fraud, theft, and other attempts to gain comparative advantage through their possession.[92] The struggles and conflicts that arose from attempts to control animals offer a great deal of insight into notions of property, law, and ownership in Ottoman Egypt.

Property disputes involving animals and other possessions frequently emerged during the execution of wills.[93] In a case from the southern city of Manfalūṭ, for example, two brothers wrangled over a cow, a donkey, and two sheep that had belonged to their late father.[94] The sale of animals from a deceased's estate was often also a cause for dispute.[95] Marriages and divorces too frequently involved the transfer of large amounts of animal capital, making them events of potential conflict over animal properties.[96]

In addition to these very common kinds of disputes in the courts of Ottoman Egypt, other more complex and rarer instances of crimes involving animals provide a more nuanced understanding of notions of property in the early modern period. One such case was the purchase of a camel by an unnamed peasant from an unnamed seller in al-Baḥayra in 1752.[97] The supposed fraud involved in this transaction came to light through the testimony of a third man, who asserted that he had entrusted the seller to watch over his horses and camels while he was occupied with another matter. This man claimed in court that the sale of the camel to the peasant was thus illegal since these animals were never legally possessed by the seller.[98] In the court's ruling, however, the judge instructed that the peasant could keep the camel he purchased because he had solid legal proof—usually either the testimony of two qualified male witnesses or some sort of written legal instrument—stating that the camel was indeed the property of the seller.[99] Thus it was the sale, not the purchase, of the animal that was deemed illegal.

Other cases of animal theft offer further insight into how the ownership of animals was established and protected by rural Egyptian courts.[100] Animal theft in rural Ottoman Egypt was frequently carried out by Bedouin[101] or roaming groups of bandits.[102] These outsiders to village communities most often stole animals for their economic value and for the food and other essentials their bodies provided.[103] Villagers stealing animals from fellow villagers was not as common an occurrence. The sheer size and clumsiness of large domesticated animals made it difficult to steal them in the first place and then to conceal them once stolen. Since the majority of peasants lived in villages where almost everyone knew everyone else and had some sense of their property holdings and animals, a theft was a very obvious act; it would be clear when a peasant no longer had a particular animal and even clearer when someone else did. Nevertheless, when thefts did occur and—on the even rarer occasion—when these cases were adjudicated in a local Islamic law court rather than through an out-of-court settlement, they show the extent to which the possession of animals was a symbol of social status and how far people were willing to go to regain their animals.[104]

In al-Baḥayra in 1787, ʿAlī ibn Aḥmad Muḥammad—described as one of the elders (*al-mashāʾikh*) of the village of al-Raḥmāniyya—accused ʿAlī ibn ʿAlī al-Ḥamd (from the same village) of stealing two steers from him.[105] According to the plaintiff's testimony, nineteen days earlier the defendant had entered his home at night and stolen one red and one piebald (*ablaq*) steer.[106] The plaintiff came to the court to demand that the defendant be made to immediately return the stolen animals. The defendant denied these accusations. To make his case, the plaintiff brought two male witnesses—sufficient legal proof—from al-Raḥmāniyya to the court. Yūsuf ibn Ḥasan ʿAṭīya and ʿAlī ibn Dāwūd al-Khawlī each testified independently of the other while in the presence of the defendant. Both witnesses claimed to have seen the defendant walking in a nighttime wedding procession (*zifāf*) in al-Raḥmāniyya with the two steers two days after the alleged theft. The witnesses added that the defendant pulled both steers from their bridles, a clear indication that he meant to display them as *his* animals. The witnesses' descriptions of the animals' colors and coat patterns matched those of the plaintiff's stolen animals. Weighing the defendant's denial against the testimony of the two witnesses, the judge ruled in favor of the plaintiff and ordered the defendant to return the two steers to their rightful owner. Upon hearing this verdict, the defendant broke down in the court and confessed to the theft. He added, however, that he had later given the steers to a man named Ḥasan Shukr, who was also present in court that day.

The crucial event in the proper adjudication of this case was the fact that the stolen steers were paraded through the village. As a very public event, a wedding procession was an opportunity for people in a village to assert their social rank and display their possessions. Important forms of capital, large animals like these two steers were key symbols of local prestige.[107] Indeed, the testimony in this case was clear: the thief held these animals by the reins as their literal leader (*huwa qāʾid*) to make sure everyone at the wedding knew they were his (even though legally they were not). The fully detailed description of the two steers—their color and coat patterns—was the key piece of information that allowed the witnesses to make the connection that led to the judge's verdict.[108]

Did the defendant really believe he could get away with parading around his village during a wedding procession with two steers he stole from a notable member of his community two days earlier? Did he perhaps know all along that the steers would be returned to their rightful owner? Did he, in other words, ignore the legal repercussions he surely knew would befall him in favor of the temporary social status that would be accorded him as the possessor of these animals? Perhaps he was put up to the task by the mysterious Ḥasan Shukr who would later take the steers for himself and who "paid" the thief for his crime by allowing him to use the steers in the wedding procession. Or maybe the defendant's claim that he later gave the steers to Ḥasan Shukr was a vain attempt to blame someone else so as to rescue his name from village shame.

Whatever the case, the defendant's motives are not clear. What is clear, however, is that the thief and the victim were from very different social classes. The plaintiff was a notable, specifically mentioned as being from among *al-mashāʾikh*, while the defendant was not. The plaintiff owned two steers and probably many more animals, while the thief likely did not (though this is unspecified). Perhaps the plaintiff even knew the judge presiding in the case and the others assembled in the courtroom. Surely most of those in the wedding procession knew the plaintiff. The disparity in the two parties' social and economic backgrounds—measured in part by their possession or lack of animals—was a key determining factor in the case's legal outcome.[109]

Animals on the Move

Because livestock were so valuable and useful as forms of property in rural Ottoman Egypt, establishing their proper ownership was an issue of the utmost seriousness and hence intense contestation. Animals' abilities to procreate and move set them apart from other forms of property. The function and logic of Islamic courts as well as wider legal, social, and economic notions of property thus had to adjust to the unique attributes of these living possessions, especially given the lack of fenced space in the Egyptian countryside.

Consider in this regard the example of Rushdān ʿAbd al-Rasūl from the village of Jalda who came to the court of Manfalūṭ to regain his donkey.[110] He alleged that the "green she-donkey" (*al-ḥimāra al-khaḍrāʾ*), now in the possession of a woman named ʿAzīza bint Ḥasan Abū Aḥmad, was actually the female offspring of one of his donkeys that had wandered off about six months earlier.[111] According to Islamic law, any offspring of an animal was the legal possession of the birthing animal's owner. For her part, ʿAzīza denied that the donkey was Rushdān's, asserting that she had legally bought the young jenny from the sale of her brother's estate a few months earlier. To counter ʿAzīza's claim and strengthen his own, Rushdān asked to have the donkey in question brought to court. Along the lines of an early modern kind of maternity test, Rushdān produced solid legal proof in the form of two witnesses who testified that this jenny was indeed his donkey's offspring and therefore his legal property.[112] Finding this evidence sufficient, the court ordered that the young donkey be returned to Rushdān.[113]

Because the offspring of one's animals was considered one's own property, an owner could sell an animal and still retain rights to its offspring. It was therefore often specified when an animal was sold whether or not the buyer was also purchasing rights to its offspring. This arrangement was in place not only to ensure legal and financial control over animals but also to help regulate breeding. By specifying the ownership of offspring and, hence, who determined breeding patterns, Egyptian peasants were able to control the breeding of their domesticated ungulates so as to select for desirable traits and diminish undesirable ones. For example, a case brought to the court of al-Baḥayra in the 1780s concerned shares in half a buffalo cow that had recently given birth to two calves.[114] The previous owner of the buffalo cow shares claimed to have sold his shares in the calves' mother but none of his shares in the offspring. He therefore came to court to assert his right to the two calves.[115]

Figure 1.5 A Camel Family. Franz Taeschner, *Alt-Stambuler Hof-und Volksleben, ein türkisches Miniaturenalbum aus dem 17. Jahrhundert* (Hannover: Orient-Buchhandlung H. Lafaire, 1925), pl. 50.

Another case about a wandering animal helps to elucidate the nature of the ownership of moving forms of property and how this ownership was established. In January 1794, ʿAlī ibn al-Marḥūm ʿUthmān Abū Bakr from the village of Sunṭīs came to the court of al-Baḥayra seeking to regain possession of his wandering cow.[116] As the plaintiff, ʿAlī claimed that he was the rightful owner of a brown cow he had legally bought from ʿAlī ʿUmar eight years earlier. About a year after this purchase, the plaintiff began to let his cow wander freely with other animals around the fields of their village community in Sunṭīs. Recently, however, it had come to the attention of ʿAlī that his cow was under the possession (*taḥt yad*) of the defendant in this case, a man named al-Sayyid ʿAbd Allāh ibn Sālim Qarqūr from the village of Zāwiyat Naʿīm, and, significantly, that she had given birth to a new calf.[117] ʿAlī thus came to court to regain from ʿAbd Allāh possession of both his cow and her new offspring.

In response to ʿAlī's claims, ʿAbd Allāh retorted that in the previous year he had rightfully purchased this cow for the price of fourteen riyāls and one buffalo cow from Ḥasan Abū ʿĀliyya of the village of Ruzzāfa, who was the representative (*al-wakīl*) of an individual named Ḥasan Abu Ḥamīda from the same village.[118] He furthermore denied that the cow had recently given birth. To strengthen his case, he brought Ḥasan Abū ʿĀliyya to court to testify that he had indeed sold this cow to ʿAbd Allāh the previous year. This witness also told the court that he had bought the cow five years earlier from a now-deceased peasant from the village of Qabr al-Umarā'. In response to this testimony, the plaintiff 'Alī produced two witnesses of his own from Sunṭīs. They swore to the court that 'Alī was the rightful owner of the cow in question and that seven years earlier he had begun to let this animal roam freely around the village. After a period of deliberation, the judge ruled that the animal was to be returned to 'Alī, adding that 'Abd Allāh was entitled to receive from Ḥasan repayment of the amount he had paid for the animal—fourteen riyāls and a buffalo cow. Once again, it was not the purchase of the animal that was deemed illegal but rather her sale, since the cow was never in the legal possession of the person who had sold her to 'Abd Allāh.

The cow in this case was allowed to roam freely around her village for seven years. From an owner's perspective, there was good reason to let one's bovine possession wander.[119] By allowing an animal to scavenge for food, an owner could avoid the costs of planting, storing, or acquiring feed.[120] Moreover, there were few feral or wild animals in the Egyptian countryside, so the threats of falling prey to another animal or of degenerate breeding were nearly nonexistent in rural Ottoman Egypt.[121] Above all, 'Alī was presumably able to keep tabs on his animal so long as she did not stray too far. As this case shows, the major threat to a wandering animal was not another animal but rather seizure by a human. Indeed, only when 'Abd Allāh attempted to claim ownership of this animal in violation of 'Alī's legal rights did this case come to court.

Beyond the letter of the law, 'Abd Allāh's seizure of 'Alī's cow also struck at the collective rights of the community in which the animal was allowed to freely wander. Over the course of seven years, peasants in the area saw her on a regular basis and knew her to be 'Alī's legal property. Moreover, they probably also used the cow for various tasks, consumed her milk, and collectively cared for her.[122] This cow, in other words, came to function as a kind of collective possession of the entire village. This in no way changed the legal reality of 'Alī's ownership, but it did mean that residents of 'Alī's village benefited from the presence of his cow and were implicated in her protection, health, and well-being.[123] The history of this cow's presence and function within the wider community existed largely outside of her strict juridical status as the property of 'Alī. Thus, 'Alī's assertion of his individual legal right to this animal was important not only to him but also to the collective (albeit legally less persuasive) rights of his village.

As an early modern agrarian empire whose calories and revenues were almost entirely based on the cultivation of foodstuffs, the Ottoman Empire was a polity that relied on the labor of millions of rural humans and animals. Animals were connected to the empire's social, economic, cultural, and architectural structures through their high value as laborers, as sources of food and drink, and as stores of energy for transport and even heat in homes during winter. This reliance on the power, productivity, and work of animals produced a rural society of a particular order and physical geography—one that no doubt would have been of a radically different shape without animals or with other kinds of animals in it. The rural Egyptian world had to be able to sustain human and animal populations in close proximity to one another to maximize the productivity of the countryside and the cultivation of foodstuffs. Ottoman-Egyptian society was thus locked into a system of intense and constructive interspecies reliance and economic production.

This rural order began to change at the end of the eighteenth century. As rural elites and government officials attempted to gain political and economic advantages over their local and imperial rivals, they began accumulating ever larger sources of wealth and power. As part of this process, they assembled large estates, initiated increasingly complicated agricultural schemes, and eventually amassed massive amounts of property. These landed estates were often run by absentee landholders and were worked by groups of peasants who had been stripped of the personal sovereignty and autonomy that had previously allowed them to maintain private property—property that consisted more of livestock than any other single item. As the next chapter shows, these political rivalries coupled with environmental and economic changes at the end of the eighteenth century would forever change how humans and domestic animals interacted in the Egyptian countryside.

2

Unleashing the Beast

Man is a tool-making animal.
—Benjamin Franklin

From a crucible of death, disease, drought, and destruction at the end of the eighteenth century, a new economic topography of the countryside began to form, one in which animals played a diminished role. Much animal labor and animal wealth would soon be wiped out, and a new energy regime would eventually emerge to replace the deficit left by a lack of animal power. The main cause of this shift away from intense human-animal interactions and animal wealth and energy was a vast reduction in domesticated animal populations at the end of the eighteenth century that coalesced with growing political efforts by local Egyptian elites to pull away from the central authority of the Ottoman bureaucracy in Istanbul. This new economic and social topography achieved by decades of rural distress and elite economic and political maneuvering set Egypt on an altered, unprecedented, and irreversible course in which human labor and land—not animals—became the keys to rural economic prestige and political influence.

The changing economic status of domesticated animals in the period between 1780 and 1820 played a central role in the transformation of Egypt from a predominantly agrarian province of the Ottoman Empire to a centralized bureaucratic proto-state with a looser affiliation with the empire. Beginning in 1780, Egypt was hit with a series of climatic and disease events that decimated both human and nonhuman animal populations. The death of so many animals in such a short period of time forced a realignment of wealth in the countryside. With a massive portion of the rural economy wiped out, rural Egyptian elites turned to other forms of wealth, the most important of which was land. In the final decades of the eighteenth century, a small group of rural leaders thus amassed a substantial amount of land and began to use that wealth as a means of political power in Egypt and elsewhere in the Ottoman Empire.[1] To make their land productive, new landholders needed effective labor to replace the many animals that had been killed by disease and drought in this period. They overcame this laboring deficit through the corvée of the Egyptian peasantry. In showing how the vital economic roles of animal and human labor changed the Egyptian rural economy, this chapter seeks to illuminate the fundamental part that animals played in the economic and energetic transformations that made possible the transition from a primarily agrarian subsistence economy to a market-driven commercial economy at the turn of the nineteenth century—a transition that was crucial to the history of all rural societies.

Indeed, between roughly 1750 and 1850, nearly all early modern agrarian societies experienced some form of intense transformation involving the commercialization and modernization of their economies. In the vast historical literature on these transitions, land, human labor, and capital accumulation are the usual protagonists and subjects of these histories. Few would disagree with John F. Richards's early twenty-first-century summary: "Human management of land everywhere has become more centralized, more intrusive, and more instrumentally effective. Large-scale capitalist forms of agriculture and resource extraction are prominent throughout the world."[2] Two of the most recent and productive attempts to tackle the problem of this particular economic transformation are the group of publications that emerged from what came to be known as the Brenner debate and newer scholarship on the question of divergence between China and Europe.[3] Much of the work that has come out of these two historiographical literatures revolves around questions of natural resources and human labor. Humans, however, were not the only living creatures driving these economic transformations. Because of their ubiquity and centrality in the early modern world, both as natural commodities and laborers, domesticated animals also played a pivotal role. As part of the transition from subsistence agriculture to commodity-based commercial economies, animal labor was replaced by human and subsequently machine labor. To fully understand the transformations of agrarian societies, we must therefore understand the historically significant causes and consequences of the shift away from a world characterized primarily by intense human-animal interactions.

Thus, in the period from 1780 to 1820, equally if not more consequential than Napoleon's invasion or the rise of Mehmet ʿAli was the massive reduction in livestock populations. Epizootics, extreme weather conditions, and environmental destruction related to higher- or lower-than-expected Nile floods killed off Egyptian animals in huge numbers at the end of the eighteenth century. Egypt's human and animal populations had of course been wiped out before by disease, environmental destruction, and war.[4] What made these late eighteenth-century demographic shifts different from previous instances of population decline and fundamentally constitutive of later periods were the several preceding decades of attempts by provincial elites in Egypt and throughout the empire to pull away from the Ottoman imperial administration. In the 1760s and 1770s,

Ottoman provincial governors, most notably ʿAlī Bey al-Kabīr and Muḥammad Bey Abū al-Dhahab, began seizing wide tracts of land to distribute to their allies and to patronize potential economic and political partners.[5] In an even clearer affront to Ottoman imperial rule—and in what would prove to be a harbinger of later attempts by Mehmet ʿAli—these provincial elites also launched several ultimately unsuccessful invasions of Ottoman territory in Greater Syria in the 1770s.[6] Mamluk grandees and other notables in Egypt in this period also created several large autonomous family estates.[7] These economic arrangements built along familial lines included large landholdings, endowments of pious foundations in Egypt and elsewhere, and strategic ties with political powers throughout the province.[8] Groups of local powerbrokers were thus already seeking to assemble or expand large landed estates by the time climatic events and diseases affected animal and human populations at the end of the eighteenth century. These population reductions thus presented these powerbrokers with the opportunity they had been waiting for to seize more land. This coupling of attempts to secure political and economic autonomy from the Ottoman Empire with demographic shifts in human and animal populations at the end of the eighteenth century made possible an irreversible phase change in the human-animal relationship in Egypt.[9] Once this transition had been achieved, Egypt was set on a new course, and there was no going back—even after animal populations had partially rebounded.[10]

The accumulation of land at the end of the eighteenth century that resulted from the political schemes of rural Egyptian elites—schemes, again, made possible by reductions in animal populations—proved central to the subsequent history of nineteenth-century Egypt, with the Egyptian state itself eventually emerging as the largest of these landed estates.[11] As estates grew in size, productivity, and output efficiency, so too did their need to irrigate and cultivate land. Market-driven demands fueling land accumulation therefore also resulted in the further centralization of both human and nonhuman labor and other vital resources.[12] In turn, this concentration and attempted monopolization of the caloric power vested in the bodies of humans and animals led to the increasing use of corvée on large estates and a decrease in the number of animals used outside of these estates. Because of corvée, the countryside's decreased overall productivity, plague, and emigration to cities and towns, there were fewer independent farmers on the land. Thus, in addition to the collection of animals on large estates, reductions in human population meant there was less of an economic imperative placed on animal populations outside of these estates. Fewer individual peasants demanded and sought out domesticated animals of their own to work alongside them on their small-scale farms. At the end of the eighteenth century, there was thus less overall demand for domesticated animals, the historic motors of the Egyptian rural economy.

As landed elites continued to expand their holdings—working to produce more food and other goods for an increasingly global market—they sought out more land, more resources, and more labor power. A key aspect of this process of emergent commercial agriculture was that estate holders undertook ever larger and more labor-intensive projects on their land, including more complicated agricultural schemes and large-scale irrigation and other public-works building projects.[13] As these rural manipulation efforts grew in intricacy, complexity, and size, and the number of potential low-cost peasant workers increased, the labor that worked on these projects shifted away from the mixed human-animal labor pools of the early modern period toward being almost completely dominated by humans. Even as decimated animal populations recovered, they were increasingly passed over for work on progressively more intricate projects that involved a great deal of capital. The fundamental social, economic, and energetic phase changes achieved in Egypt in the previous decades through political moves and the ecological forces of disease and climate thus brought animal labor closer to the brink of obsolescence than it had ever been before in Egypt's agrarian economy. Animals were, of course, still used when humans could not match their strength or stamina—to pull heavy materials or waste and to transport items to market, for example. But even these remaining economic functions would soon come to an end with the introduction of rail and coal in Egypt and other advances in mechanization and machine technology in the first half of the nineteenth century.[14] As a result of all this, the economic importance of animals decreased, leading to an overall reduction in the centrality of domesticated animals in Egypt.

This shift from animal to human and eventually machine power represented a fundamental transformation in the energy regime of Ottoman Egypt.[15] Before coal and steam, animal power—including human muscle—was the primary purveyor of usable energy in Egypt. Water power was also important, but because of the vagaries of the Nile and rural cultivators' inabilities to efficiently, easily, and usefully harness its excessive force, water was never utilized to the same degree as animal power. Wood and wind were other potential energy sources, but wood was too scarce to be burned as fuel and, apart from river transport, wind does not seem to have powered much in the countryside.[16] Thus it was animals, mostly nonhuman, that were the primary sources of energy in early modern Egypt.

Analyzing where and how societies get their sources of energy is fundamental to understanding the history of any society, culture, or economy.[17] In a place like early modern rural Egypt, or most anywhere else in the preindustrial world, animal power was a major part of this story.[18] With this in mind, the question of why so much changed in Egypt at the end of the eighteenth century can therefore be answered by analyzing the dialectical relationship that developed in this period between social changes that led to the formation of large rural estates and changes in Egyptians' demands for more and different kinds of energy sources.

Thus, the Egyptian case analyzed in this chapter offers a useful elaboration on much of the literature in both environmental and economic history that identifies the year 1800 as the crucial moment of transition between the only two energy regimes in human history: from solar energy to fossil fuels.[19] The case of Ottoman Egypt roughly follows this chronology, but is notable for showing that a critical intermediate stage in that transition was a move from animal to human labor. Thus, similar to evaluations of coal and steam machines as representing a comparative step forward over the previous caloric motors of biological power, humans in early nineteenth-century Egypt were deemed better working machines than animals.

In short, then, dwindling animal populations created a labor energy deficit, spurring a realignment of wealth in the countryside that contributed to the accumulation of land and other fixed resources by a rural elite. Seeking to expand as much as possible, these larger estates eventually demanded greater amounts of reproducible labor, discipline, efficiency, and work capacity than could be met by domesticated animals. Thus, the labor and energy output of animals that in the early modern period was wholly sufficient to achieve most tasks in the countryside was found to be inadequate in a new landscape of centralizing and expanding property. To overcome this emergent energy gap, large landholders—and eventually the Egyptian state itself—bypassed most forms of animal power for either human or, later, machine labor. The story of this chapter is thus the story of the social, economic, and political consequences of a shift away from an early modern animal energy regime.

Natural Selection

Beginning in the 1780s, several powerful outbreaks of disease, relentless food shortages, and severe weather greatly impacted cattle and other domestic animal populations in rural Egypt.[20] In the summer of 1784, a widespread plague outbreak decimated livestock in southern Egypt.[21] This plague continued into the fall of 1784[22] and then reappeared in the fall of 1785.[23] Reduced animal populations in these years, coupled with lower-than-expected Nile floods, led to food and fodder shortages throughout the countryside, as fields could not be plowed effectively.[24] In late 1784, a low flood led to humans and animals suffering intense food scarcity throughout the countryside and Cairo for over a year.[25] Many residents of Cairo were forced to comb through the city's garbage to find food—mostly melon rinds and other discarded items.[26] The situation was so dire that the starvation and death of an animal was considered an extraordinary treat for hungry humans. When the ruling elite or members of the military threw away the carcass of a diseased donkey, camel, or horse, city residents thronged to it, grabbing whatever they could from the body. The Egyptian chronicler al-Jabartī writes that people's hunger was so intense that many ate these infected animals' meat raw.[27]

In 1787, another strong epizootic spread among livestock populations in Cairo and the Delta.[28] al-Jabartī writes that "cattle collapsed in the streets and in the pastures," collecting in Cairo's lanes and alleyways and causing the countryside to be "full of the stench of decaying carcasses."[29] One wealthy cattle owner lost 160 of his cattle to the disease.[30] Evidence of the fact that livestock were, in al-Jabartī's words, "the basis of the economy"[31] for both rich and poor alike, the Egyptian chronicler notes that "peasants bewailed the loss of their cattle, realizing what a blessing its possession had been for them."[32] The summer of 1787 was especially difficult. Not only had animal populations dropped precipitously, but many humans were also dying from a new plague epidemic that had come to Egypt that spring.[33] Added to this was severe drought. Fields were parched, and the few crops that did grow were quickly consumed by rats that were, like humans, seeking out food wherever they could find it. Because so many donkeys, mules, camels, and horses had died that year, prices of the remaining animals "rose very high."[34] This epizootic persisted through the end of 1787, its virulence spiking again in October.[35]

Coincident human and animal plagues continued their unrelenting rampage during the final years of the eighteenth century.[36] The troubles of 1787 only grew worse in 1788.[37] In the spring of 1791, an extremely powerful plague outbreak followed an epizootic that had wiped out large numbers of livestock the previous year.[38] The 1791 plague was said to have killed over three hundred thousand men, women, and children.[39] This destruction of rural human and animal capital plunged Egypt into disarray (*perişanlık*).[40] By all accounts, the 1791 plague was perhaps the worst disease outbreak of the last two decades of the century.[41] It affected not just Egypt but many other parts of the eastern Mediterranean, including Istanbul, the Morea, Venice, Libyan Tripoli, Izmir, and other cities on the Anatolian coast.[42] Deadly diseases of both humans and animals continued into 1792, along with severe drought, creating a situation in Egypt in which humans, cattle, rats, and other animals were once again competing for dwindling food supplies.[43] With the flood many months away, and with no significant rainfall, fields were dry and very difficult to farm.[44] When peasants tried to plow their fields, they found only worms and rats. These circumstances benefited at least one group of animals: rat populations increased thanks to the prevalence of worms and small weeds. In fact, there were so many rats that they moved from eating worms to consuming the few fruits that were growing in the branches of trees, robbing humans of even more potential food.[45] Indeed, "whatever crops

Figure 2.1 The Alexandrian Rat at the End of the Eighteenth Century. Commission des sciences et arts d'Egypte, *Histoire naturelle,* vol. 1, pt. 3 of *Description de l'Égypte,* Mammifères, pl. 5. Beinecke Rare Book and Manuscript Library, Yale University.

were saved from the worms the rats ate."[46] Humans could not even find straw and began eating weeds. And all of this obviously affected cattle as well, since they did not have sufficient foodstuffs or herbage for themselves.

These instances of drought, disease, and food shortages for humans, animals, insects, and rodents created a particularly dire situation of competition for scarce resources. With another low Nile flood in the fall of 1792, scant amounts of grain were produced, reserves were held back from the market, shortages ensued, and prices skyrocketed.[47] Worms quickly devoured whatever piddling quantities of cereals and clover managed to grow.[48] Poor humans scavenged for dry grass and other herbage to pass off as straw for animal fodder. They either sold these bundles in Cairo for great sums or were robbed of them by soldiers and other troublemakers. In a sign of just how critical crop shortages had become, some fodder-sellers felt it necessary to sell their wares from behind locked doors to protect themselves against attack and theft.[49] Both humans and animals had nothing to eat and were literally starving to death. Animal carcasses littered the streets of Cairo to the point that "one could hardly put one's foot down without stepping on creatures lying dead in the alleyways."[50] Starved and hungry humans would immediately pounce on a fallen animal carcass, "even if it stank," wrote al-Jabartī, attempting to salvage some amount of consumable meat from it.[51] Things were so critical that people "would have eaten [human] babies."[52]

These food and fodder shortages continued throughout the last decades of the eighteenth century and the beginning years of the nineteenth. Between April 1799 and February 1800, for example, dire deficiencies in food and fodder were again exacerbated by a severe plague epidemic.[53] Things got so bad that in March

Figure 2.2 Water Buffalo in Bilbays. Commission des sciences et arts d'Egypte, *État moderne*, vol. 1, pt. 2 of *Description de l'Égypte*, Basse Égypte, pl. 75. Beinecke Rare Book and Manuscript Library, Yale University.

1800, cattle were dying of starvation because no straw, beans, barley, or dried clover could be found.[54] Some tried to sell their livestock off before they completely wasted away, but no one was willing to buy such emaciated animals even at rock-bottom prices.[55] On the rare occasions when fodder was available, it was prohibitively expensive.[56] Warfare also disrupted food supplies for both humans and animals. In late March 1800, during the French occupation, residents of Cairo faced especially trying circumstances as fighting, plundering, and looting gripped the city.[57] Ottoman and French soldiers monopolized the already low supplies of food.[58] In 1801 and 1802, the central Egyptian city of Jirja was again hit by both a human plague and cattle epizootic that claimed many lives and led to much rural distress.[59]

In addition to these human plagues, epizootics, and the many droughts and famines that hit Egypt between 1780 and 1810, the first few years of the nineteenth century were punctuated by numerous episodes of historically significant weather and wide temperature fluctuations.[60] In November 1804, for example, ominous storm clouds gathered over the city of Bilbays in the subprovince of al-Sharqiyya just to the northeast of Cairo.[61] An enormous thunderstorm ensued, and lightning strikes reportedly killed about twenty people and numerous cows and sheep.[62] Four years later, in November 1808, clouds also gathered over al-Maḥalla al-Kubrā in the subprovince of al-Gharbiyya.[63] In this weather event, hailstones "the size of hen eggs" killed numerous animals and caused widespread property damage.[64] More productively from the perspective of the area's peasants, this hailstorm also killed worms feeding on the early crops of their fields. Similar hailstones, again described as being the size of hen eggs, killed livestock in a February 1809 storm.[65] The winter of 1813 saw especially cold temperatures and even some snowfall and freezing rain in the Delta.[66] Humans, cattle, crops, and fish all died as a result of these extreme conditions. Of course, these more spectacular climatic events were interspersed with the regular destruction of the yearly flood that consistently carried away animals, crops, and often humans as well.[67]

Thus while the period from 1780 to 1810 saw Napoleon invade Egypt and Mehmet ʿAli take power in the Nile valley, it also witnessed a much more fundamental and historically consequential reduction in the number of domesticated animals in Egypt, principally as a result of disease, extreme weather conditions, and fluctuations in flood and cultivation levels. It likely took several decades for these populations to even come close to their pre-crisis numbers, if they ever did.[68] These animal population losses created a caloric energy deficit of two kinds. First, there was less meat to consume. This meant less protein for human appetites, health, and work. Second, fewer animals were available for the labor that was so essential to making the countryside function productively. This situation at the turn of the nineteenth century thus resulted in an energy regime shock to the two principal ways humans derived caloric benefit from animal bodies—by eating them and by putting them to work.

A Meaty Problem

Meat production was a realm in which animal shortages were immediately felt. By about 1815, the combination of a few decades of intermittent disease, drought, famine, flood, extreme weather, warfare, price inflations, and decreased pastureland resulted in an intense dearth of animals and thus of their meat.[69] Although eating meat was a luxury for most rural cultivators, reserved mainly for special occasions and religious festivals, most of Egypt's governmental elite ate meat as a staple of their diet. A decrease in available meat supplies was therefore a grave concern. In an attempt to gain better control of the meat market, Ottoman officials in the early nineteenth century moved to bolster their meat production capacities through the close management, centralization, and monopolization of animal supplies.[70] The results were the increased capture and commodification of the consumable energy animals provided and hence a reconfiguration of Egypt's caloric energy landscape.

Beginning in November 1816, Mehmet ʿAli's government imposed price-fixing measures in an effort to stabilize the meat supply in the face of sheep and cattle shortages.[71] These attempts, however, were largely unsuccessful, and his military and civilian administration resorted to other means to ensure a steady flow of meat to its collective stomach.[72] Given reduced sheep populations all over Egypt, Mehmet ʿAli dispatched his men to every corner of the province to collect as many sheep and other animals as possible to feed his army and governmental elite. On December 5, 1816, an enormous herd of animals arrived in Cairo from the Delta and Upper Egypt. These sheep, water buffalo, and calves were emaciated and hungry from their long and grueling journey.

In this period, Mehmet ʿAli took the further step of temporarily closing down butcher shops to ensure that all available meat was diverted to his administration. Lack of sufficient heads and locked butcher shops effectively robbed Cairenes of their meat supply. When a butcher was lucky enough to secure some meat and attempted to enter "his shop like a thief" to sell it to hungry customers, soldiers stationed in the market would take this meat away and close the shop once again.[73] Without this main source of protein, Cairo's residents were left to seek out nourishment "with great difficulty" from lentils and various other kinds of beans.[74] Chickens, as well as their eggs, were also in short supply from December 1816 through November 1817.[75]

Meat shortages grew even more extreme in the first half of 1817. So too did the government's efforts to secure meat for officials and the army. In January 1817, Mehmet ʿAli ordered an audit of all the agricultural products kāshifs (district heads) and subprovincial governors had taken from villages and peasants under their control.[76] This was an attempt to determine the availability, quantity, and location of vital foodstuffs and other supplies that were quickly diminishing. Sheep, chicken, cattle, horses, animal fodder, and eggs were particularly desired by the state in this time of want. To get a firmer grip on Cairo's scarce meat supplies, Mehmet ʿAli prohibited the slaughter of animals anywhere except in the government's central slaughter depot, which became Egypt's sole supplier of meat in the 1810s.[77] Using the recent audit as their guide, officials in Cairo directed kāshifs in January and February 1817 to buy any available livestock from peasants in their villages at very low prices and to send these animals to the central depot. Ratcheting up his demands, Mehmet ʿAli later ordered kāshifs to confiscate without payment the biggest and healthiest ram or ewe out of every ten sheep counted in a village or town in March and April. At the same time, soldiers were dispatched to monitor all roads leading into Cairo so as to seize any sheep, water buffalo, calves, or other animals brought into the city. This system of centralized seizure worked for a short time to provide enough meat for state officials and the army, with a little left over for sale to the general public.

Given these circumstances, a black market in animal meat soon emerged. Many peasants evaded attempts by kāshifs to buy animals at prices far below their market value. They and their animals would escape their villages under the cover of night and try to enter Cairo undetected to sell their animals at good prices to people willing to pay for high-quality meat. Even a group of *ḥajj* pilgrims returning home to North Africa via Cairo got in on the black-market animal trade.[78] On their way to the city from the Hijaz, they seized animals from Egyptian peasants in the countryside, set up makeshift butcher shops around Cairo, and sold meat at greatly inflated prices. Part of the attraction of this and other black-market meat was its superior quality compared to what was available in the government's central slaughterhouse.

Because of disease, inadequate fodder, general supply shortages, and the government's inexperience in dealing with so much livestock in one place, those animals that made it to the government's central meat depot were usually weak, thin, and of poor quality. And they only grew more emaciated and frail after their arrival. Even some of those animals that had died from hunger or thirst before reaching the central depot were—despite their "malodorous putrefaction"—butchered and sold for human consumption.[79] In an attempt to break the black market in animals, Mehmet ʿAli increased the number of his troops surveilling roads in and out of Cairo and inside the city more generally. Any animals found to be illegally transported, bought, or sold were confiscated and sent to the slaughterhouse. Other lawfully possessed animals were seized too.[80] In a more novel attempt to prevent cheating in the meat market, government agents pierced the nostrils of deceitful and thieving butchers and then hung meat from their noses for all to see.[81] Market restrictions on the buying and selling of animals were somewhat loosened in the spring of 1817. Authorized purchasers were allowed to buy up to nine sheep at a time from merchants who had received official licenses to bring animals into the city.[82]

The scarcity of meat was acutely felt in times of celebration and during religious feasts. During the Feast of Immolation (ʿĪd al-Aḍḥā), a celebration that customarily witnessed the slaughter of many thousands of animals by individual families, this new reality was especially visible and pronounced.[83] The feast's 1817 celebration was particularly muted. There were simply not enough sheep—or other animals—available for slaughter and consumption. In contrast to previous years when the markets "used to teem with animals," in 1817 only a very small number arrived in Cairo and, even then, just a few days before the festival.

As these accounts show, those who were able to control resources in such an environment of scarcity were clearly at a marked advantage over others. Mehmet ʿAli's government was able to control the supply of meat (and hence the caloric power it provided) by seizing animals, centralizing them in new governmental institutions, and attempting to prevent others from producing meat. In times of thin resources, states create new modes and institutions of governance. The central meat depot, road blocks, animal audits, and all of the bureaucratic machinery behind these technologies meant to maximize meat production (but that could clearly be applied in other administrative realms as well) were created by a situation of reduced animal populations in Egypt. With fewer animals, new alternatives were needed to more effectively harness and exploit the energy meat consumption provided.

This lack of animals obviously affected many other realms of life in Egypt. The second major consequence of a loss of animals at the turn of the nineteenth century was a growing deficit in the amount of animal energy available for labor. Without this established source of caloric motor power, large landholders in the countryside sought out other forms of energy, forever changing the composition and nature of the rural economy.

From Animal to Land

Egyptian peasants immediately and acutely felt the social and economic consequences of livestock reductions. After the terrible epizootics of the spring of 1787, for example, peasants found themselves lacking the energy they needed to undertake their normal agricultural tasks.[84] Without cattle, these peasants were left to seek out alternative means to thresh grain and turn waterwheels.[85] Many tried to turn these irrigation works using their own strength, but humans simply could not effectively turn waterwheels designed to be moved by the powerful muscles of water buffalo and other beasts of burden. Much of the infrastructure of the countryside was built for animals, not humans. Faced with this reality, some rural cultivators attempted to acquire donkeys, camels, and horses to replace their dead animals. But given the vast reduction in the number of animals in the countryside, these quadrupeds became prohibitively expensive.

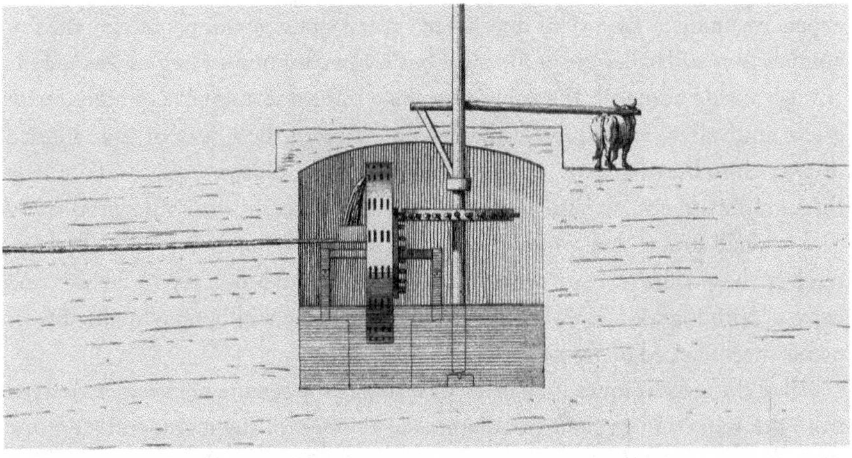

Figure 2.3 The Heavy Machinery of Egyptian Waterwheel Mechanics. Norden, *Voyage d'Égypte et de Nubie*, 2: pl. 53. Beinecke Rare Book and Manuscript Library, Yale University.

By the end of the eighteenth century, decades of diminishing animal populations had made it much more difficult and far more expensive for individual peasants to acquire beasts of burden to use in agriculture, on irrigation works, and for various other tasks only animals could do. Wealthy rural elites who secured animals through purchase, force, or theft were at a marked advantage due to their ability to amass large landholdings and expand existing ones, thereby increasing the gap between themselves and Egypt's majority population of subsistence farmers. During the plague of 1785, for example, many amirs and other wealthy individuals used the mayhem caused by the massive human death toll as an opportunity to seize large numbers of cattle from both the deceased and the living.[86] Ironically, an effort to stop another amir a few years later from similarly stealing animals in the countryside itself required the seizure of domestic animals. In preparing to move against this marauding amir in September 1802, Ottoman authorities in Cairo sought out six hundred donkeys for their soldiers.[87] Officials summoned all of the city's donkey drivers and forced them to sell their animals at dreadfully low prices. The soldiers paid five riyāls for a donkey with his harness and bridle; the going rate was fifty riyāls for a donkey without a harness. Still short of the requisite six hundred donkeys, however, soldiers began seizing animals from the general populace and the city's water carriers, who used donkeys to bring water from the central canal of Cairo (*al-Khalīj*) and the Nile to the city's various neighborhoods and districts.[88] Opportunistic soldiers who were not a part of the force preparing for battle also took advantage of this situation and began seizing animals for their own personal use. They stopped people riding donkeys and seized the animals, keeping them for themselves, extorting a ransom from their owners, or selling them at the market's elevated prices.

In the face of this stark competition for a relatively small number of now very expensive animals, Egyptians developed several strategies for protecting the few animals they still had. One of the most basic and common strategies was to hide animals inside homes.[89] The soldiers seeking out these animals countered with a few innovative techniques of their own.[90] One of these was to coax hidden donkeys into braying so as to reveal their location. Soldiers would walk quietly through the streets of Cairo and other towns until coming to a neighborhood gate or small street. They would then let out a few brays of their own—al-Jabartī renders these calls as "zarr"—in the hope of getting a donkey to answer and expose its hiding spot. These donkeys were then taken without payment, or their owners were forced to pay a ransom.

All of these techniques for seizing or resisting the seizure of livestock derived from the reduction in their populations at the end of the eighteenth century. Possessing some of the few animals remaining in Ottoman Egypt thus emerged as a primary means of cementing comparative economic and energetic advantages. Individuals or entities that could acquire animals to work in fields, turn

Figure 2.4 The Cairo Canal (*al-Khalīj*). Commission des sciences et arts d'Egypte, *État moderne*, vol. 1, pt. 2 of *Description de l'Égypte*, Le Kaire, pl. 27. Beinecke Rare Book and Manuscript Library, Yale University.

waterwheels, and transport goods and men to market or battle were in control of the most important energy sources of the period, and they used this monopolization of caloric power to great political and economic effect.

Animal losses precipitated other processes in the countryside as well. One of the most significant involved the centralized accumulation of land and other resources by rural elites who had been positioning themselves for just such an opportunity throughout the final decades of the eighteenth century. Fewer animals in Ottoman Egypt meant fewer individual peasants who owned animals. This created a class of animal-less, and hence eventually landless, peasants in need of work and income. No longer able to plow fields or move waterwheels, many of these peasants either left the countryside for towns or cities (leading to this period's urban population increases) or began to seek out work on the large centralizing estates developing around them.[91] These peasants thus represented a newly created labor pool from which large landholders could draw cheap agricultural workers. The handful of emerging elites who could take on the initial high cost of acquiring a few of the diminishing number of animals in the countryside therefore became the period's large landholders who would eventually employ many peasants on their estates—the very peasants who had lost their animals and could not replace them.[92] Animal deaths were thus a crucial factor in a shifting agricultural order that resulted in the eventual primacy of land as the foundation of wealth and prestige in Ottoman Egypt.

This realignment of rural capital to replace animals—"the basis of the economy," to repeat al-Jabartī's phrase—was the most significant outcome of the changing energy regime and political economy of animals at the turn of the nineteenth century.[93] Thus, Ottoman Egypt's economic transformation from subsistence to commercial agriculture was fundamentally based on land superseding animals as the largest and most basic source of rural wealth.[94] The decline in the most important traditional form of rural property—animals—led to the further concentration of an alternative form of wealth—land—in the hands of just a few elite families in the countryside.[95]

In the 1760s and 1770s, the rulers ʿAlī Bey al-Kabīr and Muḥammad Bey Abū al-Dhahab began seizing their enemies' land and redistributing it to their followers and other elites with whom they had connections.[96] This was a patronage system of both reward and alliance building. After the demise of these two centralizing regimes at the end of the eighteenth century, a power vacuum emerged in Egypt. Individuals who had amassed sizeable amounts of land under the rule of the two beys thus began challenging each other for the control of resources.[97] It was this class of men who most benefited from and took advantage of declining animal populations in the 1780s and 1790s. They were the ones who used these decades of disease and death to realize the transition from animals to land as the primary source of wealth in rural Egypt. In point of fact, estate inventories from the first two decades of the nineteenth century clearly show land transactions playing a far greater and animals a far smaller role than ever before in Egyptian property transactions and disputes.[98]

While land was a constant source of wealth in rural Ottoman Egypt in the early modern period, its ownership was not the basis for that wealth. Rather, until the end of the eighteenth century, the predominant rights Egyptian farmers had vis-à-vis land were usufruct rights to its products. While these rights could occasionally function as de facto ownership over land—making land open to sale, inheritance, gift, and rent—they also could very easily be taken away.[99] Indeed, the seventeenth and eighteenth centuries are replete with examples of the seizure of land by both the Ottoman state and its functionaries.[100] Thus, what Egyptian regional elites were able to accomplish at the turn of the nineteenth century, in part because of decreases in animal populations, was a move from usufruct to something more closely resembling land ownership.

As a political marker of this economic transformation, this new form of landholding would be ensconced in Egyptian administrative practice in 1812 when Mehmet ʿAli seized all of Egypt's tax farms (*iltizāms*), making the legal ownership of land the primary means of staking claims to land. This legal innovation represented the final bureaucratic instantiation of the late eighteenth-century process of the concentration of land that emerged because of elite ambitions in the countryside made realizable by the deaths of massive numbers of animals. The concentration of landed wealth by a few individuals therefore contributed to the demise of the tax farming system, which had been the dominant form of land tenure in Egypt throughout the Ottoman period.[101] What is more, the very same large landholders who had emerged in the 1780s and 1790s as the wealthy rural elite were the ones who received most of the land that was seized and then redistributed by Mehmet ʿAli; they clearly had the necessary political connections and were the only ones deemed capable of cultivating land successfully. Thus large estates grew larger, and more and more peasants were made landless and found themselves without their traditional modes of livelihood and in need of a source of income.

Given the vastly reduced number of animals available in rural Egypt between 1780 and 1810, only a few large estate holders could afford the inflated sums needed to acquire the animals necessary to overcome the minimum energy requirement for agricultural work.[102] Meeting this initial high laboring cost, however, ensured that these men would become the elite in Ottoman Egypt's changed rural economic and social landscape. In this recently remade world, newly animal-less and landless peasants were in abundance. They proved a ready labor pool from which large landholders could draw the energy supplies they needed. Over the course of this period, together with the growing economic primacy of land, there was thus a steady shift away from animals to humans as the primary energy suppliers for labor in the countryside. Animals were in short supply and expensive; human laborers were abundant and cheap.

The Most Productive Species

With humans becoming cheaper workers than animals, the corvée of human labor emerged as the most important kind of work on the expanding estates of the early nineteenth century. Not only were animals becoming far too expensive, but they were also increasingly seen as too unwieldy and too unpredictable for the regular and reproducible work demanded by the increasingly bureaucratic state in this period.[103] Thus in the first two decades of the nineteenth century, a fundamental transformation occurred in the character of labor in rural Ottoman Egypt. Whereas humans and animals had previously worked together on smaller repair projects that served their own communities, now massive numbers of forced human workers became the preferred means of infrastructural construction and manipulation. This move toward human labor took place outside of the countryside as well. In the 1820s, assigning criminals to hard labor in the Alexandria dockyards became one of the state's favorite criminal punishments.[104] In realms as varied as penal law, food production, and administrative practice, the increasing reliance on human labor meant that animals were being gradually stripped of their economic worth—indeed, that they were being actively deskilled—and that humans were emerging as the dominant component of Egypt's new, decidedly human, energy regime.[105] Whereas animal prices shot up drastically immediately after the population declines of the last decades of the eighteenth century, animals became less valued in the new economy as the infrastructure of rural labor became increasingly centered on human workers.

Corvée had certainly been utilized in Ottoman Egypt before the early nineteenth century, but not to the same degree or with the same ferocity. In the seventeenth and eighteenth centuries, corvée was most commonly organized at the very local level of the tax farm.[106] Yūsuf al-Shirbīnī, writer of a seventeenth-century satire on country living, offers a useful description of this early modern system of labor, highlighting as well the integral role of animals in village communities.

> The corvée is found only in those multazims' villages that include ūsya land [the only land over which *multazims* (tax farmers) held direct administrative control]... The multazim sends oxen, timber, plows, and whatever else is needed and appoints an agent to take charge of it and prepares a place for the timber and animals belonging to it... He also delegates someone to spend money on the upkeep of the animals, etc., and to keep careful accounts... In some villages the corvée applies to a number of men, fixed by household, for example. Thus they say, "From such and such a household one man is to go, and from such and such two" according to the quota set for them in the distant or more recent past.[107]

Because corvée in this earlier period was a very local affair, organized and overseen by those who had long resided and were immediately recognized within village communities, without the full weight of a state bureaucracy behind it, it was mostly small scale and temporary. And even then, early modern corvées were usually undertaken only on the small minority of lands over which tax farmers exercised direct usufruct control (about 10 percent of Egypt's total arable land).[108] Peasants were forced to clear mud from canals or to bring in crops—relatively easy tasks compared to the massive infrastructural projects that would come later. Many peasants, however, not surprisingly, found even this local form of forced labor objectionable and attempted to escape from it.[109]

The term most often used for this sort of corvée in the early modern period was *al-ʿauna*, deriving from an Arabic root related to "assistance" or "help."[110] In contrast, the term most often given to corvée in the nineteenth century was *al-sukhra*, from the Arabic root for words related to "derision," "subservience," "servitude," and "exploitation."[111] This semantic shift in the description of human labor is a telling marker of the difference between these two forms of corvée. While coercion was clearly integral to the nature of the practice itself, it is important to understand that in the earlier period human and animal laborers worked directly to improve their local communities; in the later period, by contrast, peasants were much more solidly controlled and made to serve under the forceful power of a strong and faceless state. Indeed, between the sixteenth and the nineteenth centuries, as a general phenomenon, small-scale peasant labor moved from being under the local supervision and utilization of village tax farmers to being controlled by the centralized, more distant, and more abstract bureaucracy of Mehmet ʿAli.[112] This shift away from the local to the centrally bureaucratic is embodied in the response of a group of peasants in May 1814—two years after the abolition of the tax farming system—to their local tax farmer's demands of labor from them: "Your days are over, and we have become the pasha's [Mehmet ʿAli's] peasants."[113]

Indeed they had. And the pasha put his peasants to work. Throughout the first few decades of the nineteenth century, a new and decidedly imbalanced preference for human over animal labor emerged in rural Egypt. Some of the best examples of the shift from animal to human labor are the numerous repair, expansion, and building projects undertaken on Egypt's irrigation network. At the beginning of Mehmet ʿAli's rule in 1805, the total length of Egypt's canals was approximately 514 miles. New construction efforts initiated by Mehmet ʿAli's administration more than doubled this total length, to roughly 1200 miles by the end of his reign in 1847.[114] The new waterways led to an increase of about 18 percent in the amount of cultivatable land in Egypt between 1813 and the 1840s.[115] Keeping these canals clean was an enormous task. In the Delta alone, approximately 20,730,118 cubic meters of silt had to be dredged from canals every year.[116] These projects of canal construction, expansion, and maintenance began very early in Mehmet ʿAli's reign and were soon occurring all over Egypt. Some of the largest canal projects undertaken by Mehmet ʿAli's government were al-Faraʿūniyya in the central Delta, Shibīn in the Delta, al-Zaʿfarāniyya and al-Sharqāwiyya in al-Qalyūbiyya, al-Būhiyya in al-Daqahliyya, al-Wādī in al-Sharqiyya, al-Fashn in al-Minyā, and al-Shunhūriyya in Qinā.[117] It was humans, not animals, that dug these canals, kept them clean, hauled away excess dirt, and lifted the heavy tools and baskets needed to carry out this work.

As the intense and widespread canal work of the first quarter of the nineteenth century shows, corvée was not simply a temporary measure to replace those animals that had been removed from the labor pool between 1780 and 1810. Even when animals became available again, humans remained the preferred form of rural labor. The economy had fundamentally shifted away from animal to human labor, and not even a recovery in the availability of domestic animals could push it back to its former character.

The permanence of this shift was marked, in part, by the criminal legislation Mehmet ʿAli's government undertook in 1829 and 1830.[118] Laws enacted in these years made corvée an integral feature of punishment in Egypt. For example, periods of forced labor in the Alexandria dockyards were linked to the financial scale of crimes; they ranged from one year of labor for stolen property valued under 1,000 piasters to four years for an amount over 60,000 piasters.[119]

Because corvée was so different from the human-animal laboring nexus that had existed before the nineteenth century, and because it was used so extensively throughout Egypt, it had massive effects on the social and economic structures of rural people's everyday lives.[120] Consider the movement of peasant laborers entailed by corvée. The physical relocation of human caloric energy to fuel projects that did not benefit these humans' own families or communities meant that rural people were further alienated from their own lands and that it was ever more difficult to stop the march toward the concentration of land and energy resources in just a few hands.

The details of this new system of human labor are telling. Peasants in the early nineteenth century contributed about sixty days of corvée labor every year, excluding the amount of time it took to travel to and from a worksite.[121] Workers were collected and delivered to a worksite by the heads of their villages and were assigned tasks as a village unit so that peasants from the same community worked side by side on construction projects.[122] Over the course of one year, an average of four hundred thousand men could be forcibly moved to work.[123] To put this number in perspective, the population of Cairo in 1821 was 218,560 and that of Egypt as a whole was around 4.5 million in 1800 and 5 million in 1830.[124] The demographic effects of this enormous annual movement of people—about 8 or 9 percent of Egypt's total population—were in actuality somewhat greater, since workers frequently brought their families with them to construction sites. Moreover, workers were often charged with bringing their own food and tools. In the case of very large repair jobs, however, the government often provided food—and sometimes even a daily wage.[125]

Specific examples of corvées from all regions of Egypt in the early nineteenth century further illustrate the magnitude and wide reach of this forced labor system. The maintenance of the Shibīn Canal, which ran past al-Maḥalla al-Kubrā in the central Delta, required a yearly corvée of fifty to sixty thousand men from villages throughout the Delta.[126] The order for a corvée between 1817 and 1821 to rebuild the Ra's al-Wādī Canal, which flowed east from al-Zaqāzīq to Lake Timsāḥ, demanded the labor of eighty thousand men. Within just eight days, this enormous number of peasants had been brought to the worksite from villages throughout the subprovince of al-Sharqiyya in the northeast of the Delta.[127] In

Figure 2.5 THE MAḤMŪDIYYA CANAL AT THE END OF THE EIGHTEENTH CENTURY. Commission des sciences et arts d'Egypte, *État moderne*, vol. 2, pt. 2 of *Description de l'Égypte*, Alexandrie, pl. 99. Beinecke Rare Book and Manuscript Library, Yale University.

1829, 32,300 men were brought to dig a new canal in the northwestern subprovince of al-Gharbiyya.[128] And in 1838, Mehmet ʿAli ordered twenty thousand of his own troops to work on the Zaʿfarānī Canal, which fed water to Cairo.[129]

An order sent by Mehmet ʿAli to Upper Egypt in December 1835 for twenty-four thousand men to clean and dredge canals and reinforce embankments makes clear the state's violently paternalistic treatment of the Egyptian peasantry used in corvée.[130] "If you say it upsets the fellahin [peasants] when there is no need," he wrote to his functionaries, "then I say the boy does not willingly go to school but is forced by his parents until he grows older and knows the value of learning, so driving all the men to dykes and canals is difficult for them but is necessary."[131] In this period, soldiers were instructed to forcibly (*qahran wa jabran*) move unwilling peasants to canal sites. Mehmet ʿAli continued, "If land is *sharaqi* [unwatered] because Jisr [the canal of] Banu Khalid is not well shored up there is no punishment but death."[132] He added, "If we see one *qirat* of land unwatered we will bury you in it."[133]

Perhaps the most famous and deadliest instance of corvée in the early nineteenth century was the reconstruction from 1817 to 1820 of the Maḥmūdiyya Canal between Alexandria and the Rosetta branch of the Nile.[134] Estimates for the labor forcibly moved to work on this canal range from 315,000 to 360,000 individuals.[135] So many men were put to work on the canal that it caused labor shortages elsewhere in Egypt. In August 1818, for example, the *muḥtasib* (market overseer) of Cairo, Muṣṭafā Aghā, ordered certain city lanes, alleys, and dead-end streets cleaned of accumulated dirt and debris that had made them impassable.[136] The men who would have normally undertaken this work were away at the Maḥmūdiyya construction site, so shop owners and other urban residents were forced to do the work themselves.[137] One of the fragilities of a system such as corvée was that it was based almost entirely on only one form of labor. Devoting hundreds of thousands of human hands to the Maḥmūdiyya clearly meant shortages elsewhere.

Of the over three hundred thousand workers brought to dig and reinforce the canal, nearly a third of them—one hundred thousand—died.[138] By all accounts, work conditions on the canal were very harsh. Indeed, Mehmet ʿAli's earlier promise to bury unproductive peasant laborers in the dirt seems to have been fulfilled. According to al-Jabartī's description of forced human labor on the canal in August 1819:

> He [Mehmet ʿAli] ordered the governors of the rural districts to assemble the peasants for work, and this command was executed. They were roped together and delivered by boats, thus missing the cultivation of sorghum, which is their sustenance. This time they suffered hardship over and above what they had originally suffered. Many died from cold and fatigue. Dirt from the excavation was dumped on every peasant who fell, even if he were still alive. When the peasants had been sent back to their villages for the harvest, money was demanded from them plus a camel-load of straw for every *faddān*, and a *kayl* each of wheat and beans. They had to sell their grain at a low price but at a full measure. No sooner had they done this than they were called back to work on the canal in order to drain the extremely saline water which continued to spring from the ground. The first time they had suffered from extreme cold; now, from extreme heat and scarcity of potable water.[139]

All of these instances of corvée in the first third of the nineteenth century were the end result of a transition in the energy regime of Ottoman Egypt from animal to human labor. These examples illustrate the massively transformative social power of the changing political economy of animals in the decades around the turn of the nineteenth century. An enormous reduction in the number of animals in the countryside led to the emergence of land as the primary source of wealth in rural Egypt. As small-scale rural cultivators lost their animals and were unable to acquire other forms of property, they found themselves increasingly excluded from an economy now shifting away from animals to land. Large property owners were also severely impacted by animal deaths at the end of the eighteenth century, but unlike poorer farmers, they were able to replace their lost animals with some of the growing number of cheap peasant laborers. Thus, as the energy regime shifted toward humans, animals lost the centrality they had enjoyed in rural Egyptian society for millennia, and humans emerged as the primary form of labor in the new rural economy.

From Animal to Rail

If humans largely came to replace animals as sources of labor in the Egyptian countryside, so too did rail replace animals as the primary form of transport in the nineteenth-century Egyptian economy. In the early modern period, domestic animals were central to the transportation network of Ottoman Egypt. They were the long-distance carriers of goods, people, information, and ideas across Egypt and throughout the Ottoman Empire. Camels, for example, were the most essential transporters of materials and merchants between Cairo and Suez.[140] Those eighty miles of desert were a hinge of world trade in the early modern period. Goods in transit between the Indian Ocean and the Mediterranean necessarily had to traverse this stretch of land—there was no Suez Canal, of course—and this movement could only be accomplished through massive amounts of intense animal labor.[141] Animal transportation was similarly crucial to the smooth operation of the yearly pilgrimage to Mecca and Medina. This intercontinental movement of people, goods, and other living and nonliving things necessitated a huge number of ready and robust animals every year.[142] As in other instances when Ottoman authorities in Egypt needed large numbers of animals, they often took to forcibly seizing beasts of burden for the annual pilgrimage caravan.[143] Needless to say, these instances of transcontinental animal transport stood alongside the more common, everyday uses of animals to move goods and people between villages and towns in Egypt.

Machines, however, soon came to replace animals as means of transport. Mehmet ʿAli made the improvement of Egypt's transportation network one of his top priorities in the nineteenth century. This began with modest attempts to improve roads by making them wider and clearing them of excess dirt and trash to facilitate the easy movement of people, animals, carts, and the like.[144] Very quickly, though, Mehmet ʿAli sought out the newest and most promising technology of conveyance—railroads. Egypt's first rail lines were laid down in 1851, but beginning in the 1830s, Mehmet ʿAli and his hired European advisors engaged in intense conversations about building rail connections between Cairo and Alexandria and Suez—the world's first rail links outside of North America and Western Europe.[145] The effort and time put toward the question of rail in Egypt by Egyptians, Europeans, and others reflected the country's growing inclusion and importance in the global capitalist economy.[146] External market forces pushing for rail in Egypt thus shaped rural labor and the role of animals in the countryside in fundamental ways.[147] In the words of John Galloway, the most

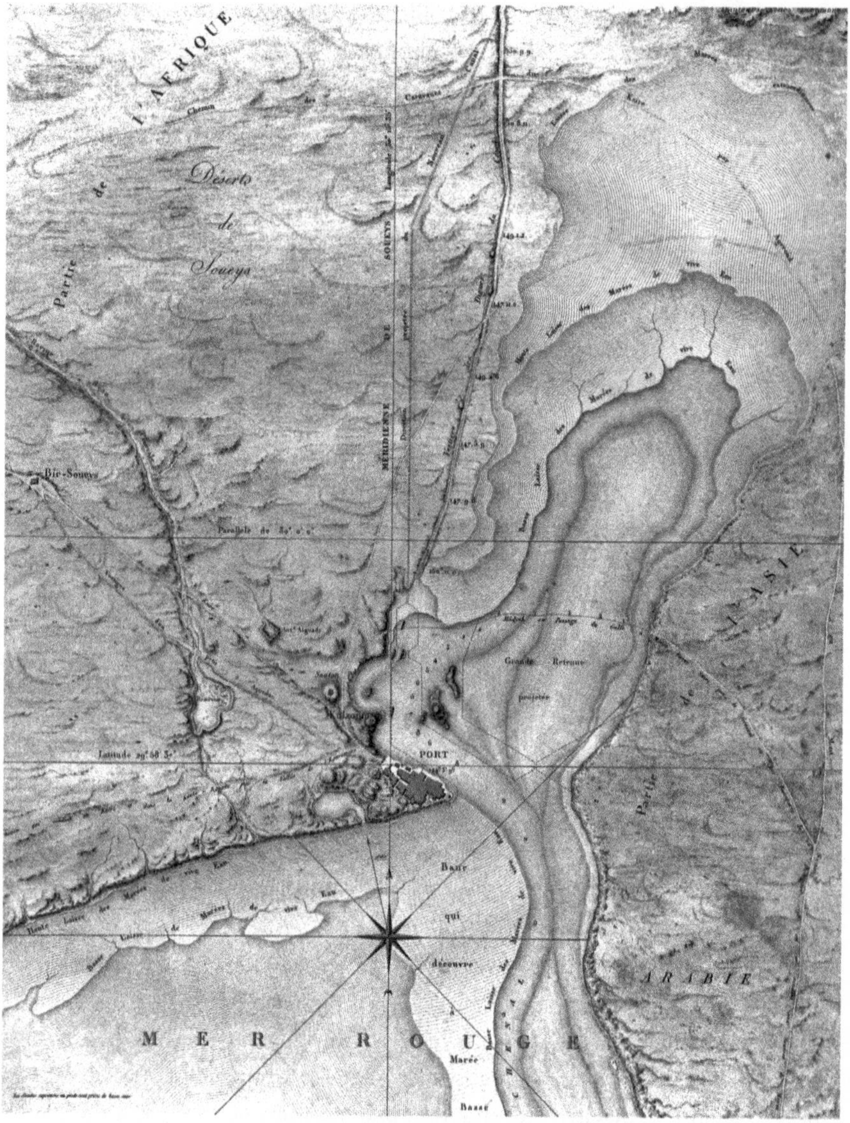

Figure 2.6 EARLY NINETEENTH-CENTURY SUEZ. Commission des sciences et arts d'Egypte, *État moderne*, vol. 1, pt. 2 of *Description de l'Égypte*, Isthme de Soueys, pl. 11. Beinecke Rare Book and Manuscript Library, Yale University.

important British rail consultant hired by Mehmet ʿAli to study the feasibility of a rail link between Cairo and Suez, Egypt was "the direct highway between Europe and India."[148] Rail would "enable quicker returns," "afford sales at lower rates," increase "the consumption of articles," and generally extend "the advantages of trade and civilization to a larger number of our fellow-creatures."[149]

Post was also of paramount interest in discussions about rail construction in Egypt.[150] In the 1840s, mail arrived to Britain from India, via Egypt and France, only four times a year. With a rail connection between the Red Sea and Nile—again, these eighty miles were one of the world's most crucial overland chokepoints—it was hoped that communications between Britain and the Mediterranean on the one hand, and Britain and India and China on the other, would be vastly improved.[151] All of this was a modern British imperial twist on the very old reality of Egypt's role as the middleman between the Indian Ocean and Mediterranean.[152] It nevertheless reflected nineteenth-century ideas undergirding European colonial expansion and the perceived "interests" of Europe in "the civilization of the East."[153]

In discussions about the feasibility of rail and whether or not it was to be preferred to the construction of a canal either between the Mediterranean and Red Seas or between Cairo and Suez, animals represented the very antithesis of the speed, efficiency, capability, and comfort of modern technologies.[154] Animals' transport capacities were disparaged as remnants of a bygone era. Just as humans came to be preferred as the most efficient and effective means of cultivating food or dredging a canal, railcars and train tracks came to be seen as more desirable than camels, donkeys, and horses for the transportation of goods and people.[155] Like the move from animal to human labor, the move from animal muscle (both human and nonhuman) to steam power represented a fundamental reworking of the energy regime of Ottoman Egypt.[156] Two results of this new energy regime, in which machine trumped animal, were the increasing separation of the animal and human realms and a vast reduction in the perceived productive social and economic functions of animals in Egypt. Reflecting Mehmet ʿAli's government's preferences for rail, Galloway wrote:

> The animals employed in the transport, *viz.* the camel, the horse, and the donkey, have been used for ages, and their powers and habits are well known, and have long been used to their utmost extent. The high temperature of the climate must always prevent rapid travelling by animal transport. These circumstances most clearly demonstrate the great difficulty of making any material improvement as to *speed* in this mode of conveyance. Again, the *cost* to passengers is very great, as the transit company are necessarily obliged to keep a very extensive and expensive establishment of servants and cattle all the year round.[157]

The italicization of "speed" and "cost" in this quotation is indicative of the focus of the Egyptian bureaucracy in this period. Transportation and movement had to be efficient, quick, and cheap. Animals were simply too slow, too unreliable, too dangerous, too fragile, and too animated to fulfill these requirements, and they therefore had to give way to machines.[158] Galloway terms rail "the triumph of science, humanity, and intelligence."[159] His use of the term "humanity" of course evokes his belief in the power of human knowledge, spirit, intellect, and will to create and use technology. This word should also be read in a more literal sense: that which is not animal. As the philosopher Yi-Fu Tuan usefully and provocatively explicates, the problem with an animate being—whether human or animal—is that "it moves and breathes, it has a biological rhythm to which it must defer, and it has a will of its own which can be made to cower but never can be totally defeated."[160] All "civilizations," Tuan continues, have therefore in varying degrees and in very different ways attempted to reduce humans and animals to "inanimate nature—to mechanical things."[161] In nineteenth-century Egypt, nature (animal) was replaced by a mechanical thing (rail), relegated to a position of irrelevance, and indeed—following from Tuan—consigned to the status of uncivilization. Unable to turn animals into usable inanimate things, the Egyptian state shunted them aside for machines.[162]

The emergence of rail in Egypt usefully elaborates the specific mechanisms that made animal labor obsolete in various sectors of the Egyptian economy. Conventionally, rail and steam have been portrayed as the crucial causes that led to a move away from animals as traditional sources of solar energy, power, and transport in rural economies. What happened in Egypt between 1780 and 1830 roughly follows the generally accepted chronology of the transition from a solar energy regime to one based on fossil fuels around 1800.[163] There was, however, a crucial intermediary step between animals and rail: humans. Indeed, for nearly fifty years, humans were the preferred source of energy in Ottoman Egypt.

Animals were thus already losing their productive economic and social functions decades before discussions in the 1830s and 1840s about the importation of rail and steam to Egypt. Had rail not come to Egypt in the middle of the nineteenth century, animals would likely still have remained outside of many sectors of the rural economy, given their already diminished economic and social status by the early years of the century. While the general notion of an energy regime shift from solar energy to fossil fuels still holds, the decisive factor that transitioned the Egyptian economy away from being one run on animals was not the importation of the new technology of steam power but rather the emergence of new political elites in Egypt at the end of the eighteenth century. These elites used human labor—first by necessity and then by choice—to build their large estates and bases of wealth. Their actions, in conjunction with epizootics and other environmental forces, were the primary factors in the large-scale removal of domestic animals from the Egyptian economy, a phenomenon that fundamentally reshaped Egypt's energy landscape. Rail was simply a later—not a causal—step in a process already set in motion.

As the chapters in Part I show, animals were central to the massive economic, social, and energetic transformations Egypt experienced between 1780 and 1830. These nonhumans' histories help to explain the formation of large landed estates at the end of the eighteenth century, shifts in the character and makeup of rural labor, mechanization, the emergence of new technologies of rule, and the early nineteenth-century administrative restructuring of Egypt. Because domesticated animals were so enmeshed in Egypt's rural economy, society, and energy regime, the colossal reduction in their numbers at the end of the eighteenth century led to changes in all parts of rural life. The waning of animals' centrality in the rural economy was thus one of the most crucial factors in Ottoman Egypt's transition from an early modern agrarian subsistence economy to one based on commercial agriculture, large landholdings, and intensive human labor. Understanding the changing political economy of animals thus helps to explain the turn of the nineteenth century as whole—a period that saw the first years of a century and a half of rule by Mehmet ʿAli and his family and that witnessed the unleashing of the immense force and power of the nineteenth-century Egyptian bureaucracy.

What happened in rural Ottoman Egypt as the eighteenth century gave way to the nineteenth also occurred in other early modern agrarian economies between the sixteenth and the nineteenth centuries. From Qing China to southern England, as land became concentrated in fewer and fewer hands, rural economies moved from subsistence to commercial agriculture.[164] One of the factors that made this global transition possible was a reconstitution of the energy regime of rural labor. Animal power became increasingly marginal as massive numbers of human laborers took over as the central motors of rural economies. All forms of caloric power would, furthermore, continue to lose out over the following decades as rail and steam superseded both animals and humans in various sectors of the economy. The extinguishment of animal energy fundamentally reconfigured Ottoman Egypt.[165]

PART TWO

BARK AND BITE

3

In-Between

> In the lands of the infidel Franks, the so-called Europeans, every dog has an owner. These poor animals are paraded on the streets with chains around their necks, they're fettered like the most miserable of slaves and dragged around in isolation. These Franks force the poor beasts into their homes and even into their beds. Dogs aren't permitted to walk with one another, let alone sniff and frolic together. In that despicable state, in chains, they can do nothing but gaze forlornly at each other from a distance when they pass on the street. Dogs who roam the streets of Istanbul freely in packs and communities, the way we do, dogs who threaten people if necessary, who can curl up in a warm corner or stretch out in the shade and sleep peacefully, and who can shit wherever they want and bite whomever they want, such dogs are beyond the infidels' conception.
>
> —Orhan Pamuk

When humans have written about other animals, they have more often than not written about dogs.[1] Dog domestication and human history have indeed marched in lockstep, fusing together the fates of both species.[2] Biologists, historians, anthropologists, ethologists, and others have long argued about whether humans made dogs or dogs made humans. Yes, we domesticated them for hunting, protection, and companionship, but they have also clearly managed to control us to their benefit by making us feed, support, and provide for them. "So who then," as Michael Pollan has usefully asked with regard to certain plants, "is really domesticating whom?"[3] Humans can survive without dogs, but domesticated dogs cannot survive without humans. As the history of bulldog breeding shows, many domesticated dogs bred for characteristics such as loyalty, cuteness, pedigree, size, coat color, and playfulness literally cannot be brought into this world without humans (caesarean rates for bulldogs are over 80 percent).[4] Given this imbalance, it is clear that dogs have succeeded in forcing us to keep them around. Rousseau had a dog named (interestingly enough for us) Sultan.[5] In his last days, Hitler purportedly only trusted his dog Blondie, who—along with Eva Braun—was killed at the Führer's side as the Russians entered Berlin.[6]

Ronald Reagan loved his Lucky and Rex. Toto, Lassie, and Rin Tin Tin are all household names in the United States and in many other countries as well.[7] The examples of humans and dogs together are nearly infinite. Why all this attention to canines?

This chapter argues that dogs' material and symbolic existence in-between the worlds of the human and the animal is what has kept them tied to humans for so many millennia. When so many humans have looked into so many dogs' eyes, they swear they have seen the human. Yet when those very same dogs kill, snarl, and shit, the very same humans are quickly reminded of the animal in the dog.[8] Humans have been drawn to dogs precisely because they usefully call into question the supposed dividing line between human and animal, wild and domestic, nature and culture, civilization and instinct, kingdom and dogdom. Dogs are unique crossers of all sorts of boundaries that illuminate just what those boundaries are, how they came to be, why some are so invested in maintaining them, and in which directions they might be moving.

The history of dogs in the Muslim world has generally been misunderstood. Contrary to the conventional thinking that they were always considered ritually impure in Islam and, therefore, that Muslims and dogs have been forever locked in an antagonistic relationship, the actual historical record of Muslims' writings about and interactions with dogs is on the whole much more positive.[9] Dogs were more often than not seen as productive species in human communities.[10] They herded livestock, were commonly involved in hunts, provided their masters with companionship and protection, exemplified ideals of loyalty and trustworthiness, fought in wars, and were living consumers of urban and rural waste. Dogs in Ottoman Egypt, as elsewhere, were therefore economically, socially, culturally, and ecologically productive historical actors.[11] Ottoman officials indeed actively encouraged the increase and maintenance of dog populations by providing them with food and water and by punishing those who committed violence against them.

The history of human interactions with dogs in Egypt over the last five hundred years is built upon a set of ideas about dogs in Islamic societies that dates back to the time of the Prophet. The enormous corpus of religious, legal, medical, and cultural texts that served as the backdrop for ideas and practices connected to dogs in the early modern period makes clear that Muslims and others in Islamicate societies were not wholly sure how to deal with the many dogs in their midst. For some, dogs were indeed animals to be feared because of their smelly saliva, life on the streets, or perceived cursed nature. For others, dogs were the most trusted living companions a human could have.[12] The story of dogs in Egypt in the sixteenth, seventeenth, and eighteenth centuries continued much of the tradition of these mixed ideas about the functional in-betweenness of dogs in human society.

Against the classical literature about dogs, the actual historical record of the animal in Ottoman Egypt brings to life an even more positive sketch of the human-dog relationship. Like livestock, dogs were everywhere in Ottoman Egypt and ultimately had to be dealt with in culturally, economically, politically, and socially serious and productive ways. For the first few centuries of Ottoman rule in Egypt, human-dog interactions were mutually constructive. Dogs protected, provided companionship, participated in military campaigns, ate trash, and helped in the hunt. In return, humans provided food and water for them, prohibited killing them, and wrote a great deal about them.

At the turn of the nineteenth century, this relationship—like so much else in Ottoman Egypt—drastically changed. Dogs in the early nineteenth century were no longer considered productive and constructive members of society. They came to be seen primarily as noise polluters, competitors for urban space, potential disease vectors, and useless sources of filth. They were therefore deemed economically negligible and ultimately culturally ignorable. For these and other economic, demographic, and epidemiological reasons, dogs became increasingly expendable in the early nineteenth century and eventually became targets of eradication campaigns. Their stories reveal how human relationships with canines changed more over the fifty years between 1780 and 1830 than they had for millennia before that. Their stories also help to explain changing notions of disease and health at the turn of the nineteenth century, reforms in urban sanitation and politics, and the growing power of the Ottoman-Egyptian state to exact interspecies violence. The history of changes in human-dog relations in Ottoman Egypt is thus part and parcel of the history of the enormous transformations that forged Egypt at the turn of the nineteenth century.

The Prophet's Puppies

The constructive relationship of mutual reliance and productivity between dogs and humans in the early modern period was the product of centuries of ideas about dogs in Muslim societies. Much of the classical Islamic religious and legal literature about dogs turns on the question of the animal's saliva.[13] The origins of this debate can be traced to a Prophetic report (*ḥadīth*) citing the need to wash a container several times—even cleaning it with a shower of dust, according to one variant of this report—after a dog's licking of the vessel. The various versions of this *ḥadīth* instruct that the container be washed different numbers of times (one, three, five, or seven); sometimes the directive to use dust is included and sometimes it is not. Another report states that if a dog touches one's garment, the garment is to be rubbed forcefully if dry and washed with clean water if the dog left it moist.[14] Other *ḥadīth* claim that the Prophet instructed that any place where a dog had lain down was to be washed for fear that some of his saliva had dripped on the ground. In short, it was thought that a dog's saliva was impure and could void a Muslim's ritual purity.[15] The fixation in the Prophetic tradition on canine saliva has precedents in the Quran itself: "So his [the unbeliever's] likeness is as the likeness of a dog. If you attack it, it lolls its tongue out; if you leave it, it lolls its tongue out."[16]

Many commentaries written about dog saliva cite deference to the authoritative *ḥadīth* insisting on the washing of a vessel touched by a dog as the primary justification for the judgment that the animal and its saliva were impure. Thus dogs, like pigs, were to be taken as impure, despite any evidence to the contrary.[17] Many jurists, however, took issue with this opinion. A number of Mālikī legal scholars began from the premise that everything in nature must be considered pure, unless proven otherwise through lived experience or by the authority of a textual tradition. To argue against the stated position that dogs were impure, these scholars therefore attacked the reliability and authenticity of many of the texts about dogs. Other writers took a different tack and posited that edicts to wash a vessel licked by a dog were aimed at preventing the spread of disease among the community of believers. They argued that a container only had to be cleaned if it could be firmly established that the dog that licked it was known to be infected with a disease.[18] If there was no proof of infection, then the dog's purity had to be assumed and accepted. A different set of jurists made a distinction between rural and urban dogs. They claimed that only the latter were impure because they consumed human and other garbage. Using a similar logic, others argued that domestic dogs were pure—thanks to their human masters who fed, housed, cleaned, and cared for them—while wild or feral canines or street dogs were impure. These arguments for and against the notion of the impurity of dogs, and particularly of their saliva, not only reflect the depth and complexity of religious, legal, and cultural thinking on the subject but also show that dogs were constant members of human society demanding engagement and cultural understanding.

A similar but less-discussed issue related to dogs and ritual purity was whether or not the mere sight of the animal during prayer would nullify that prayer.[19] This idea was part of a larger tradition claiming that donkeys, pigs, and other animals—as well as women and sometimes non-Muslims—would all void the prayer of a pure (male) Muslim if they passed in front of him as he prayed. As with the issue of canine saliva, this idea also had its skeptics. Some *ḥadīth* unequivocally claimed that the Prophet himself prayed while dogs played near him.[20] In general, it seems that the Prophet and his companions had a fairly positive view of dogs.[21] Indeed, no less an authority than the Prophet's wife Aisha threw the notion that canines and women were nullifiers of prayer into doubt when she argued that the association of dogs with women was demeaning to

Figure 3.1 Woman's Best Friend in Ottoman Egypt. Commission des sciences et arts d'Egypte, *État moderne*, vol. 1, pt. 2 of *Description de l'Égypte*, Basse Égypte, pl. 77. Beinecke Rare Book and Manuscript Library, Yale University.

the latter and had no basis in the teachings and actions of the Prophet. Given that this necessary constitutive condition of the *ḥadīth* (the connection between animals and women) was in question, jurists ruled that the entire report was inauthentic, thus invalidating the idea that the sight of women or of dogs, pigs, and other animals voided prayer. The equivalence drawn between these various "problematic" classes of creation constitutes what the modern legal scholar Khaled Abou El Fadl calls "a symbolic nexus between marginalized elements of society."[22] He explains that the ambivalence of early Islamic legal scholars on the question of dogs was a product of the fact that "discourses on dogs played a symbolic role in the attempts of pre-modern societies to explore the boundaries that differentiated human beings from animals. In that sense, the debates about dogs acted as a forum for negotiating not just the nature of dogs but also the nature of human beings." Seen through this wider lens, both the regularity and wide divergence of opinions about dogs stand as elements in a long debate within and without Islamic thought over the nature of humanity and God's creation. For these Muslim writers, dogs were a vehicle for part of this argumentation, rather than a specific end in and of themselves.

Thus, very little in the authoritative classical literature of Islam suggests that dogs were explicitly impure, unwanted, or dangerous. To the contrary, there is quite a bit of evidence advocating a much more open relationship with the animal. In addition to *ḥadīth* about the Prophet himself praying in the presence of dogs, other reports relate that the Prophet's younger cousins and some of his companions owned and raised puppies.[23] Dogs were known to roam freely around Medina, and some were even reported inside the Prophet's mosque. A prostitute—in some versions of the story it is an elderly woman or a sinning man—secured her place in heaven by giving water to a dog dying of thirst in the desert.[24] The protective dog of the Men of the Cave that "stretches out its paws on the threshold" of their abode was likewise assured a place in paradise.[25] Only the Quranic (not the Christian) version of this story includes a dog. There is even some debate in Islamic scholarship as to the dog's name, a clear indication of his importance.[26] Whatever his name, he had the ability to speak and was purported by some to be the spiritual leader of the Men of the Cave and by others to be the reincarnation of a human.[27] Another long tradition in Islamic scholarship identifies the canine endurance of wounds as a desirable attribute which military and political leaders should seek to emulate.[28]

That *ḥadīth* and other texts in the Islamic legal and religious corpus mention dogs is not at all surprising. The animal was everywhere in the world in which these thinkers were writing. Perhaps what is surprising to some is that there is no definitive answer to the question of the purity or impurity of dogs and, moreover, that there are indeed many statements expressing explicit support for the animal by and for Muslims.[29]

Toward a Dogma

These early Prophetic reports and other religious texts suggesting a mixed view of dogs in Islamic society were superseded by the writings of the famous ninth-century theologian and scholar al-Jāḥiẓ. Along with several later writers, al-Jāḥiẓ in many ways settled the debate over whether dogs should be identified with their meritorious or malignant qualities.[30] Dogs were useful and important. Although a great many of al-Jāḥiẓ's works describe the characteristics of different animals, address numerous aspects of animal life, and sketch various facets of the human-animal relationship, his most sustained study of these topics is the aptly titled *The Book of Animals* (*Kitāb al-Ḥayawān*).[31] This mid-ninth-century book includes a massive compendium of Quranic references, *ḥadīth*, and other religious writings about animals; citations to Greek scientific texts, especially the works of Aristotle and Galen; observations of Bedouin tribes near al-Jāḥiẓ's homes in Basra and Baghdad; discussions of relevant classical Arabic prose and verse; and engagement with other literary, scientific, religious, and cultural sources.[32] Both its form and content make it an important precursor to similar later texts by the likes of Qazvīnī and al-Damīrī.[33] al-Jāḥiẓ's book is best conceived of as both a work of Aristotelian zoology in which he is chiefly concerned with the physical forms, behavioral characteristics, and personalities of animals, and as a theological treatise striving to evidence the perfection of God's creation.

Figure 3.2 Dog from Ottoman Translation of Qazvīnī's *ʿAcaʾib ül-Mahlukat*. Walters Art Museum, Ms. W.659, 113a. Used by permission of Images for Academic Publishing.

Among walking animals—al-Jāḥiẓ divides creatures into four classes, those who walk, fly, swim, and crawl—dogs figure quite prominently.

In a fictitious debate in *The Book of Animals* between a supporter of the dog and a supporter of the cock, al-Jāḥiẓ makes a strong case for why dogs were especially praiseworthy and superior to other animals. For al-Jāḥiẓ, the great intelligence, compassion, and skill dogs possessed proved the wisdom and design of God's creation. As animals that shared much of the inner nature of humans, dogs held particular significance for understanding the natural world and humans' roles within it. In addition to their proximity to man in temperament and their abilities to reason and feel, dogs also served society as guardians of flocks, companions for the lonely, and aids in hunting.[34] Because of its detailed accounting of various aspects of dogs, al-Jāḥiẓ's work is also significant as one of the earliest texts to give a sense of the different dog varieties that lived in his part of the Middle East (primarily Iraq, Greater Syria, and the northern Arabian Peninsula).[35] Among the most common dogs of his day were the greyhound (*al-salūqī*), Kurdish sheepdog (*al-kurdī*), Pekinese (*al-ṣīnī*), and basset sheepdog (*al-zīnī*).

One particularly instructive anecdote from al-Jāḥiẓ's work about both the closeness of dog and human and God's divine order was that of a dog that suckled a human infant.[36] Plague had ravaged a certain household, killing all of its members and leaving their home empty. A few months passed and one of the family's heirs went to retrieve something from the house. When he entered the home's courtyard he was startled to find an infant child playing with some young puppies. Surprised, afraid, and intrigued, the man continued watching the human child until the mother of the puppies appeared. The human baby then went over to the bitch and started suckling her dugs, which she freely gave to him.[37] al-Jāḥiẓ relates this story to make several points.[38] The first is a lesson about humanity's ineptitude and cruelty. How could this house have been boarded up and abandoned with a human child left inside? Are humans that cruel, careless, and unaware? The more important point for al-Jāḥiẓ is about the divine order scripted by God for his earthly creation. God endowed the bitch with the natural ability to feed the child and the child with the instinct to feed from the canine mother. He also ordained that this particular dog would be in the house to allow the child to survive with no other humans around him. "Divine direction," al-Jāḥiẓ remarks, is "everywhere in nature."[39] Finally, this story is a very clear example of the intimate, innate, imperative, instinctual, and in-between shared natures of humans and dogs.[40] Domestic dogs need humans for food, shelter, and some amount of protection, and humans—as this story clearly illustrates—in turn need dogs.

Written in the century following al-Jāḥiẓ's death, a text known as *The Superiority of Dogs over Many of Those who Wear Clothes* by the Iraqi writer Ibn al-Marzubān is the longest Arabic treatise devoted entirely to dogs and their virtues for human societies.[41] Ibn al-Marzubān's text, like al-Jāḥiẓ's, represents a sort of compendium of stories and verse about dogs from various cultural traditions—Persian, Indian, Greek, Arab—and therefore serves as a snapshot of the thinking about dogs in the tenth-century Muslim world. The dog and the human exist in this text as just two parts of a much larger spectrum of living things. Like all other creatures, both think, feel, act, and choose. This text's fluid interspeciality and seamless slippage between the human and the animal comes through in Ibn al-Marzubān's identification of two categories of dog—"the dog of humans (*kalb al-nās*)" and "the dog of dogs (*kalb al-kilāb*)," the former being much more dangerous than the latter.[42] With this formulation, Ibn al-Marzubān means to say that both dogs and humans could domesticate and hence be masters of dogs. The human-possessed creatures, no doubt because of their interactions with people, were much more vile, violent, and aggressive than those dogs who remained exclusively among their canine kin. In Ibn al-Marzubān's words, "A dog of dogs does no harm to those around him; but you are constantly tortured by a dog of humans."[43] This and Ibn al-Marzubān's many other vignettes about the human-dog relationship are a clear statement that he considered dogs more trustworthy, useful, loyal, and loving than humans or any other sentient being.

As an illustration of canine superiority, consider the story related by Ibn al-Marzubān of a certain unnamed king who enjoyed hunting and traveling with a dog he himself had reared.[44] The dog never left his king's side, and the king always shared his lunch and dinner with the animal. This dog even had a name, something of a rarity in the human-dog relationship as it appears in the Islamic literature of this period.[45] During a particular hunting trip, the king ordered his cook to make a milky rice pudding (*tharīda*) for the evening meal at the end of a day's hunt. The cook prepared the pudding and then left it uncovered as he began to make something else. Unbeknownst to the chef, a viper entered the cooking area through a crack in the wall, ate a small portion of the pudding, and spat poison into the dish. Two living beings saw what the snake had done to the pudding: the king's trusted dog and a mute elderly servant woman also in the king's company.[46] Upon his return from the hunt, the hungry king promptly asked for his pudding. As he sat down to the table, the old woman made a gesture to warn the king about the poison in his meal.[47] She was ignored. The king then grabbed his spoon and his dog began to bark. Assuming that his faithful companion was hungry as well, the king threw some of his dinner to the animal, but the dog avoided it entirely and continued to howl. Thinking that something more serious was wrong with his dog, the king ordered his men to take the animal away so he could finish his dinner in peace. The king then dipped his spoon into the pudding. Seeing that his master was about to consume the poisoned pudding, the dog wiggled free from the guards, leapt onto the table, overturned the king's dish, and—in a sacrificial show of loyalty—lapped up the spilled pudding. Initially angered by what his dog had done, the king soon noticed the animal becoming weak. In a matter of moments, the dog fell dead with his skin peeling away and his flesh disintegrating. Realizing what had just happened, the king thanked his best friend, praised the animal's loyalty and sacrifice, buried the dog between his own father and mother, and built an ornate mausoleum over the animal's grave.

In juxtaposing the mute old woman with the loyal dog, this story creates a kind of hierarchy of being. The only two creatures who see the viper poison the pudding and who are therefore responsible for saving the king's life are the dog and the physically impaired elderly woman. Despite both creatures being unable to speak—a human subaltern who literally cannot speak and a canine that can only bark—the story makes clear that the dog was obviously a much more constructive member of this community than the woman, a human endowed with social weakness by both her sex and physical compromise.[48] Ibn al-Marzubān's message is clear: the human is weak and incapable of saving the king; the dog is strong and not only able but also willing to sacrifice his own life to save the king's.

Of course, the woman is not the only inadequate human in this story. The foolish cook who forgot to cover the pudding, the careless architect or groundskeeper who allowed for a crack to develop in the kitchen wall, and the king's many other servants all fail to save the life of their sovereign. Of all the beings in this story—human, canine, and viper—it is thus the dog that emerges as the most heroic of God's creatures, his sacrifice securing for him a place between the king's parents and, implicitly, in heaven for all eternity.[49] Unlike earlier accounts from the Prophetic tradition that questioned whether or not dogs prevented humans from fulfilling certain religious or social obligations, this story clearly shows a dog enabling the continuation of human political and social functions, as embodied in the person of the sovereign.

Several other stories in Ibn al-Marzubān's text also pit dog against snake. The two creatures indeed represent opposite ends of the animal spectrum in Islamic literature—one a loyal, trustworthy, productive member of the society of the living; the other a slithering, venomous killer of the innocent.[50] Stories of dogs, humans, and snakes help to highlight the canine's privileged position as a creature existing somewhere between the human and animal realms. Their in-betweenness—at once recognizable and familiar, yet animal and other—is part of their appeal for humans. That said, juxtaposed to the snake, the dog is obviously much closer to the human than to this debased member of the animal world. Instinctually perceptive enough to sense the dangers posed by snakes, dogs in Ibn al-Marzubān's text are forever vigilant and loyal and always at the ready to protect their human masters—even at times becoming martyrs to save them.[51] Above all, Ibn al-Marzubān's dogs understand their productive social roles as protectors and companions. Even when his master's judgment is impaired by alcohol, for example, the dog is ever present to watch over the human and prevent the snake from taking advantage of his master's inebriation.[52]

The quintessential tale of human, dog, and snake is the one that ends Ibn al-Marzubān's text. It is a story common to many traditions and seems to have first appeared in the sixth-century BCE Sanskrit collection known as the *Pañchatantra*.[53] In the sixth century CE, a Sassanian prince had the work translated into Pahlavi. After the Arab conquest of Iran in 652, the work was translated into Arabic and included in perhaps the most famous collection of allegorical tales about animals in the Muslim world, *Kalīla wa Dimna*.[54] From the Arabic, a Greek translation was completed in the late eleventh century and a Hebrew translation in the middle of the thirteenth century. Between 1263 and 1278, the Jewish convert to Christianity John of Capua translated the Hebrew text into Latin, giving it the telling title *Directorium Humanae Vitae*. This translation was, however, not the story's first entrance into Latin Christendom, as it had also existed in French peasant legend for some time before the thirteenth century.[55]

In Ibn al-Marzubān's version of the story, a male widower left both his young son and dog home one day as he went out on an errand.[56] He returned home a few hours later to find his dog waiting outside the front door with blood dripping from his muzzle. Assuming the worst, the man killed the dog in a fit of rage before entering his home to collect his son's body. When he approached the child's cradle, he was startled to find his son safe and sound asleep. Next to the cradle lay the chewed-up remains of an enormous viper. Realizing that his dog had saved his son's life and that he had wrongfully killed the animal, the man was filled with remorse and gave his dog a proper burial.[57]

Here again the dog appears as martyr. Wrongfully accused and killed, the dog is vindicated only in death and recognized as a hero with an honorable burial. Unlike the tale of the king and the poisoned pudding, undergirding this story of dog and viper is a lingering anxiety about the wild nature inherent in the animal. Although the dog was presumably domesticated, the assumption is that the animal—when left to its own devices—will succumb to its wild instincts and kill the defenseless child. It is precisely the perceived wild nature of the dog-turned-beast—the story is purposefully explicit about the fact that blood dripped from his muzzle—that necessitates that the creature be killed to protect human society. The line between the wild and the civilized, between wolf and trusted domesticated dog is all too thin. That the dog is not wild is, of course, the eventual irony and crucial moral lesson of the story. In the end, the dog is found to be the protector of the domestic human space from the wild savagery embodied by the viper. The multiple dualities imbedded in the now-dead body of the dog—wild and domestic, animal and human, ferocious and protective, worldly and divine—instantly and jarringly flip as the man enters his home to discover the mistake and truth of his actions. Ultimately, the dog is vindicated as an integral member of human society, in both life and death.

Here Boy

From the time of the Prophet until the tenth century, writings about dogs in the Muslim world moved from a focus on debates over the impurity and character of the dog toward a more confident stance about the animal as a constructive, productive, and integral component of human society that fulfilled necessary social, familial, and political roles.[58] This earlier tradition was crucial in informing engagements between humans and dogs in early modern Ottoman Egypt. Two different source bases bring to life the significance of dogs as important members of Egyptian society until the end of the eighteenth century: religious, allegorical, and legal treatments of dogs, and the social historical record of humans' interactions with canines.

One of the most sustained defenses of dogs in Ottoman Egypt was penned by a religious scholar writing in Cairo between the middle of the sixteenth and the middle of the seventeenth century. Nūr al-Dīn Abū al-Irshād ʿAlī ibn Muḥammad Zayn al-ʿAbidīn ibn ʿAbd al-Raḥmān al-Ajhūrī was born in the village of al-Ajhūr in the subprovince of al-Qalyūbiyya in 1560.[59] He was a Mālikī jurisprudent (*faqīh*) who spent most of his intellectual career at al-Azhar in Cairo writing on various legal topics, including the permissibility of coffee-drinking and smoking.[60] He was blinded very late in life by an angry student who hit him over the head with an extremely heavy book.[61] He died in

1656. His opinions and writings on the general acceptability and ritual cleanliness of the dog (*ṭahārat al-kalb*) survive in a text composed as a hypothetical debate between different schools of religious law.[62] al-Ajhūrī represented his own Mālikī tradition against the positions of an unnamed, and possibly imaginary, Shāfiʿī scholar. The debate consisted of eight points of contention. In each of these eight sections, al-Ajhūrī first presented a reason why dogs were clean and productive beings in society; his opinion was then countered by the Shāfiʿī scholar defending the position of the ritual impurity (*najāsa*) of dogs; and al-Ajhūrī then responded to these critiques.[63] Portions of the debate revisit some of the classical disputes over the cleanliness of the dog—the issue of saliva being the most central. In stark contrast to earlier debates, however, the position advocating the impurity of dogs was no longer tenable. Thus by the seventeenth century, as al-Ajhūrī demonstrates, an evolution had occurred in the thinking about human-dog relations in Egypt, making the impurity of the animal a moot point.

For example, al-Ajhūrī writes that dogs entered the mosque of the Prophet in Medina and were cared for by those in attendance.[64] Even though the animals rested their muzzles on the carpeted floor of the mosque, leaving some saliva on the sacred ground, they were not banished from the hallowed space, and there was no indication that these areas had to be washed or cleaned in any special way. Indeed, the Prophet himself, al-Ajhūrī writes, allowed dogs to remain in the mosque and was seemingly not bothered by their presence or saliva.[65] al-Ajhūrī's interlocutor offers up the rather weak retort that perhaps the dogs in question had dry mouths and therefore left no moisture on the mosque floor.[66] al-Ajhūrī is quick to respond that this is highly unlikely given the propensity of dogs to pant and slobber.[67] More importantly, al-Ajhūrī adds, because there is no specific mention of the dogs' wet mouths, it must be assumed that their saliva was accepted as an unproblematic part of their recognized nature and was hence not considered impure.[68] The feeble critique offered up by al-Ajhūrī's Shāfiʿī debater and the Mālikī's strong response suggest both the tiredness of these arguments against the dog by the early modern period and the ascendency of the position supporting the purity of the animal and its saliva.

Putting to rest any lingering doubts on the subject, al-Ajhūrī writes that there is no danger of impurity in the human consumption of animals collected in the mouths of dogs during a hunt, given "the ritual cleanliness of the dog's saliva (*ṭahārat rīqihi*)."[69] This idea taps into an older one that even some Shāfiʿī scholars accepted, namely that even if dogs are ritually impure, God made their saliva clean as a special dispensation (*rukhṣa*) to human hunters who needed dogs for their own sustenance.[70] In a final example of the purity of the dog, al-Ajhūrī cites a *ḥadīth* about a man who one day came upon a desperately parched canine eating moist earth in an attempt to squeeze some water from the dirt.[71] Taking pity on the animal, the man took off his shoe and used it to ladle water from a nearby well into the dog's mouth until he was no longer thirsty. In return for this good deed, God is said to have granted this man entrance into paradise. The Shāfiʿī responded to this *ḥadīth* by claiming that the man likely first poured water from his shoe into a vessel and only then gave the water to the dog *from the vessel*, thereby protecting and preserving the purity of his shoe from the dog's saliva.[72] al-Ajhūrī responds to this challenge by making the very simple point that if such a vessel had been available, the man would surely have filled it with water from the well and given it directly to the dog rather than use his shoe as an intermediary container.[73] It had to be assumed, in other words, first that the man's shoe was the only remotely suitable container available, and second—and most significantly—that there was no legal objection to the dog's saliva touching the man's shoe and presumably then his foot.

During his youth in the countryside and later as an adult in Cairo, al-Ajhūrī was likely quite accustomed to seeing dogs all around him. Although he approached his writings on the animal through his training and expertise in religious law, his opinions about dogs both derived from and were constitutive of a very intimate relationship between Egyptians and dogs in the early modern period. Canines were everywhere in this most lucrative of Ottoman provinces, and their many

Figure 3.3 Dog Existing Happily and Unproblematically in the Social Fabric of Ottoman Cairo. Commission des sciences et arts d'Egypte, *État moderne*, vol. 2, pt. 2 of *Description de l'Égypte*, Arts et métiers, pl. 16. Beinecke Rare Book and Manuscript Library, Yale University.

social roles thus deeply affected the cultural, political, environmental, economic, and—as al-Ajhūrī shows—legal histories of Ottoman Egypt.[74] In addition to religious and literary texts, historical chronicles and other narrative and archival sources further elucidate the essential and productive social and economic roles of dogs in human societies.

By all accounts, Cairo in the sixteenth, seventeenth, and eighteenth centuries was a city full of dogs. Antonius Gonzales, a Franciscan Recollect from the southern Netherlands who served as chaplain to the French consul in Cairo in 1665 and 1666, wrote that the city had innumerable dogs.[75] Literally every street in Cairo was home to large groups of ownerless dogs who lived by eating rubbish and thereby helped keep the city clean.[76] There were so many dogs that it was not uncommon to see packs of twenty or thirty of them following people as they walked.[77] From the time one opened one's doors in the morning until closing them at night, it was a constant battle to prevent these street dogs from getting inside one's home. These large numbers of dogs were not simply the accident of a bustling urban setting; rather, they were purposely maintained and carefully cultivated by Ottoman authorities to keep streets clean of refuse.[78] Killing these animals was illegal. Any person found guilty of violence against a dog or of killing the animal (or a cat) was strongly punished.[79] This aggressive punishment of those who would harm dogs was likened to the censure of those who used force against the elderly or physically impaired—another parallel between dogs and those humans judged socially or physically weak.[80]

Figure 3.4 Dog Being Watered in Ottoman Cairo. Commission des sciences et arts d'Egypte, *État moderne*, vol. 1, pt. 2 of *Description de l'Égypte*, Le Kaire, Citadelle, pl. 71. Beinecke Rare Book and Manuscript Library, Yale University.

Figure 3.5 Dogs Playing in Ottoman Cairo. Commission des sciences et arts d'Egypte, *État moderne*, vol. 1, pt. 2 of *Description de l'Égypte*, Le Kaire, pl. 39. Beinecke Rare Book and Manuscript Library, Yale University.

Various institutions in Cairo existed to serve, protect, and maintain the city's canine population.[81] Feeding bins and watering troughs were placed throughout the city to provide sustenance for these animals.[82] Dogs were also regularly fed in mosques, many of which maintained large stone water basins at their entrances for dogs, mules, and other animals.[83] Butchers, fishmongers, and various kinds of shop owners put dogs to good use as cleaning agents, guards, and helpers of various sorts.[84] Dogs also kept unwanted vermin like rats, hares, and wild pigs out of the city.[85]

Dogs were, in short, integral actors in the urban fabric of Ottoman Cairo. They served many useful social and economic functions, and their vast numbers were noted by all who came to the city. The daily constructive interactions of Cairenes with dogs meant that the religious, historical, and literary writings of this period were heavily infused with often quite positive takes on the animal.[86] A work written at the end of the seventeenth century exemplifies how certain features of the traditional literature concerning dogs were usefully recast as these ideas were filtered through Egyptians' experiences and understandings of dogs and humans in the early modern period.

Unlike earlier stories—the man watering the thirsty dog with his shoe, for example—in a late seventeenth-century anthrozoological reversal, it was the dog that gave sustenance to a poor and hungry man. A formerly rich man had been overcome by debt and was left penniless.[87] Hoping to regain some of his lost riches, he left his family and set off to seek a new fortune. He eventually arrived on the outskirts of a town that seemed to have many wealthy merchants and estates. The man sat for a moment to rest before continuing on into the town. Another man soon came by with a group of four hunting dogs dressed in ornate silks and brocades with gold collars and silver chains around their necks. He tied his dogs and left to get them food, soon returning with a golden dish of sumptuous fare for each. The dog owner then left again to allow his animals to enjoy their meals. The poor and increasingly famished man hungrily eyed the dogs' meals, but his remaining pride prevented him from making a move toward their food.

One of the four dogs, however, recognized the man's abject hunger and motioned toward him as if to say, according to the account, "Come and take some of this food."[88] The man hesitantly approached, and the dog freely gave of his food. The man eagerly ate until he was satiated. He then rose to take his leave, but the dog, in another in-between move, again motioned to the man indicating that he should take with him the remainder of the food and even the gold dish if he liked. Careful to make sure no human eyes saw him, the man put the dish in his sleeve and left the town for a different one where he promptly sold the gold dish for a great sum of money. This sale proved to be the turning point in the man's fortunes. He bought many goods, started a business, and made enough money to repay his debts. He soon returned to his home village, where his recent spoils allowed him and his family to ease back into their generally comfortable former existence.

After some time, the man felt compelled to return to the town to thank and repay the dog and his master for the gift of their dish. The man set off for the town. As he approached it, he saw that the entire city had been deserted and had completely deteriorated to nothing but "crumbling ruins and cawing ravens."[89] As he roamed the desolate town, he came upon a decrepit old man, who asked him why anyone would come to such a godforsaken place. The new arrival told the old man his story and that he had come back to repay those who had helped him so long ago. The old man guffawed, incredulously mocking the idea that a dog had knowingly given the man a gift. Frustrated, the traveling man left the city and returned home with the following verse on his tongue: "Gone are the men and the dogs together / So to the men and dogs alike, farewell."[90]

As in previous stories, the dog again sacrifices for the betterment and longevity of man. By giving up his food and the gold dish, the dog allowed the man to regain his economic and social station. This is lost on the old man at the story's end. Weakened by age and poverty, he was ill-equipped to grasp the generosity of the dog. The man and the dog in this story effectively communicate and even converse as if sharing the same language. The aphoristic final verse suggests an equality between human and dog, making the point that both are capable of empathy, care, and moral rectitude. The dog helps the man in the same way that masters provide for their dogs. The social world sketched in this story is one in which humans and animals engage in intimate and cooperative relationships of social reliance and sustenance—relationships in which man ultimately relies on dog, not the other way around.

Herds, Vomit, Hunt, Warfare, Affection

Dogs played numerous other productive social roles in Ottoman Egypt. One of these was acting as caretakers and protectors of flocks of sheep and herds of goats and other animals.[91] As a learned peripatetic holy man told a group of followers to prove to them the existence of dogs in heaven, "I saw that the Tenth Heaven was full of flocks of sheep and goats, and as you know flocks need dogs, which they are never without, and the shepherd has to have a dog to guard his flocks."[92] Dogs additionally played constructive and instructive medicinal roles for human communities, directing people on various issues of disease and treatment. For example, Egyptians learned through the observation of dogs which plants aided in the purging of the body.[93] Dogs were well-known gluttons and had developed a strategy of consuming certain plants to induce vomiting.[94] Medical practitioners in Egypt, from antiquity to the Ottoman period, thus learned from dogs which plants and herbs could be used to treat human stomach ailments.[95]

There is also a long history of hunting with dogs in Egypt and other parts of the Muslim world.[96] Rabbits, gazelles, hares, and other animals were some of the most common prey caught by dogs in Ottoman Egypt.[97] Dogs were also used to hunt stags in Lower Egypt in the Ottoman period.[98] The most common breed employed for both herding and hunting was the Egyptian greyhound and its close relative the *salūqī*. Depictions of greyhounds and *salūqī*s on the hunt exist from as early as the Middle Kingdom period (2134–1785 BCE).[99] They are often shown running alongside a hunter's chariot in pursuit of foxes, hyenas, onagers, and other desert creatures.[100] Sometimes they are depicted moving in packs on leashes. The Egyptian pharaoh Ramses IX (r. 1131–12 BCE) loved his hunting greyhound so much that he took the unusual step of being entombed with him. Egyptian greyhounds and *salūqī*s were so renowned for their hunting abilities that they were traded around the Mediterranean very early on, most likely during the Minoan period.[101] They are thought to have reached southern Europe in Greco-Roman times.[102]

Into the Islamic period, the *salūqī* continued to hold pride of place for hunters and herders alike. Like other writers in the classical Islamic tradition, Abu Nuwas (d. 810s) describes the use of greyhounds in nearly half of his poems about hunting.[103] The ʿAbbasid caliphs were avid hunters and regularly imported *salūqī*s from Yemen to Iraq for the task. These prized creatures were housed in structures built solely for their care, had strictly monitored diets, and received veterinary treatment for injuries and disease.[104] The Muslim empires of the early modern period greatly valued hunting dogs and participated in an active global dog trade and network of gift exchange. In the Ottoman, Safavid, and Mughal courts, hunting dogs were often found in very close proximity to the sovereign. During his visit to Iran in the 1670s, Jean Chardin noted that it was only court officials and nobles who were able to afford dogs, which they proudly celebrated and displayed as symbols of status, wealth, and prestige.[105] The Mughal court also regularly sought out hunting dogs—tellingly known as the Persian hound (*sag-i tāzī*).[106] In the early seventeenth century, Mughal emperor Jahangir asked his Safavid counterpart Shah ʿAbbas to kindly send him a group of European hunting dogs.[107] ʿAbbas sent nine mastiffs. Jahangir later also implored the British envoy Sir Thomas Roe to procure for him more mastiffs, Irish greyhounds, and "such other Dogges as hunt in your Lands."[108]

The Ottoman court too imported dogs over vast distances. Mastiffs were regularly captured in Poland, Russia, and Moldavia and then brought to Istanbul.[109] Some dogs even came from as far away as China. A merchant visiting the Ottoman Empire in the early sixteenth century wrote that "in the exalted courts of the [Muslim] rulers toward the land of the Sultan of Rūm [the Ottoman sultan], there are such dogs which the Rūmīs call the 'Sasanid dog' but [which] by origin is the Tibetan dog. And these Tibetan dogs are found in the mountains of China and it is from there that one acquires these dogs."[110] Even before the early modern period, Ottoman sultans regularly sought out and celebrated dogs. Sultan Murad I (r. 1362–89) was known to lavish ornate silver collars on his beloved and prized hunting dogs.[111] The many luxuries and precautions afforded hunting dogs in the Muslim world, their importation over vast distances, and the wealth and prestige they commanded were almost always in service of the hunt. As a poet cited by the seventeenth-century Egyptian writer al-Shirbīnī succinctly puts it, "Hunter and dogs cannot live apart!"[112]

To maintain their corps of hunting dogs, the Ottomans developed a very sophisticated and expensive regime of canine training, feeding, and exercise. The two main dogs used in the hunt were the greyhound (*tazı*) and the spaniel (*zağar*), whose keepers were known as *tazıcılar* and *zağarcılar*, respectively.[113] Divided by breed, the dogs were housed in separate quarters in Üsküdar.[114] In each of two very spacious rooms, up to sixty dogs were kept on extremely slack chains fixed to the wall. Both rooms were heated by fireplaces at each end. Sofas lined all four walls, and there were also sheepskins strewn about for the dogs to sleep and lounge on. Each dog was walked and groomed by his handlers every morning and evening.[115] The dogs were so well fed and immaculately kept that a French visitor to their lodgings in the early seventeenth century described the canines as "marvelously polished."[116]

Dogs were also regularly used for military purposes in Ottoman Egypt.[117] The most common cadre of dog soldier used by Ottoman armies throughout their early imperial conquests and by various military factions stationed in Egypt were a group of mercenary dog keepers known in Ottoman Egypt as the *saymāniyya*.[118] These soldiers were originally the keepers of dogs used as advance forces in military campaigns.[119] The animals would be sent in as a first strike team to maul enemies and scare them from their positions.[120] The *saymāniyya* were just one of many factions within the Ottoman janissary corps devoted to the upkeep and care of imperial hunting dogs.[121] Others included the *ṭurnacıbaşı*, *ṣamsuncubaşı*, and *zaġārcıbaşı*.[122] Still, the *saymāniyya* were the most prominent of these canine fighters and were regularly used to great effect. In a battle between the Egyptian notables Ghīṭās Bey and Ismāʿīl Bey ibn ʿAwwaḍ Bey in 1714, for example, each side used dogs to attack the other. Ghīṭās Bey's men finally triumphed, thanks to a combined force of sixty *saymāniyya*, muskets, cannons, and a group of hired Bedouin troops.[123]

Even Cairo's seemingly endless numbers of street dogs were used in warfare.[124] In 1711, a soldier named Muḥammad Bey employed a regiment of street dogs to aid him in attacking the ʿAzab barracks near Rumayla Square in the vicinity of the Cairo Citadel. He sent out his men to collect twenty dogs from the area around the square. These men were then instructed to tie a wick to each of the dogs' tails. The animals would be kept in a storehouse in the area until time came to spring the attack. The wicks on their tails would then be lit, and they would be released to run toward the ʿAzab barracks. Cannon and musket fire behind them would both provide cover and scare the dogs into running faster. Confused, surprised, and terrified by this onslaught of enflamed canines—the plan went—those held up in the barracks would fire on the dogs, thereby exhausting their ammunition. In the unlikely event that a dog actually reached the barracks alive, this would be all the better from the perspective of Muḥammad Bey and his attacking forces. According to al-Damurdāshī's account, in the end, this attack plan failed and the barracks were successfully defended.[125] Later that evening after the battle, a soldier walking in Rumayla found one of the canine combatants howling from the pain of his burned tail. He picked up the injured dog, took it to his barracks, and attended to the animal's wounds.

In the previous stories of dogs sacrificing for humans—the dog who kills himself by eating the king's poisoned pudding or the dog who defends the baby by killing the viper—dogs give freely of their well-being and even lives to save humans from destitution, danger, and death. A crucial aspect of these stories is that the humans recognize the canine sacrifice that has saved them only after it is too late. The feelings of guilt and remorse that result are meant to impart a moral lesson. In the story of the soldier's compassion for the burned dog, however, the animal's sacrifice and the human's remorse are of a very different order. The animal does not willfully sacrifice himself, but is indeed sacrificed by the humans. The human's remorse is for an act of violence committed knowingly, consciously, and deliberately, not for violence done in error as in the earlier stories. Thus, by the eighteenth century dogs acting compassionately, willfully, and selflessly on behalf of human society were replaced by humans exercising a monopoly over social agency and using animals as tools for selfish and worldly purposes, warfare foremost among them.

Furthermore, the soldier's show of compassion toward the dog in this final story is one of only a handful of examples pointing to an affective relationship between humans and dogs in the early modern period. It stands as an exception that indeed proves an early modern rule. As I have shown in this chapter, although there were myriad constructive and productive relationships between dogs and humans in early modern Ottoman Egypt, none of these indicate a strong or widespread affective relationship between Egyptians and dogs (and certainly not general human sensitivity or concern for animal emotion or pain). The soldier did tend to the hurt battle dog, and hunters too seem to have cherished their dogs, but these shows of affection derived not from compassion but concern and respect for the social and economic utility dogs provided for humans. While present-day human-dog interactions are principally mediated through companionate relationships of affection, love, comfort, and emotion, in the early modern period the human-dog relationship was based primarily on loyalty, security, aid, protection, productivity, the prestige derived from hunting, and the utility of the canine consumption of waste. Dogs were, in other words, useful and beneficial for human communities primarily because of their economic, social, political, military, and pharmacological attributes and abilities. Their value was not built on any emotional or affective basis.

Despite the virtual absence of an affective mode of interaction between humans and dogs in early modern Ottoman Egypt, dogs were nevertheless conceptualized as existing somewhere between the human and animal worlds. They possessed admirable and desirable human traits—loyalty, economic and social productivity, keen perception, modesty, a willingness to sacrifice—often much more so than many humans. They were thus kept close to human communities. In various realms of life—from medicine to warfare to morality—humans and dogs were engaged in cooperative and constructive—though, again, not affective—relationships throughout the Ottoman period. Dogs kept Cairo's streets clean, aided in military exercises, taught humans how to heal their own bodies, caught game for political prestige, and helped maintain supplies of sheep and goat meat. Recognizing the utility and importance of dogs to their multiple human worlds, Egyptians and Ottoman urban authorities sought to support and safeguard these animals—to provide them with food, shelter, and protection. There were even *waqfs* (pious foundations) to support and protect canine welfare in early modern Istanbul.[126] Humans who perpetrated violence against urban dogs were regularly and severely punished. Even during times of plague in Egypt, special attention was devoted to dogs and cats to ensure their safety.[127] Dogs were thus recognized and valued for their unique and indispensible contributions to many realms of early modern life and were intertwined with humans through established modes of social, economic, military, and medicinal interaction.

At the same time, dogs were still set apart from humans as wild and animalistic. They ate garbage, were ultimately instinctual, and shat and copulated on the street. No matter how intertwined they were with humans, dogs were always infinitely distant from the human realm. It was the maintenance of this in-betweenness of dogs—ever-present yet forever other, human and nonhuman—that made them so productive for early modern Egyptians. They could be used to keep the city clean *and* to mark a moral and existential difference between humans and other animals. Needless to say, this multifunctionality of dogs could only obtain in a situation of intense dog-human proximity and regular interaction like the one that existed in early modern Ottoman Egypt.

While this set of circumstances continued more or less throughout the early modern period, it would drastically and dramatically change in the early nineteenth century and is thus a crucial indicator of the enormous shifts that marked human-animal relations at the turn of the century. The social role of dogs would change more in the decades around 1800 than it had for millennia. This transformation would have monumental consequences for both humans and dogs and for Egyptian society more broadly. As the next chapter shows, new notions of public sanitation, hygiene, social organization, public health, and governance would eventually render dogs' former productive functions unnecessary, meaningless, and then obsolete. Without a constructive and productive role to play in human societies, dogs emerged primarily as sources of annoyance, disease, pollution, and danger. As a result, it would soon come to be deemed necessary to remove them from the realms of human habitation. Thus, paradoxically, as affective bonds between humans and dogs began to emerge in the nineteenth century partly to replace older social, economic, and ecological roles for dogs in human society, widespread violence against the animal also increased. In the early nineteenth century, dogs were no longer valued for loyalty, productivity, utility, or security. They were certainly no longer used to draw equivalences between the human and the animal. Dogs were indeed no longer anything but animal. Dogs were no longer in-between.

4

Evolution in the Streets

> The first reason people kept a dog was to acquire an ally on the hunt, a friend at night. Then it was to maintain an avenue to animality, as our own nearness began to recede. But as we lose our awareness of all animals, dogs are becoming a bridge to nowhere. We can only pity their fate.
>
> —Edward Hoagland

In the early decades of the nineteenth century, dogs' in-betweenness became a problem in Ottoman Egypt. Canine in-betweenness posed a direct threat to the increasingly rapacious and powerful Egyptian state bureaucracy of this period. Dogs challenged its ideas and practices of order, its strict definitions of spaces and social roles, its disciplinary control and modes of policing, and its attempts to forge a legible society and economy.[1] As the Ottoman-Egyptian state endeavored to make all of these governing ideals reality, it faced not only a canine challenge but also the pressures of Cairo's rapidly increasing population and its demands for space and work. The main strategy the state administration employed to deal with the dog challenge was to attempt to remove the animal from the city entirely. This removal was primarily accomplished by pushing Cairo's trash outside the city's walls. With this social, economic, and sanitary function taken away from dogs in the first third of the nineteenth century, they were given new roles in Egypt—disease vectors, noise polluters, sources of filth, and menaces to social order. These emergent ideas about dogs went hand in hand with new notions of disease etiologies, hygiene, urban sanitation, and governance—all of which reinforced the imperative to remove dogs from the city.[2] Some humans did try to invent other novel and more accommodating roles for dogs in Egyptian society at the time—chiefly as companionate species—but these efforts were overwhelmed by the dominant wave of recently imagined and more troublesome roles and attributes ascribed to dogs.[3]

These shifts in the human-dog relationship in the early nineteenth century had many enduring consequences. Attempts to remove dogs from Egypt's cities resulted in interspecies violence on an unprecedented scale. The radical alteration of Cairo's urban environment separated dogs from urban waste for the first time in millennia, creating a divergent evolutionary pathway for dog species in Egypt.[4] The Egyptian example therefore shows how, in the environmental historian Edmund Russell's words, "the state has been an evolutionary force."[5]

What happened to dogs in this period forever changed understandings of governance in Egypt and shaped the relationship of both the human and the nonhuman populace to the state. In the early modern period, the loyalty of the animal was to his individual master, and the master had both a monopoly of control over his animal and a set of responsibilities to that creature. In the nineteenth century, the singular master in Ottoman Egypt became the state. Loyalty, allegiance, and deference—canine and otherwise—were to be given exclusively to the state. The state, rather than families, households, or individuals, was to be the sole and final arbiter and caretaker of life, economy, and society. These understandings and practices of governance affected the lives of Cairo's street dogs before they targeted Egypt's humans. The independent and sovereign dog master or, even more problematic, the unattached street dog with seemingly no interest in any human connection, could not continue to exist under Egypt's new state regime of order, fixity, and control.[6]

Dogs' social, political, and economic fortunes were thus forever reshaped in this period. So too were their biological futures fundamentally set on a new course. The history of dogs and their many relationships at the turn of the nineteenth century reveals that nearly every aspect of interspecies interactions, from the place of dogs in Cairo to canine evolutionary biology, was altered by the epochal transformations in Egyptian society in this period. The forceful consequences of these new modes of rule in Egypt that forever changed dogs' lives and genes would also soon come to echo in the lives of humans and other creatures.

On the Prowl

Dogs were removed not just from Cairo but from rural Egypt as well. As domesticated animal populations declined at the end of the eighteenth century and human laborers came to replace them as the preferred tools of rural agricultural work in the early nineteenth century, dogs became less important as guards and night watchmen over other animals.[7] Fewer livestock in the countryside meant dogs began losing their roles as overseers of herds, historically one of their principal functions in human societies.

What replaced dogs as guardians of personal property in the Egyptian countryside were law and the police. The most important new instrument created by Mehmet ʿAli's state to order, monitor, manage, and administer the countryside was the 1830 Law of Agriculture (*Qānūn al-Filāḥa*), which unintentionally but directly dealt a severe blow to the roles of dogs in rural Egypt.[8] The fifty-five individual statutes of the law have a lot to say about domesticated animals and tellingly nothing to say about dogs.[9] The protection of livestock was completely given over to the work of bureaucrats dispatched to the countryside. These legal and policing agents were tasked with ensuring that animals were not wrongfully taken from their owners, and if they were, that the thieves were adequately punished. Instead of a dog's snarl, growl, or howl, bureaucrats, legal statutes, and administrative proceedings now protected individuals' personal property. Just as peasants replaced animals as the bulk of Egypt's rural labor regime, so too did human representatives of the state and the law replace dogs as guardians of domesticated animal populations.

For example, statute thirty-four of the 1830 law served to fulfill dogs' former function of protecting fields from the intrusion, destruction, and consumption of other animals.[10] The new law stated that if a domestic animal that ate or otherwise damaged a farmer's crops was determined, after investigation, to have escaped because of its owner's negligence or been intentionally let loose, the owner had to financially compensate the victim for the damaged goods and was also subject to fifty lashes.[11] If it was determined that the offending animal's owner was not at fault and that the animal acted of its own volition, the owner was still held responsible but only had to pay the price of the damaged crops. Whereas dogs used to defend fields from the threats of wandering forage animals, the logic of the centralizing Egyptian state in the early nineteenth century determined legal statute to be a superior and more desirable form of property and crop protection than dogs.

Another clause in the 1830 law charged the village qaimmaqam or shaykh with the responsibility of preventing animal thefts in the countryside.[12] If such a theft occurred and was prosecuted, the thief would be required to pay the animal's rent to the owner and would also be punished with twenty-five whips of the kurbaj. Here again, humans and law replaced dogs in their former role as guardians of domesticated animal herds. Dogs used to try to scare off thieves; even when they could not prevent a theft from occurring outright, they were crucial in helping to alert others. In the early nineteenth century, the responsibility for discovering and deterring thefts in rural Ottoman Egypt was transferred from dogs to village qaimmaqams. Other relevant sections of the 1830 law outlined punishments for those who killed another person's domestic animals,[13] stipulations concerning slaughter,[14] and penalties for the illegal sale of animals.[15]

Taken together, these new regulations were a key aspect of Mehmet 'Ali's attempts to institute a novel regime of order in the Egyptian countryside.[16] Part of this effort included replacing what were deemed ineffective procedures and unadministratable actors with the ordering, rationalizing, and legibilizing control of a centralized bureaucracy and legal code.[17] While perhaps not intentionally aimed at stripping dogs of their productive social and economic functions in the countryside, the 1830 agricultural law was one of the most important legal instantiations of a new rural reality that shaped, and was shaped by, larger processes affecting the human-animal relationship in early nineteenth-century Egypt. This new reality involved increasingly interventionist governing techniques, capitalist market relations, and fewer dogs.

Kill the Dogs

As in the countryside, in Cairo human police also came to replace dogs as means of security. The removal and replacement of dogs in urban security first emerged under the occupying regime of Napoleon's army during its three-year incursion.[18] On the night of November 30, 1798, French military forces undertook a campaign to rid Cairo of the many dogs that constantly harassed them during their nighttime security marches.[19] Engaging in the security function they had upheld for centuries, these dogs harangued the strange and unknown French troops who patrolled Cairo's streets, barking at them and chasing them from lanes and alleyways.[20] To rid the city of what was to them a nuisance and security risk, these soldiers walked through Cairo's streets that November night with baskets of poisoned meat that they fed to as many dogs as they could find.[21] By morning Cairo was filled with dead dogs.[22] Men were hired to remove the dog carcasses to dumps outside the city where the bodies were in all likelihood burned.[23] This massive dog eradication effort was the first documented episode of systematic and sweeping dog killing in Egypt.[24]

Almost two decades later, there was a similar instance of violence against Cairo's dogs that specifically invoked the earlier French incident. On September 10, 1817, the Egyptian pilgrimage caravan left Cairo for the Hijaz. That year there was a shortage of boats to transport pilgrims across the Red Sea from Suez to Mecca and Medina. Many of these pilgrims, some of whom had traveled great distances from North Africa or southeastern Europe, had no choice but to return to Cairo disappointed that they were unable to complete their journey. This massive influx of people led to enormous congestion in the city, which was doubly magnified by the chaos caused by the many infrastructural projects Mehmet ʿAli was undertaking in Cairo at the time. Not only was the city crowded with people, but many lanes and alleys were also clogged with raw materials, merchandise, foodstuffs, wood, and dirt from construction sites. Horses, donkeys, and camels were also jostling for space, since they were used to carry building materials and to remove stone, dirt, and rubble from construction sites.[25] The

Figure 4.1 Dogs and Other Animals in the French Encampment in Cairo. Commission des sciences et arts d'Egypte, *État moderne*, vol. 1, pt. 2 of *Description de l'Égypte*, Le Kaire, pl. 40. Beinecke Rare Book and Manuscript Library, Yale University.

city was, in other words, packed with people, animals, and goods. al-Jabartī goes on to describe this situation in 1817:

> Add to all this the packs of dogs, sometimes as many as 50 in one street, continuously barking and howling at passersby and one another, disturbing everyone especially at night and making sleep impossible. The French did well in killing these dogs. Once settled in Cairo they saw that these large packs served no need or purpose [*min ghayri ḥāja wa lā manfaʿa*] except barking and baying at them in particular since they were strangers. Therefore a party went around the city with poisoned meat and by morning all the streets were littered with dead dogs. Adults and boys dragged them by ropes into vacant lots outside the city; thus the earth and its inhabitants (*al-arḍ wa man fīhā minhā*) were rid of them.[26]

As the praise and invocation of earlier French anti-dog efforts suggest, Egyptians were more than happy to borrow the ideas and practices of large-scale dog eradication measures from the French. Like other techniques of Egyptian governance in the early nineteenth century, the state's treatment of dogs largely derived from Mehmet ʿAli's perception and desired emulation of European ideas and models of rule. His employment of European advisors and consultants to assist him in creating various industries and an administrative infrastructure for military training and education is a well-known phenomenon.[27] Less well-known is that this emulation of colonial European politics and violence also played out outside of the formal institutions of the state.

In contrast to the earlier French violence, the dog killings of 1817 were not based on any explicit security or military concerns. Rather, issues of public order and urban density—such as noise, pollution, annoyance, and social necessity—were, by the 1810s, the most exigent concerns for those seeking to rid Cairo of its dogs.[28] Other dog eradication attempts by Mehmet ʿAli's government in this period also invoked these anxieties as justifications for the removal of dogs from Egypt's cities. In one such operation, Mehmet ʿAli sent his men out to round up as many dogs as possible from Cairo and Alexandria.[29] These animals were then boarded onto a ship in the harbor of Alexandria, and the vessel was sailed out to sea and sunk so as to rid Egypt's streets of these canines in one fell swoop. At roughly the same time (in the late 1820s) for roughly the same set of reasons, Ottoman sultan Mahmud II similarly attempted to rid Istanbul of its street dogs by rounding them up and sending them to an island in the Sea of Marmara.[30] The vessel carrying these dogs, however, capsized near the city's shore, and the unwanted canines swam back to Istanbul.[31] Issues of public order, urban health, sanitation, annoyance, and disease control were thus the primary factors shaping the human-dog relationship in urban Egypt (and elsewhere, of course) in the first half of the nineteenth century.[32]

This sort of mass violence against dogs was new to Egypt in the early nineteenth century and was largely unthinkable a few decades earlier. How to explain this gigantic shift in the human perception and treatment of animals? Where did all this violence come from? Part of the answer to these questions lies in the clear distinction al-Jabartī makes in his account between useless street dogs that barked and prevented sleep and the implicitly more civilized and socially valuable "earth and its inhabitants."[33] This distinction evidences the new and growing gap being forged in Egypt in this period between the human and the animal.[34] This gap emerged in the countryside through shifting labor regimes that made human labor vastly more important than animal labor, thus separating the work of the two species as never before. In Cairo and elsewhere, the founding and expansion of numerous governmental, educational, military, and legal establishments came to define the human realm much more clearly in this period as a regulated and controlled space of learning, health, law, policing, and bureaucratic productivity. Dogs blurred the boundaries of these increasingly protected and policed spaces, literally and figuratively crossing into them when they were not supposed to—even shitting in them—and therefore presented a direct challenge to the growing authority of the state's bureaucracy. Like the human social world, the realm of the animal was becoming much more starkly defined, closely

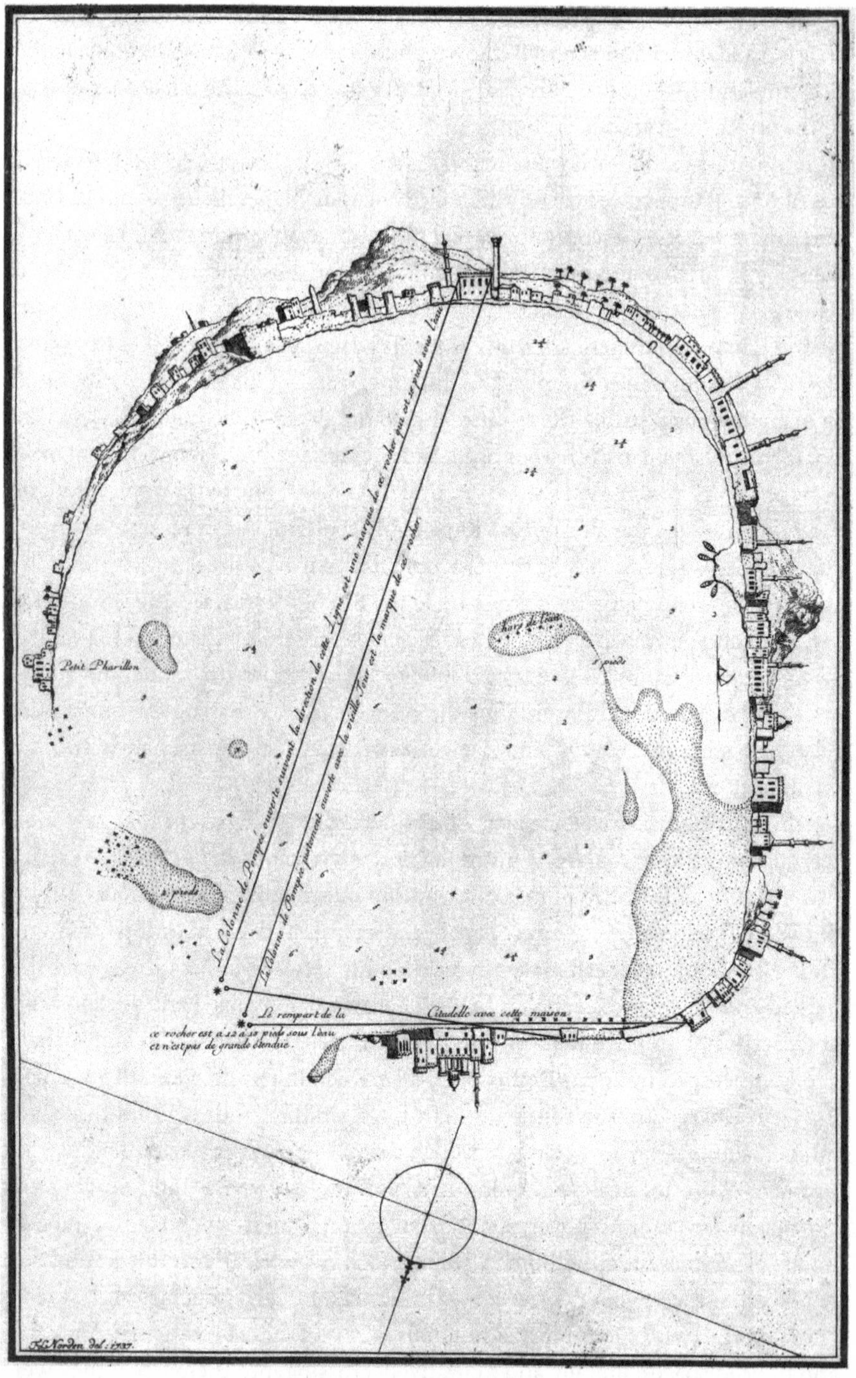

Figure 4.2 Alexandria Harbor. Norden, *Voyage d'Égypte et de Nubie,* 2: pl. 2. Beinecke Rare Book and Manuscript Library, Yale University.

managed, and spatially cordoned-off in the early nineteenth century. This widening gap between dogs and humans made it much more palatable, even desirable and obligatory, to enact violence against dogs and eventually attempt to do away with them altogether. If a parallel human world could easily exist without the need for dogs to undertake productive social functions—humans or other social actors having taken over these tasks—then why keep dogs around at all? Dogs thus became a problem in Cairo not through any action or fault of their own but because of the changing anthrozoological state around them. One of the solutions to this problem was to kill the dogs.

A Dirty Job, but Somebody's Got to Do It

The changing relationship between Cairo and its dogs was a phenomenon found in cities across the Ottoman Empire and throughout the world in this period. From Istanbul to Seattle, dogs in cities since the nineteenth century have been markers of both civilization and uncivilization.[35] In modern America, most sophisticated urbanites want dogs, but only if secured on leashes and if their feces are collected by their owners.[36] Dogs on the loose or without the proper vaccinations are a telltale sign of urban disorder.[37] Similarly, in late nineteenth-century Istanbul, reformers debated what the presence of so many street dogs in the city meant for the city's status as a modern ordered metropolis and what, if anything, should and could be done to remove them.[38]

In Cairo, the first half of the nineteenth century represented a time of reordering, sanitizing, clearing, and building in the context of a massive influx of people to the city.[39] Streets were cleared of dirt, garbage, and debris.[40] Quarantine measures were instituted as a means of controlling disease.[41] Lakes, canals, and other urban bodies of water were drained and filled in to remove what were thought to be problematic sources of disease and to provide more land for urban expansion.[42] All of this, of course, had important consequences for Cairo's street dogs.[43]

For Mehmet ʿAli's government, street dogs were bothersome pests, potential disease vectors, and dirty beasts that had to be expunged from the city. Mehmet ʿAli's plans to clean and reorder the city thus included a campaign to remove dogs from the spaces they had historically inhabited and to take from them their essential role in urban life as consumers of the city's garbage.[44] Intertwining efforts in the early nineteenth century to cleanse cities of canines and to reorder urban trash removal thus altered the human-dog relationship more profoundly in a few decades than had centuries of interspecies relations before that.[45]

In the seventeenth and eighteenth centuries, Cairo's street dogs drew their sustenance from the food and water provided by Ottoman authorities, from what they could forage on the city's bustling streets, and most importantly from Cairo's many trash mounds. Cairo was renowned for its garbage mounds, so much so that these heaps of trash and the many dogs they supported made Cairo a point of reference for cities across the globe. When visiting China in the middle of the nineteenth century, for example, the British traveler George Fleming was reminded of what he had seen in Cairo years earlier: "Closely allied to the Pariah dog of India, the savage pests of Cairo and Egypt generally, those of Syria, and those snarling droves which we have been so often obliged to pelt off with stones by moonlight, in the narrow streets of Stamboul,—the Pariah dog of North China is, like them, allowed to breed and to infest the towns and villages free from disturbance, to congregate on the plains or in the field during the day, or to kennel in the graveyards; while at night they prowl about the streets like our scavengers at home, sweeping off the quantities of filth and trash that strew the thoroughfares."[46] Thus Cairo's dogs joined those of other cities in the Ottoman Empire as the globe's archetypal street dog.[47]

The garbage mounds these canines ate from were often the first sights visitors saw as they approached the outskirts of Cairo from the Nile in the west, and they also served as elevated vantage points allowing visitors to view the city off in the distance.[48] These extramural mounds were the result of centuries of the city's residents disposing of their garbage by throwing it over the walls away from view.[49] Over the years, there thus accumulated "an almost continuous band of high mounds which virtually surrounded the city on all sides."[50] These mounds even came to serve a protective function by reinforcing the city's walls, and they were also some of the highest available lookout points for guarding the city. During the French occupation, Napoleon's soldiers took advantage of the elevated position provided by these mounds to build watchtowers and defensive positions on top of them all around the city.[51] For Cairo's dogs, the city's mounds were especially inviting. They would comb through the debris to find anything of consumable interest and leave the rest.[52] The mounds inside the city walls were less impressive than those outside but were nonetheless noticeable to all who visited Cairo. These intramural mounds were much smaller than their extramural counterparts primarily because the city's dogs kept them in check. They had learned over the years that the freshest and best food scraps and other waste were to be had inside the city, and there was thus fierce competition among Cairo's street dogs for the choicest intramural morsels.[53]

At the end of the 1820s and the beginning of the 1830s, Mehmet ʿAli and his son Ibrahim took to clearing these rubbish mounds.[54] From their perspective this was a crucial step in solving some of the city's major infrastructural, demographic, and planning problems. They sought to remove these mounds to make room for more construction to serve an ever expanding population and to rid the city of what they took to be stinky sources of disease.[55] These cleared garbage mounds could then be put to good use filling in the city's many swamps and lakes, another

Figure 4.3 Mounds, Dogs, and Humans inside Cairo's Walls. Commission des sciences et arts d'Egypte, *État moderne*, vol. 1, pt. 2 of *Description de l'Égypte*, Le Kaire, Citadelle, pl. 69. Beinecke Rare Book and Manuscript Library, Yale University.

major component of urban reform in this period.[56] A British traveler visiting Cairo in December 1832 observed that Cairo's dirt and garbage, "when carried out of the city, is not thrown, as formerly, into heaps, but is employed in filling up the pits, hollows, and inequalities which are found in the environs. In the meanwhile all the old mounds of rubbish are clearing away at a vast expense, and the land thus gained is laid out in gardens and olive plantations."[57] Throughout the early 1830s, Ibrahim used refuse to fill in and level land across Cairo and its vicinity. Garbage mounds were cleared to make room for building on the site that would become Garden City; soon thereafter, Ibrahim ordered trees to be planted and roads to be constructed there.[58] The trash from this area was taken to fill in the lake of Birkat Qasim Bey in the southern part of the city. This newly reclaimed land was then also developed. Later in the 1830s, other rubbish heaps were cleared from the northern and northwestern sections of the city to fill in lowland areas around the new road between Būlāq and Azbakiyya.

While the removal of garbage mounds increased the amount of urban land available for human construction, the process was detrimental to Cairo's dogs.[59] It deprived them of important sources of food and places of congregation. Many accounts of the work undertaken to clear these trash mounds note the intense barking and "savagery" of the dogs watching the removal of the mounds they used to frequent for sustenance.[60] Dogs clearly understood that the world around them was drastically changing for the worse.[61] From the perspective of the humans who shared the city with these dogs, removing Cairo's garbage mounds meant the animals were becoming increasingly irrelevant. What good were dogs in Cairo if they no longer helped to keep the city clean? It was not simply that dogs were no longer useful and productive for human communities, but they were also judged to be directly detrimental to urban life. As ideas about disease changed in the early nineteenth century and sickness became understood as a function of the physical and natural environments around human bodies, large groups of mangy, smelly, dirty dogs were deemed increasingly undesirable in the midst of large human populations.[62] Thus the issue of Cairo's garbage and the dog populations it supported crystallized three major problems related to the governance of the city: waste removal, disease, and the availability of usable space.[63]

The first problem was what to do with massive amounts of human and human-generated urban waste. With more people producing more garbage and more pressure for housing space, the solution to this problem was to either move waste outside the city altogether or collect it to fill in Cairo's urban lakes and ponds.[64] The second related issue was how an expanding urban population impacted conceptions of disease in Cairo. Urban proximity—between individual humans and between humans and other animals—created anxieties about how diseases developed, moved, spread, and were cured.[65] Mehmet ʿAli and most of his governmental officials—in contrast to the majority of the European medical community in Egypt—were contagionists.[66] They believed that physical proximity to the sick and to filth made the healthy more vulnerable and susceptible to disease. Smelly piles of garbage, dank bodies of water, and scraggly dogs all came to be seen as potential disease vectors that needed to be expunged from society.[67] As the historian Khaled Fahmy observes about this period, "It was the concern about the city's smell, informed as these concerns were by the dominant miasmatic theory of the spread of diseases, that informed most of the authorities' policies."[68] These new ideas about the relationships among disease, place, and the human body made dogs objects of medical knowledge, government action, and urban policing in unprecedented ways in Egypt.[69] Finally, anxieties about disease and Cairo's garbage impacted human-dog relations by pitting humans against dogs in a situation of increased pressure on and competition for scarce resources and shrinking urban spaces.[70] In a landscape in which space was at a premium and in which dogs no longer possessed productive social and economic functions, a novel, adversarial relationship developed between the two species. In this newly competitive and hostile interspecies arena, it was quite obvious that human populations had the upper hand and would soon use that advantage to violently push dogs and other creatures out of Cairo as they took over the spaces these animals used to inhabit and control.

Man's Worst Friend

Another major consequence of the decline in dog populations in Cairo was a change in human attitudes toward dogs. They came to be seen much more negatively in the early nineteenth century and with more revulsion, fear, and violence than had been the case for at least the previous millennium. The perceived negative characteristics of dogs—some old, some new—emerged as the most dominant human view of the animal: ritual impurity, annoyance, danger, the potential for disease, fecal waste. Edward William Lane, a British orientalist resident in Egypt at various points during the late 1820s and 1830s, identified the shifting attitude of Egyptians toward animals:

> In my earlier intercourse with the people of Egypt, I was much pleased at observing their humanity to dumb animals... Murders, burglaries, and other atrocious and violent crimes, were then very rare among them. Now, however, I find the generality of the Egyptians very much changed for the worse, with respect to their humanity to brutes and to their fellow-creatures. The increased severity of the government seems, as might be expected, to have engendered tyranny, and an increase of every crime, in the people: but I am inclined to think that the conduct of Europeans has greatly conduced to produce this effect; for I do not remember to have seen acts of cruelty to dumb animals except in places where Franks either reside or are frequent visiters, as Alexandria, Cairo, and Thebes [Luxor].[71]

Lane clearly blamed this shift from the humane to the violent treatment of "dumb animals" on European influence—the French precedent of dog culls bears this out—and also linked it to the "tyranny" precipitated by new government regulations and administrative practices. Not only was Mehmet ʿAli's state becoming more exacting, intolerant, and cruel in its treatment of Egyptian animals, but it was also increasingly perpetrating this violence against Egyptian humans, leading them, Lane opines, to become more violent toward one another. It would thus seem that Egyptians' disgust of dogs—and increasingly of certain humans—and the notion that canines were only impure (*najas*) with no redeeming productive social or moral capacities were ideas cemented in this period.[72] Unsurprisingly, as Lane confirms, this was also the period in which violence against dogs first began to be perpetrated on a wide scale in Egypt. For the first time in Egypt's history, dogs were now frequently beaten on Cairo's streets for nothing more than "mere wantonness."[73]

Another outcome of Egypt's shifting interspecies terrain was the emergence of the potential for a very different kind of relationship between humans and dogs, one that would fully develop only later. Since dogs were no longer primarily consumers of urban trash and were not, of course, entirely removed from Cairo, some of them came to play new roles in Egyptian society as companion animals. For elite Egyptians, dogs would become widespread as pets only in the late nineteenth century.[74] There are some earlier glimpses of the developing affective relationships between humans and dogs, but these examples largely stand as exceptions that prove the rule.[75]

In the 1830s, a lonely Cairene woman's dog died.[76] With "neither husband nor child nor friend," this woman's most trusted and beloved companion was her dog.[77] When he died, she determined, against observed practice, to honor him with a proper Muslim burial. Rather than a quiet and unceremonious interment in a solemn burial site, she resolved to have her dog buried in one of the most sacred and important burial grounds in Cairo, the cemetery housing the tomb of al-Imām al-Shāfiʿī. She washed the body as prescribed for a proper burial, wrapped the corpse in the appropriate shroud, and prepared a bier on which to mount the body.[78] She then hired reciters of the Quran, chanters, and wailing women to perform the appointed lamentations and final rites for the body as it was processed through the woman's neighborhood to the cemetery.[79] All of this was done with the ceremony appropriate to any proper human burial.[80] As the procession moved through the city, many of the woman's neighbors whispered among themselves, unsure as to who exactly had died since all assumed the woman had lived alone and had no immediate kin. Since no one knew this reclusive woman well, none ventured to ask her about the identity of the corpse on the bier.

After some time, someone eventually did question the woman about who had died, and she answered, "It is my poor child."[81] Hearing this, a group of her female neighbors accused her of lying since it was well-known that the woman did not have any children. In an attempt to keep her secret from spreading any further, the woman confessed to these neighbors that it was her dog who had died and begged them not to tell anyone. Before long, however, word got out to the assembled crowd that the funeral procession they were a part of was for a dog.[82] The ceremony immediately came to a halt, groups of men gathered and screamed at the woman for her insolence and disrespect, the hired chanters and

Quran reciters cursed the women for embarrassing them, and the police quickly assembled to protect the woman and prevent a melee.

The love and compassion of this woman for her dog—the only real companion she had in the world, her "child" as she put it herself—is perhaps unsurprising to us today. She knew that giving her dog a proper Muslim burial in a prestigious cemetery near the tomb of al-Imām al-Shāfiʿī was entirely inappropriate according to established Islamic ritual and observed Egyptian practice, but she nevertheless accepted the risk of her actions to honor her beloved companion, trying all the while to hide the identity of the body underneath the shroud.[83] The violent and intense reaction of the processional crowd at the discovery that this ceremony was for a dog—and at the realization of such a close affective relationship between a human and a dog—belies the shifting early nineteenth-century Egyptian attitude toward the animal. It was clearly unacceptable in this period to allow a canine to participate in a ritual designed to commemorate the human dead.[84]

In the early nineteenth century, the vast majority of Egyptians likely had no idea that Muslims and dogs had been in intense communion for centuries since the time of the Prophet, let alone that, according to some *ḥadīth*, the Prophet himself sometimes prayed in the company of dogs.[85] Earlier complex arguments that had been developed and debated by Muslim scholars for over a millennium in nuanced and careful treatises about the positive nature of dogs—their loyalty, protective capacities, intellect, and productive social and economic functions in human societies—were beside the point. Instances of compassion toward the animal, of learning about the medicinal properties of various plants by observing his behavior, or of giving him a proper burial among human graves (a phenomenon with many precedents as we have seen) were all immaterial. A new human-dog order was being forged, one that separated the two creatures into two distinct realms. This was likely the period in which the modern conventional notion that Muslims always considered dogs ritually impure (*najas*) came to take hold as the most common understanding of the human-dog relationship in the Muslim world. Treating a dog like a human—whether in life or death—was deemed a social and interspecies transgression that almost no Egyptian could tolerate in the early nineteenth century.

Thus, the story of dogs in Cairo during this period suggests something of the incongruous contradictions of the history of humans' relationships with animals. Dogs' smell, movement, barks, and waste were rarely problems for Egyptians before roughly 1815, but in the span of only a few decades, they emerged as the conceptual pillars of a project to remove dogs from Egypt.[86] And yet even as dogs were being set apart from many parts of the human realm, some Egyptians would eventually bring them into their homes to develop affective relationships with them. These companionate relationships were not widely recognized or respected in the early nineteenth century and would only become widespread in the twentieth. Dogs thus elicited and held in productive tension both human desire for their work and affection and human aversion to their animality and instinct.[87] They existed on the razor's edge between the human and the animal—intimately close in emotion yet infinitely distant in biology. After living closer to the human side of the human-animal line for millennia, or at least productively straddling both sides of the line, dogs in early nineteenth-century Egypt were marked as exclusively animal. The history of the emergence of this razor's edge between the species is a mix of human choice, unintended evolutionary causes and consequences, and dog agency.

Stinky Evolution

The history of Cairo's changing relationship with its garbage in the first third of the nineteenth century, and the consequences of this history for the city's dogs, is a story of evolutionary proportions. The leveling of the city's garbage mounds ended one of the most important historic and evolutionarily conditioned roles of dogs in human communities. Canine garbage consumption was no longer a trait selected by humans, and for many decades in the nineteenth century, humans would *not* select for other dog traits. Only later in the century would they come to seek out other specific characteristics for their canines such as cuteness or breed purity.[88] In the middle decades of the nineteenth century, dogs were chiefly pariahs with no perceived productive social or economic functions. Garbage had been taken away from them, and they were not yet pets.

While contemporary interactions between humans and dogs revolve primarily around petkeeping and affective bonds, for the vast majority of the thousands of years of dog-human relations these interactions were built around the canine consumption of garbage. Dog domestication itself is a story of human trash, which is why the Egyptian case elucidates how robbing dogs of their roles as consumers of urban waste was a process with evolutionary consequences.[89] Ten thousand years ago, *Homo sapiens*—only about ninety thousand years old at the time—moved from being a predominantly nomadic species to one that started settling down into sedentary communities. These fixed settlements began to accumulate large amounts of waste, which early humans usually disposed of very close to their living spaces. Some of this waste included excess food and animal products that these *Homo sapiens* chose to avoid. Scavenging wolves eventually stumbled upon this free and relatively easy source of food. Some wolves were probably too scared or too put off by the sight or smell of these early humans to approach the discarded food. Some others though did come and take it. This was the first in a series of selections. Wolves who were braver and less fearful of human communities became consumers of their food waste. This early trash thus brought humans and wolves into closer proximity than they had ever been before and was the first step in a long process of domestication.[90]

Over generations and centuries, many humans began to recognize the benefit these wolves offered and came to tolerate their presence on the outskirts of their settled communities. Of course, humans tolerated the calmer wolves most, and perhaps some even developed constructive working relationships with tamer wolf pups, realizing that they could be put to good use for hunting and other productive activities. At a certain point, because of direct human intervention, the wolves' own selected physical proximity to one another, or some combination of these and other factors, tamer wolves began breeding with one another. Breeding among those wolves who had come into close contact with humans genetically instantiated behavioral differences. Thus began a gradual separation of the genetic lines of wolves who had been interacting with humans for generations from those who had stayed away from humans.[91] Over many generations, tamer and tamer wolves eventually emerged with physical and behavioral characteristics quite distinct from their wild wolf ancestors.[92] At the same time, humans killed, drove off, or ran away from those wolves and other animals for which they did not have any desire or use. The result was that only those animals that would readily submit to human actions and could be easily controlled remained in the gene pool available for wolf domestication. These wolves thus came to be selectively bred through interspecies interactions and choices for certain characteristics humans wanted or found attractive and useful. After thousands of years of this process, it became clear that the wolves with which humans were interacting were actually no longer really wolves at all, but distinct early versions of many dog species.[93] Thus human preference and selection for utility, tameness, and general agreeability; wolves' attraction to human settlements and their choices to stay close to those settlements; and of course a good dose of chance, serendipity, and unintended consequences made dogs out of the wolves who first came to forage in human garbage about ten thousand years ago.

In the years between this first period of domestication and the early modern centuries, human preferences, dog behavior, and the human-dog relationship developed in many disparate and undetectable directions. Along the many twists and turns of this evolutionary history, one important trait was consistently desired—the human preference for dogs that consume garbage. For much of human history, dogs were thus one of the preferred and principal means of dealing with the problem of what to do with the things humans no longer wanted and that caused them revulsion.

Dogs stopped eating garbage in Egypt in the 1830s. Only in the early twentieth century would Egyptians start using dogs for affective purposes and as signs of class distinction and urbane domesticity. Dogs were therefore in a kind of

Figure 4.4 DOG NOVELTY IN EARLY TWENTIETH-CENTURY EGYPT. Pyramid village, trained animal show, dog and baboon. © 1927 Museum of Fine Arts, Boston. Used by permission.

evolutionary holding pattern from roughly 1840 to 1920—between a period when one of their traits that had been desired for thousands of years was no longer being selected for, indeed was actively being destroyed, and a new epoch in which affective bonds would emerge as the most important form of interaction between dogs and humans.

Although it may initially seem surprising, a few decades is ample time for an evolutionary divergence to take place. As Dmitry K. Belyaev's famous experiments with tame and wild foxes in the 1950s and 1960s show, selecting for certain behavioral traits can biologically change an animal's genome in just fifty years.[94] The enormous changes that occurred in Cairo at the turn of the nineteenth century, changes that altered what behavioral traits dogs were (and were not) selected for, gave them a new behavioral evolutionary niche and therefore physically modified their genes.[95] Following from Edmund Russell's claim that "the state has been an evolutionary force," the case of Cairo's dogs shows how the state could operate at the level of both the urban built environment and the genome.[96]

These colossal biopolitical shifts in human-dog relationships and in the dog genome around the turn of the nineteenth century are just two examples of the epochal changes historians and scientists are increasingly identifying as characteristic of an era termed the Anthropocene. This is the name given to the period from roughly the last half of the eighteenth century until the present day, a period during which human actions have come to affect the earth, oceans, and atmosphere on a global and most likely irreversible scale.[97] Humans in the early twenty-first century directly impact over 60 percent of the world's land surface; they shape 41 percent of the world's marine environments; over two-thirds of fisheries have been depleted, exploited, or overexploited because of human harvesting; humans have caused the extinction of over a quarter of bird species; humans consume 40 percent of the plants grown in any given year; and human activities result in the emission of 160 tons of atmospheric sulfur dioxide per year, more than twice the amount of the earth's naturally produced emissions.[98] Since the late eighteenth century, humans have, in other words, become a global geological, atmospheric, and ecological force as never before.[99]

The term *Anthropocene* was first coined by the Nobel Prize-winning Dutch atmospheric chemist Paul J. Crutzen in 2002.[100] Interestingly, the periodization he offers and that scientists and historians generally accept neatly maps onto the period of transition analyzed in this book. Crutzen writes, "The Anthropocene could be said to have started in the latter part of the eighteenth century, when analyses of air trapped in polar ice showed the beginning of growing global concentrations of carbon dioxide and methane. This date also happens to coincide with James Watt's design of the steam engine in 1784."[101] This was also the year the Laki Fissure erupted in Iceland contributing to drought, famine, and disease in Egypt.[102] It is clear that the last three decades of the eighteenth century represented a fundamental moment of atmospheric, geological, ecological, and therefore human flux and transition for Egypt and the entire globe. In trying to understand the changes that occurred at the turn of the nineteenth century in Egypt and throughout the Ottoman Empire, historians must account for the transformations in energy, atmosphere, biota, and human connections to nature captured under the rubric of the Anthropocene. Are Selim III, Mahmud II, Napoleon, and Mehmet ʿAli products of the Anthropocene? It seems unlikely that the massive political, social, and economic changes that reformed the Ottoman Empire around the year 1800 are unrelated to these more fundamental global shifts.[103]

This chapter maps just one of the manifestations of this global transition from Holocene to Anthropocene—the violent and sweeping reworking of physical spaces and of the environments that shape the evolution of species.[104] Urban dogs' environments in Egypt were radically altered in the first half of the nineteenth century, beginning a process that set them on a slightly divergent evolutionary track. Evolution is, of course, a constant process, but that does not put it outside of history. Indeed, the history of dogs in Ottoman Egypt represents an empirical story of one particular turn in the evolutionary history and historical evolution of a specific time and place.

The period from 1770 to 1840 was a wrenching one for dogs and humans. Dogs' roles in human societies, their urban environments, the way humans came to interact with them, and indeed their very biology changed more in these few

Figure 4.5 Dogs in Cairo Zoo, 2010. Photograph by author.

decades than they had for millennia. Behind these social, political, economic, and evolutionary transformations were changing notions of disease, urban sanitation, population management, and governance. The Egyptian state had become the only viable master of both dog and human.

There are few street dogs in Cairo today. Garbage removal remains a challenge in the city, but dogs are generally no longer considered part of the solution.[105] Some in the Egyptian bourgeoisie—Muslims and non-Muslims alike—keep dogs as pets in their homes and apartments. As a sign of just how distant humans and canines have become in Egypt, dogs are also—curiously to some—found in the Cairo zoo. Except for the few who can afford the food, space, and vaccinations required, *Canis familiaris* has become anything but familiar in Cairo.

PART THREE

CHARISMA AND CAPITAL

5

Enchantment

I always used to wonder how a rhinoceros and an elephant would act if they were brought face-to-face.

—Babur

The Ottomans never conquered Rome. But Rome's fate was far from clear at the turn of the sixteenth century, with the Ottomans having gained Constantinople only fifty years earlier and with their military in Otranto in 1480. The portended unification of the two Romes under an Ottoman flag inspired various interpretations of the political and theological consequences of such a victory. For the Ottomans, bringing Rome under their control would have made them the undisputed military and political force in early modern Europe and the Mediterranean and would have brought Christendom under the banner of the greatest power in the Muslim world. Romans, Venetians, and others interpreted this threat quite differently. If the Ottoman sultan had come to Rome, the story goes, he would have converted to Christianity and thereby reunited the western and eastern halves of Christianity. Neither scenario ever came to fruition.

Nevertheless, in a replay of Mehmed II's entrance into Constantinople in 1453, when he made the Hagia Sophia the first building he entered in the newly captured capital, several Venetian accounts of the imagined Ottoman conquest of Rome have the sultan riding triumphantly into the city after the cessation of violence and going directly to the church of St. Peter. Then, according to one account, "The Turk will have his horse eat on top of the altar that is the first shepherd's. Our glorious Lady will show a miracle: the horse, now meek, will immediately kneel. When the Turk sees this miracle sent by God he will immediately be baptized and will repent his error."[1] In this geopolitical tale of continental conquest, the horse is the key actor that enacts both Ottoman surrender to Rome and a Muslim's conversion—elements that together secure the unity of the world through Christianity.[2] The horse, powerful military creature that he was, symbolized the assumed fierce nature of the Ottoman sultan himself.[3] The Roman Christian God thus tames both the horse and the Ottomans.

Throughout the sixteenth century, such stories were regularly told by the Romans, the Ottomans, and many other powers to mark their claims of universal kingship and political supremacy. And while Rome did remain on the empire's agenda throughout this period, Ottoman concerns were mostly with rivals to the east. The main threat came from the rise of the Safavids in Iran in 1501.[4] The ascendency of the Mughals in South Asia and the entrance of the Portuguese into the Indian Ocean in the early modern period also directed Ottoman energies east.[5] What was at stake for the Ottomans in all of their military and rhetorical ventures in the early modern world—whether against Rome or Iran—were claims of universality. Which sovereign could rightly claim to be the strongest in the Mediterranean or the true caliph of the Muslim world?[6] And how would such worldwide power be registered and recognized by populations from Spain to China? As the apocryphal story of the sultan's horse in Rome suggests, animals played significant roles in the symbolic, military, and economic competition for ascendancy in the early modern world.

This chapter examines the role of large charismatic animals in the Ottoman Empire and Egypt in the context of rival claims to universal sovereignty in the early modern Muslim world. Understanding this history and, more generally, the relationship between humans and charismatic megafauna, requires analyzing the major institutions that mediated their interactions. In Egypt and the Ottoman Empire, these included the menagerie, the imperial procession, the lion house, and the hunt. Because charismatic megafauna were mostly used to project sultanic authority, this chapter will focus on the role of these animals in Egypt and the Ottoman capital. Istanbul was one of the primary destinations for animals sent from Egypt. More importantly, the animal institutions that existed in both places were representative of a wider early modern human-animal relationship of trade and animal display.

In examining various animal institutions and modes of presentation, this chapter emphasizes three aspects of the early modern animal world of wondrous creatures. First, this was a world made possible by Indian Ocean trading networks that served to move these animals to—and more rarely, from—the Ottoman Empire.[7] Second, these animals were transported to and from the empire as part of established forms of gift-giving and protocols of exchange whereby sovereigns used animals to project their authority against rival imperial powers and among their own populations.[8] Finally, the menagerie, dynastic procession, and hunt were imperial institutions to be enjoyed by royalty and the wealthy elite of the empire. These animals were seen and enjoyed by the masses on very few and controlled occasions. As the next chapter shows, the roles charismatic megafauna played in the early modern Ottoman Empire were very different than those they would play in the nineteenth century, when newly established institutions would come to arbitrate the human-animal relationship.

The various modes of managing the multispecies relationships decribed in this chapter and the next were central to notions of sovereignty and to the projection of that sovereignty in Egypt and throughout the Ottoman world. Unlike domesticated animals and dogs, charismatic megafauna were almost always the exclusive domain of imperial power and were strategically and successfully used by Ottoman and Egyptian authorities to display that power.

Animals on the Ocean

Egypt had been part of a globalized economy in charismatic megafauna long before the early modern period, and the Ottomans were neither the first nor the last imperial power to create a collection of animals. Indeed, the earliest known evidence of a sovereign cultivating and maintaining a menagerie for political purposes comes from ancient Egypt around 2500 BCE in the form of inscriptions at the Saqqara cemetery depicting collections of gazelles, oryx, addax, ibis, falcons, cheetah, hyenas, and mongooses.[9] Hatshepsut's (r. 1490–70 BCE) renowned menagerie in Thebes (Luxor) included elephants and leopards imported from India and monkeys and a giraffe from Somalia; a later wall painting depicts the delivery of antelope, a giraffe, cheetahs, and monkeys to the pharaoh.[10] And Egyptian religious beliefs about the afterlife meant that animals were often mummified and entombed alongside and even independently of humans.[11]

Other states in the ancient period also collected animals. Evidence from the Indus River valley dating to 2000 BCE shows that the kings of Ur kept lions in pits and cages.[12] Similarly, around 1150 BCE, the Chinese emperor Wen Wang constructed a walled nine-hundred-acre menagerie that housed deer, birds, and many ponds with enormous numbers and varieties of fish.[13] The Greeks, Ptolemies, Assyrians, and Romans all kept large menageries for various purposes.[14] In Egypt, the Ptolemies created one of the ancient world's most extensive zoological collections and are purported to have organized the largest animal procession ever to that date: ninety-six elephants, twenty-four lions, fourteen leopards, fourteen oryx, sixteen cheetahs, fourteen wild asses, and hundreds of domesticated animals.[15]

Much later in the thirteenth century, Marco Polo wrote of Kublai Khan's royal menagerie at Shang-Tu.[16] In 1235, the Holy Roman Emperor Frederick II is purported to have established the first major menagerie in Western Europe.[17] Henry III of England transferred Henry I's collection of animals, adding a few of his own, to the Tower of London in 1252.[18] The Chinese Ming emperor Yung-Lo maintained an impressive collection of zebra and other African animals. In the 1410s, Yung-Lo's envoy Zheng He brought back to the Ming court two giraffes, one that originated in Kenya but was picked up by Zheng He in Bengal as a gift to China and another that was transported directly from the eastern coast of Africa.[19] On the other side of the globe, when Hernando Cortés reached Tenochtitlan in 1521, he found Montezuma's exquisite collection of birds, reptiles, and other wild predatory animals.[20] In the early modern period, due to Europe's increasing connections to various parts of the world, many European royal families began collecting animals. Urban menageries were sponsored in Frankfurt late in the fourteenth century, in Angers in the middle of the fifteenth century, in The Hague at the end of the sixteenth century, and at Versailles in 1665.[21] Even the Vatican had a menagerie. The prized creature among the lions, civets, monkeys, birds, leopards, and other animals that made up Pope Leo X's famous collection was a diminutive albino elephant named Hanno, who died in 1516.[22]

The Ottoman menagerie in early modern Istanbul should be understood in the context of these earlier and contemporary zoological institutions. By all accounts, the Ottoman animal collection was one of the world's largest and most diverse.[23] One historian has gone so far as to say that "the best royal zoo in early modern times was found in the Ottoman capital."[24] At any given time, the menagerie near the grounds of the Topkapı Palace was home to elephants, lions, leopards, tigers, rhinoceroses, dogs, horses, birds, and many more creatures. Over the course of the sixteenth century, the number of animals in the menagerie ranged between thirty-five and two hundred.[25] Visitors were always quick to comment on what the sultan kept on his grounds.[26] An accountant for the British Levant Company named John Sanderson, who lived in Istanbul at the end of the sixteenth century, was floored by the mysterious beasts he regularly saw in the city. Of one of them that had come to the capital from Ethiopia via Egypt he wrote, "The admirablest and fairest beast that ever I sawe was a jarraff, as tame as a domesticale deere and of a reddish deere colour, whitebrested and clovenfooted. He was of a very great hieth; his forelegs longer then the hinder; a very longe necke; and headed like a cambel, except two stumps of horne one his head."[27] When Sanderson watched Sultan Mehmed III lead his armies out of Istanbul on the Hungarian campaign of 1595, he wrote, "The jarraff before spoken of, beinge prince of all the beasts, was ledd by three chaines of three sondry men stalkinge before him. For it is the custome that, the Great Turke in person goinge on warefare, most or all in generall the cheefe men and beasts attend him out of the citie."[28]

Other visitors to Istanbul also noted that the city's charismatic megafauna largely came from elsewhere—chiefly from Egypt, India, Iran, and East Africa. In the words of Thomas Gainsford, a British diplomat who joined the Istanbul embassy in 1607, "Under the [palace's] walls are stables for seahorses called *Hippopotami*, which is a monstrous beast taken in *Nilus*, Elephants, Tygres, and Dolphines: sometimes they have Crocadiles and Rhinoceros: within are

Roebuckes, white Partridges, and Turtles, the bird of *Arabia,* and many beasts and fowles of Affrica and India."[29]

The global trade in charismatics was largely an Indian Ocean affair centered around animals exported from the Indian subcontinent.[30] India and Egypt had been linked through trade for millennia.[31] They were the two final nodes of the Indian Ocean world, with Egypt playing an important role as one of the primary way stations for goods moving from India and points farther east to Europe and the Mediterranean. In the early modern period of Ottoman rule in Egypt, the province's role in this commercial network expanded as European and Ottoman trade with South Asia and other parts of the Indian Ocean world greatly increased to feed growing tastes throughout the Mediterranean for Indian spices, textiles, and much more.[32] Many Egyptian merchants in the Ottoman period made their fortunes by tapping into the Indian Ocean trade, either directly or through intermediaries who purchased goods for the Egyptian market in Jidda, Aden, or elsewhere on the Red Sea coast.[33] Although trade obviously moved in both directions—high returns were secured, for example, from the traffic in Arabian and Persian horses from the Middle East to South Asia—the bulk of this commerce moved west from the subcontinent.[34] In Egypt, as most everywhere else, robust demand existed for Indian imports—spices, finished goods, textiles, and animals.[35] During their three-year incursion, the French quickly recognized the centrality of the Indian Ocean trade to Egypt's economy and attempted to take advantage of it.[36] Moreover, although he did much to reorientate the bulk of Egypt's commercial interests toward the Mediterranean, Mehmet ʿAli nevertheless recognized the importance of Egypt's eastern trade. He regularly stationed government trade and purchasing agents in the subcontinent to facilitate the importation of Indian goods to Egypt. For instance, in the fall of 1816, he sent to India one of his representatives, a man named Ḥasan al-Maḥrūqī, with the impressive sum of 500,000 French riyāls to purchase items for the Egyptian market.[37] In June of 1817, Mehmet ʿAli personally made the eighty-mile overland journey to Suez with one of al-Maḥrūqī's relatives to see what his envoy had brought back.[38] Mehmet ʿAli furthermore tried to regulate the domestic Egyptian market in Indian products by instituting various price controls on these goods and by periodically limiting their availability in the market.[39]

It was in the context of this Indian Ocean commercial sphere that India functioned as the primary supplier of elephants, tigers, rhinoceroses, and monkeys to Egypt, Iran, the Arabian Peninsula, and from these places to Istanbul and Europe.[40] Even before the Portuguese rounded the Cape of Good Hope at the end of the fifteenth century, many of these Indian animals found their way to European courts. As with most living and nonliving things from South Asia, these creatures moved to Europe through the Middle East.[41] For example, in the summer of 802, the famed Hārūn al-Rashīd facilitated the movement of an Indian elephant—later named Abul Abaz—to the residence of Emperor Charlemagne.[42] As trade between India and the Ottoman Empire and Europe intensified in the sixteenth century after the 1517 Ottoman conquest of Egypt, an increasing volume of animals was in transit between South Asia and the Ottoman Empire. A historic hinge of trade between Asia and Europe, Egypt was an overland pinch-point for many of these animals.[43] Egyptians were thus regularly exposed to the fantastic spectacle of these creatures as they moved between the Red and Mediterranean Seas.

The Five-Legged Sultan

The Ottoman conquest of Egypt in 1517 threw the empire into the Indian Ocean for the first time in its history, leading to growing rivalry between the Ottomans and the Portuguese for control of trade to and from India. The role of Egypt and Aden in this imperial competition for influence in the Indian Ocean was made manifest in late 1547 and early 1548 when the Ottoman navy set out from Suez to recapture Aden from the Portuguese.[44] Principally through its Red Sea links to Aden, Egypt had always been connected to the Indian Ocean trading world of charismatics, spices, and other goods. This was no different once the Ottomans conquered Egypt. Indeed, the new province became the Ottomans' very lifeline to the Indian Ocean after 1517. Beginning in the sixteenth century, these and other similar connections of commerce between the three major empires of the early modern Muslim world—the Ottomans, Safavids, and Mughals—facilitated the movement of charismatic megafauna and other animals of symbolic and economic importance from South Asia through Egypt and Iran and then on to Istanbul and other parts of the Ottoman Empire.[45]

The history of South Asian elephants is a prime example of the circulation of prized living creatures among the three empires.[46] In both Istanbul and Cairo—the two largest cities in the Ottoman Empire—Indian elephants played an active role as instigators of imagination, projections of imperial authority, and symbols of Ottoman attempts to manage and come to terms with nature. In these and other important ways, the history of elephants in these two Ottoman cities reflects wider early modern uses of the animal in the Islamic world, South Asia, Europe, and beyond. One of the principal factors spurring demand for Indian elephants in both the Ottoman Empire and Safavid Iran was their dual use as diplomatic gifts and symbols of imperial power.[47] For example, in 1738, the post-Safavid leader of Iran, Nadir Shah, sent the Ottoman Sultan Mahmud I an Indian elephant as part of his attempts at rapprochement with the Ottoman Empire.[48] The gifting of elephants was a common practice in South Asia and

Figure 5.1 Elephant from Ottoman Translation of Qazvînî's *ʿAca'ib ül-Mahlukat*. Walters Art Museum, Ms. W.659, 109a. Used by permission of Images for Academic Publishing.

one likely exported from the subcontinent to the Ottoman and Safavid worlds.[49] Indeed, there is a long tradition in South Asia of gifting and using elephants as signs of power and prestige,[50] and—as evidenced by the major accounts of the reigns of several Mughal rulers—elephants continued to play a significant role in South Asian society and politics in the early modern period.[51] The symbolic power of elephants was surely not lost on Ottoman and Safavid visitors to the Mughal court.[52] For example, when the Ottoman admiral Seydi ʿAli Reis arrived at the Mughal court in October 1555, Humayun greeted him with a retinue of four hundred elephants and several thousand men.[53] By adopting the largely South and Central Asian practice of staging and gifting elephants, the Ottomans and Safavids attempted to appropriate the near-universal prestige of the animal to produce their own symbolic capital.[54]

In 1531, envoys of Charles V to Süleyman's court were especially awed by the sight of two elephants and a number of beautiful horses in the first courtyard of the palace.[55] Charles V himself also used elephants to good effect in the imperial competition for status and power that characterized the early modern Mediterranean. When Charles V's nephew Maximilian II was sent from Spain to take up residence in Vienna in the early 1550s, he was given an elephant to celebrate his new appointment.[56] This South Asian pachyderm originated, most likely, in either Kerala or Sri Lanka and was sailed first to Portugal and then to the Spanish capital Valladolid in 1551. In the letter accompanying the elephant, another of Maximilian's uncles, John III of Portugal, suggested: "I think you should name the animal and give it the name of the deadly enemy of the West as well as of your splendid home, namely Sultan Suleyman. In this way he will become your slave and will be properly humiliated. As your show animal, he will move to your residence in Vienna, named after the Sultan who wanted to destroy it!"[57] In this rather direct statement of early modern rivalry played out through the possession of an elephant, Süleyman the pachyderm was both a wonderous delight for the city's residents and a symbolic reminder of the pacification of the Ottoman enemy on the other side of the city's gates.[58]

Another envoy to Süleyman's court, Ogier Ghiselin de Busbecq, a Flemish diplomat who served as Holy Roman Emperor Ferdinand's ambassador to Istanbul and who penned a well-known account of the empire in the sixteenth century, was thoroughly enamored of an elephant he saw perform in the palace.[59] "When this *Elephant* was bid to dance, he did so caper or quaver with his whole Body, and interchangeably move his Feet, that he seem'd to represent a kind of Jig; and as for playing at Ball, he very prettily took up the Ball in his Trunk, and sent it packing therewith, as we do with the Palm of the Hand."[60] No doubt these imposing creatures undertaking such nimble, agile, and intricate contortions impressed upon the sultan's visitors the gravity of their impending audience with a sovereign who could make ferocious creatures perform such wondrous light-footed acts. Similarly, a Habsburg representative sent to Edirne in 1628 took special care to note his marvel at seeing two elephants perform for him during his meal with officials of the Ottoman court.[61] These elephants were possibly those that had been sent to Istanbul in November 1620 as a gift from the Safavid Shah ʿAbbas I. This gift package included four Indian elephants, two large tigers, and a rhinoceros.[62] These diplomatic gifts from the Ottomans' rival were thus reused to great effect by the sultan in Istanbul to awe visitors to his own realms. The shadow of God on earth obviously wanted to convince his rival sovereigns that his shadow spread over much more than just the earth's human elements.

Elephants had other public uses in the Ottoman capital as well.[63] Illustrations of the circumcision ceremony of Ahmet III's sons in 1720 show several elaborately adorned elephants participating in the festivities.[64] Whether or not these miniature depictions of elephants mean that living elephants were actually present during the two-week celebration is debated. Some of these creatures were clearly fantastical. One elephant is, for instance, shown shooting fireworks from his nostrils.[65] At the very least, the image of the elephant—if not the living, breathing being himself—was quite familiar to city residents in the early eighteenth century.[66] This was obviously a result of the many gifts of elephants made to sultans over the preceding two hundred years, gifts that these Ottoman sovereigns usefully deployed for purposes of public pageantry. It was also partly a result of the knowledge of the elephant form as seen in other guises—most importantly as depicted in many well-known Mughal miniature paintings and in elephant figurines that were popular in early modern Europe and often given as gifts to the Ottomans in the sixteenth and seventeenth centuries.[67] The ambassador Busbecq, for example, brought to Süleyman from Emperor Ferdnand the gift of a clock mounted on top of an elephant.[68] A large number of these clocks have survived, and they seem to have been quite common imperial gifts throughout the Mediterranean in the sixteenth century.[69] Moreover, the elephant clock is part of a much older tradition of such machines in the Islamic world. Device manuals from various periods discuss elephant clocks. Perhaps the most famous of these texts, al-Jazarī's *Book of the Knowledge of Mechanical Devices*, completed in 1206, outlines in great depth and detail the construction and proper function of an elephant clock.[70]

Beginning in the late 1730s, the sultan consistently had at least one elephant in his menagerie in Istanbul.[71] In a firman and accounting book from March 1739, the sultan writes that the daily stipend to be given to elephant keepers (*filbanan*) to cover both their salaries and the maintenance of the animals in their charge was 510 akçes.[72] This money was to be taken from the customs revenues of Istanbul and was further specified as totaling 15,300 akçes a month.[73] Later cases confirm that these allowances held steady at these levels (510 akçes per day for a total of 15,300 akçes per month) through March 1740 at least.[74] Most of this money went to feeding the elephants.[75] Every day the palace kitchen provided each elephant with one kıyye (1.28 kg) of sugar, one kıyye of butterfat, and fourteen kıyye of bread.[76]

By all accounts, being an elephant attendant was quite a comfortable life. At the beginning of 1742, there were fourteen men put in charge of a single elephant.[77] The head elephant keeper (*filcibaşı*) was given a monthly salary of 7.5 guruş, and the other thirteen elephant keepers (*filbanlar*) were each paid 4 guruş per month.[78] The number of elephant keepers was reduced from thirteen to eleven in October 1742,[79] and was reduced again in December of that year to ten.[80] In addition to their normal salaries, these men also received daily rations of salt, onions, chickpeas, rice, bread, firewood, wax, and other necessities.[81] The total amount earmarked for the elephant handlers' provisions was forty-five akçes a day.[82] As before, all of these monies were taken from Istanbul's customs revenues and distributed by the head of the palace stables (*mirahor ağa*).[83]

There is no evidence that Ottoman elephant handlers received any formal training in how to care for Istanbul's elephants. If the surviving pictorial evidence is any indication, elephant handlers indeed seem to have been rather lacking in their abilities to ride, direct, and control the animal.[84] Beyond serving

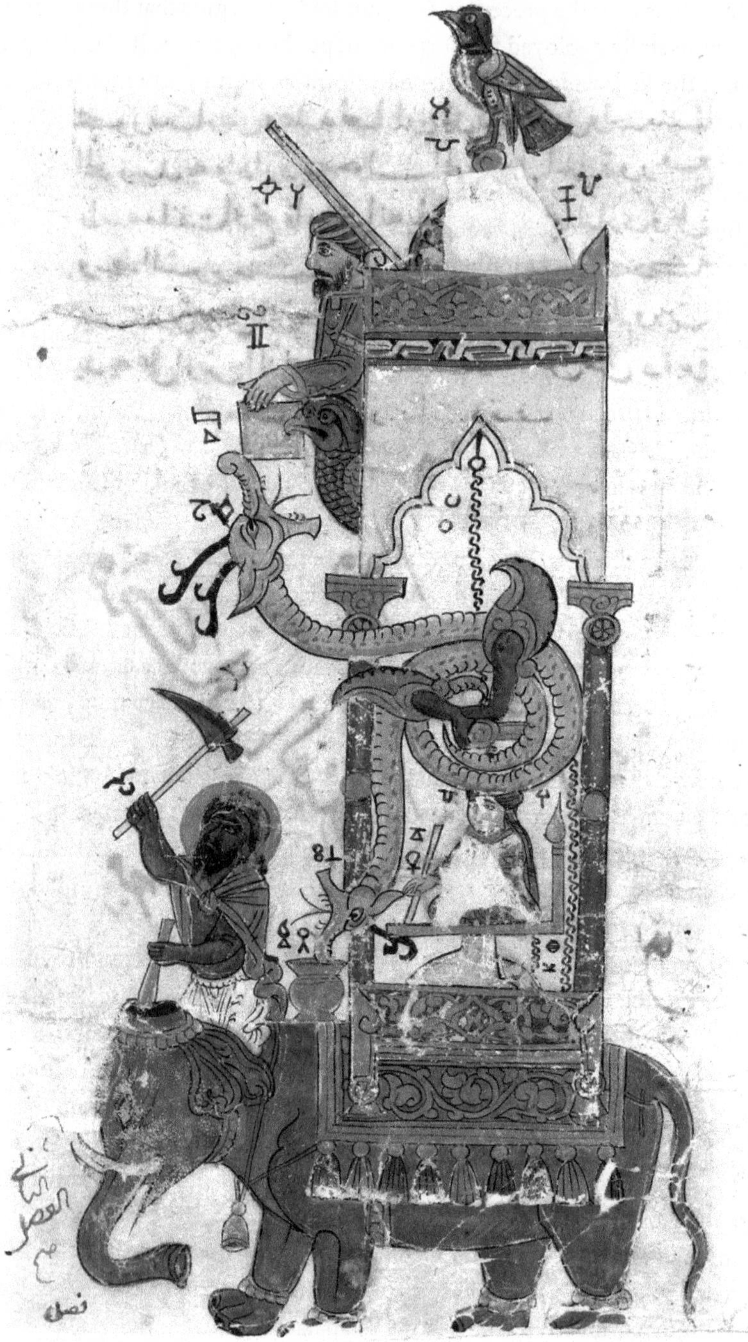

Figure 5.2 AL-JAZARĪ'S ELEPHANT CLOCK. Abī al-ʿIzz Ismāʿīl al-Jazarī, *Book of the Knowledge of Ingenious Mechanical Devices*, Metropolitan Museum of Art, Bequest of Cora Timken Burnett, 1956 (57.51.23). Used by permission of Images for Academic Publishing.

Figure 5.3 ELEPHANT RIDER IN SEVENTEENTH-CENTURY ISTANBUL. Taeschner, *Alt-Stambuler Hof-und Volksleben*, pl. 27.

as objects of wonder and awe, elephants were sometimes used to pull wagons, but this seems to have been rather rare. And unlike in South Asia, they were never used for warfare in the Ottoman world.

It seems reasonable to assume that a mahout (elephant handler) accompanied elephants on their journey from the subcontinent and then remained in Istanbul for at least some time to care for the creatures in their new home and to teach others how to properly tend to them.[85] This was often the case with Indian elephants that came to Egypt in the eighteenth century.[86] There is, moreover, some evidence suggesting that elephant handlers were in demand in Istanbul, since Ottoman authorities occasionally sought them out and paid for their travel to and maintenance in the city.[87]

There were numerous other imports of elephants to the Ottoman capital throughout the eighteenth century. In December 1778, an elephant was sent to Istanbul as a gift from the ruler of Jalbār.[88] For reasons unstated, Sultan Abdülhamid I decided to send this elephant to Edirne.[89] Perhaps he was content with the elephants currently in his menagerie; perhaps he felt that the imperial hunting residence in Edirne could have used an elephant; or perhaps he simply did not want to deal with the expense and bureaucracy of keeping another elephant in the capital. Whatever the case may have been, this creature was sent to Edirne in the care of a low-ranking imperial official known as the *kışlak emini,* who was charged with ensuring that the elephant comfortably settled into his new home and had sufficient amounts of food and drink.[90] Five days after the elephant was sent to Edirne, the sultan authorized fifty guruş to be sent to the former Ottoman capital to support the animal's upkeep.[91] In the spring of 1818, another group of Indian elephants came to the Ottoman capital—this time as a gift from the Qajar court—and a new cadre of five elephant handlers was hired to look after the creatures.[92] The same year a new group of lions was brought to the city, indicating strong interest in expanding and cultivating the imperial animal collection in Istanbul during those years. As a young man taking the throne in a period when there were many pressures on the sultanate, bolstering an institution like the imperial menagerie was perhaps part of Sultan Mahmud II's efforts to publicly pronounce and project his authority for all to recognize and respect.[93]

Cairo was the other major destination for elephants coming to the Ottoman Empire. The itineraries of these animals to the empire's two largest cities traced the major land and sea trade routes to the east. Many of the elephants that came to Istanbul as gifts from India or Iran traveled entirely overland to the imperial capital. Others took the sea route across the Indian Ocean and up the Red Sea, finally docking to again set foot on land in Suez or another of Egypt's Red Sea ports.[94] While some of these creatures continued on through Egypt only to get on other ships in the Mediterranean to travel to Istanbul and elsewhere, many of them stayed in Egypt—almost always in Cairo—and came to play important roles in the cultural and political life of the city in the early modern period.

In 1512—just a few short years before the Ottoman conquest of Egypt—four elephants were dispatched from India to Cairo as a gift to the Mamluk Sultan Qanṣūh al-Ghawrī from an unnamed "king of India" (*malik al-hind*).[95] When the ship carrying the animals arrived on the Egyptian Red Sea coast in November of that year, it was found that only two of the elephants had survived the trip. These two survivors were immediately taken from Suez to al-Ghawrī's court and entered Cairo with great pomp and circumstance.[96] All of Cairo came to watch the elephants as they were paraded through the city's gates and then through the center of the Mamluk capital accompanied by a stream of drummers and horn players. On their backs they wore red velvet sashes, and their trunks were overlaid with ornate metalwork. It was said to be a day of great delight as the city celebrated these "two wondrous creatures."

This celebratory moment was a welcome respite from the general sense of foreboding hanging over Cairo in the 1510s.[97] The Ottomans had already made clear their desire to capture Egypt, and al-Ghawrī was busy trying to thwart these Ottoman threats through a possible alliance with the Safavid shah.[98] As a show of friendship and in an attempt to placate the Ottoman sultan, al-Ghawrī sent one of his two recent elephantine acquisitions as a gift to Selim I.[99] He dispatched the animal, along with a few other delicacies from Egypt, to Istanbul with one of Selim's envoys who had come to Cairo bearing a rather less cheerful gift for al-Ghawrī—the head of his vassal ʿAlā' al-Dawla, whom the Ottomans had just defeated in Syria, and a fethname (declaration of war) formally proclaiming Ottoman intentions to conquer the Mamluks.[100] Selim received the elephant in Istanbul, where it occasioned great wonder and excitement among the city's residents, just as it had in Cairo.[101] The four elephants in this account—two of whom died while at sea, one of whom stayed in Cairo, and the other who went on to Istanbul—moved along established trade networks that had long circulated animals, spices, and other goods in the Indian Ocean.[102] Elephants played important symbolic roles in careful political maneuverings and negotiations—in this case those on the eve of the Ottoman conquests of Egypt and Syria. al-Ghawrī likely sought to elicit from Selim the same kind of awe and wonder that residents of Istanbul and Cairo experienced upon seeing elephants, in the hopes that Selim would at least reconsider his military plans. Unluckily for al-Ghawrī, his elephant gift did nothing to stop Ottoman armies from taking his territory.

Beyond these tales of imperial conquest, the two South Asian elephants that arrived alive in Egypt help to illuminate how early modern Ottoman subjects (and those soon to be subjects) experienced the few charismatic megafauna they encountered over the course of their lives. The epic procession of elephants through Cairo or their inclusion in imperial ceremonies like circumcision festivals or celebrations of military victories were rare occasions for Ottoman urbanites. The sight of elephants during these events exposed residents of various Ottoman cities to marvels of creation they had never before seen. Elephants allowed Cairenes and Istanbulites to experience what must have seemed like a freakish anomaly of nature and also to engage with a living creature that had come from oceans away. These animals thus expanded the ecological, fanciful, and geographic horizons of these humans' imaginations. They also performed important recreational functions for Cairenes. The history of an elephant sent from India to Egypt in 1776 suggests some of these elephantine recreational roles.

> During this year came a group of Indians with a small elephant which they brought to al-ʿAynī Palace. They put him into the big barn. The people rushed to this spectacle. At the gates servants stood who took money from the sightseers. Likewise the Indian grooms of the elephant collected a great deal of money. The people would bring him pastries and sugar cane and they would amuse themselves observing him

> sucking the cane and eating it with his trunk. The Indians would address him in their language and he understood their words. Whenever they presented him to a grandee, they would say something to him and the elephant would go down on his knees and salute with his trunk.[103]

As this passage illustrates, Indian mahouts rather than Ottoman or Egyptian merchants in South Asia were the ones who most often brought elephants to Egypt.[104] Thus, it was typically locals who controlled the export and transport of charismatic animals. This points both to the importance of the animal trade to local economies throughout the Indian Ocean and to the circulation of people between places like Egypt and India.[105] The presence of Indian mahouts in Egypt further suggests that Egyptians, much like Istanbul's elephant keepers, did not have much expertise in handling elephants and therefore relied on the know-how of Indian specialists who accompanied the animals.[106]

Most importantly, this passage relates the experiences of wonder, amazement, and spectacle Egyptians enjoyed with this elephant.[107] They brought him food, were dazzled by the way his handlers controlled him, and were impressed by the respect he showed Egyptian elites.[108] Unlike camels or donkeys in the countryside or dogs in the city, elephants inspired awe, provided spectacle, and changed Egyptians' sense of the world around them.[109] The early modern economy of elephants and other charismatic creatures was not based on everyday use or laboring productivity but rather uniqueness, splendor, novelty, imagination, and celebration.

The history of elephants in Ottoman Cairo and Istanbul thus points to a certain early modern economics and aesthetics of the human-animal world as well as to the role of animals in the politics of early modern imperial rivalries. Recognizing the political potential of megafauna charisma, Ottoman officials in Egypt and Istanbul consciously deployed animals like elephants to impress upon their subjects their imperial power and ability to tame and display nature. Unlike later institutions of animal enclosure such as the zoo or veterinary hospital, these means of displaying animals were not meant to illustrate the sovereign's animalistic conquest and subjugation of nature, but rather his capacity to stage the wonders of the natural world in controlled spectacles like processions or imperial ceremonies. The sovereign was the sponsor of this marvel, not a brute attempting to dominate it. Fantastic creatures like elephants were thus symbols of how human political authority could harmoniously harness and reconcile the awesome wonders and random violence of the natural world. While humans succeeded in bringing elephants halfway around the world from India to the Ottoman Empire, only elephants could arrest the imagination of an entire city and expand its residents' worldviews beyond anything they could have ever envisioned.

Istanbul Roars

Like elephants, lions and other large cats were also effectively used to portray the sovereignty of the Ottoman state.[110] The story of lions in the Ottoman Empire is one that largely takes place in Istanbul. It is nevertheless essential to consider these cats to fully understand the history of human-animal relations in Egypt, because lions further illuminate the early modern economy of animal trade and other aspects of how the Ottomans utilized charismatics throughout their realms. Lions were usually either gifted to Istanbul from the empire's North African provinces or from Safavid Iran.[111] In 1582, in celebration of the circumcision of Sultan Murad III's son (the future Sultan Mehmed III), the former provincial governor of Algiers, Haydar Pasha, sent a gift package that included two lions, two tigers, a caracal, a ram, lavish textiles, and a collection of silver vessels.[112] In 1620, a similar collection of animals was sent from the Safavid court to the Ottoman palace as a diplomatic gift.[113] In May 1743, Sultan Mahmud I promulgated a firman ordering that two lions and two tigers (s. *kaplan*) be delivered from western Tripoli to Istanbul;[114] twenty or so years later another firman was issued to send more lions and tigers from Algeria;[115] lions and tigers were also sent from Algeria to Istanbul in late 1806;[116] and another firman in 1816 ordered still more lions to be sent from Algeria.[117]

Figure 5.4 Lion from Ottoman Translation of Qazvīnī's *ʿAcaʾib ül-Mahlukat*. Walters Art Museum, Ms. W.659, 80a. Used by permission of Images for Academic Publishing.

The palace was thus regularly in the business of first finding and then transporting lions and other large cats to the Ottoman capital. The sultan and his retinue understood that these felines were a rarity in the Ottoman world and that, like elephants, they could be used by the Ottoman state as a potent symbol of its power and grandeur. More than elephants however, Ottoman leaders actively sought out lions to bring to Istanbul, seeing them as perhaps the best animal analogy for the message of rule they sought to impart to both their subjects and rivals—a statement that only Ottoman imperial power could harmoniously reconcile nature's violently competing forces.

Sultans always made sure their guests saw the empire's collection of large cats during their stays in Istanbul. For their parts, European visitors to the Ottoman capital were always quick to note how impressive they found the city's captive lions and other felines. For instance, in September 1608, Ahmet I was entertaining a German delegation at the palace.[118] After their official audience with the sultan, the guests were escorted to the hall of ambassadors (*elçi hanı*) where a full dinner was awaiting them, along with that evening's entertainment—two chained lions and a group of imperial musicians. In their accounts of the visit, the sultan's guests made special note of their amazement at this animal show.

European visitors to Istanbul were always quite interested in the imperial lion house (*arslanhane*), an institution the Ottoman bureaucracy wholly supported and carefully cultivated as a tool of political pageantry.[119] The lion house was built in the remnants of the tenth-century Church of Christ Chalkites, built near the historic Chalke Gate of the Great Palace of Byzantine Constantinople.[120] As late as the early seventeenth century, a Polish-Armenian visitor to Istanbul noted that frescos and mosaics of saints from the old church could still be seen on the inside walls of the enclosure.[121] In 1588, a Habsburg ambassador to Istanbul noted eight lions in the lion house. He was impressed with how tame they were and wrote that their handlers were able to play with them freely, "as if they were large dogs."[122] He later describes them being as docile as sheep. In a similar vein, a French visitor to Istanbul in 1672 and 1673 commented that the leopards he saw in the lion house were so tame and peaceful in their richly adorned fabrics that their keepers were able to carry them on horseback.[123] A mid-seventeenth-century Russian visitor to the lion house wrote that it contained four lions (one of which was specifically mentioned as coming from Africa), snow leopards, a hyena, a fox, three wolves, and a jackal.[124] As these accounts suggest, the lion house seems to have functioned as a kind of auxiliary zoological collection, housing those animals considered unsafe to allow to roam freely in the capital's menagerie.

The lion house had several other functions as well. Both this seventeenth-century Russian visitor and Ivan Luk'ianov, a Muscovite priest resident in Istanbul between 1701 and 1703, commented on the presence of a crocodile skin, a giraffe skeleton, the head of a rhino, and the head of an elephant with an enormous trunk.[125] The lion house therefore seems to have served an additional function as a kind of cabinet of animal curiosities in the Ottoman capital.[126] Other visitors to Istanbul noted that the lion house was also home to a set of administrative offices around the cats' main pen.[127] Given its strategic location near Istanbul's centers of power, it is not surprising that the lion house served as a kind of spillover space for the palace's important bureaucratic functions. For example, offices of designers, draftsmen, and dyers in the lion house supported Ahmet III's imperial library.

A story related by Ivan Luk'ianov demonstrates both his excitement at the novel prospect of seeing a live lion and also some of the public functions of the lion house.[128] After handing over his entry fee of one para, Luk'ianov was led by an attendant to the cage of a sleeping lion. Wanting to see the animal in all its ferocious glory, Luk'ianov implored his Turkish guide to awaken the beast, but the handler refused, citing his desire not to disturb or anger the recently fed and tired animal. Frustrated and wanting the show he had paid for, Luk'ianov, a priest lest we forget, picked up a piece of wood near the cage and hurled it at the beast to wake him up. Nothing happened. The fierce king of the jungle turned out to be nothing more than the empty hide of a lion stuffed with straw. Feeling duped and cheated, Luk'ianov demanded his money back, but soon gave up on his request when shown some of the lion house's many other wonders.

In 1802, portions of the lion house burned down in one of Istanbul's many fires.[129] After the conflagration, sections of the enclosure's charred remnants were subsequently torn down to allow for the extension of the imperial armory (*cebehane*).[130] Other parts of the lion house were, however, quickly returned to their former use. In late 1802, five damaged lion cages were rebuilt and two brand new ones were constructed, all seven at a cost of 300 guruş each.[131] The sultan then issued a firman in the summer of 1803 authorizing the complete reconstruction of the lion house (albeit with slightly smaller dimensions) and also funding the purchase of necessary building materials and other supplies.[132] These repairs were followed up in 1806 with the restoration of parts of the floor and roof rafters.[133] Further work on the lion house was carried out in 1815 and 1816 when stones and bricks from an old building near the Hippodrome were brought to patch up parts of the structure.[134] Iron bars were also used to close off a group of arches to prevent the lions from escaping, and new eaves, windows, floors, and doors were added. The total cost of these refurbishment efforts was 3330.5 guruş. The new lion house was well worth the high price tag, and it was soon put to good use. During the middle of the 1810s, Istanbul received the gift of a new lion from its provincial governor in Algiers.[135] As part of the budgeting for the repairs of 1815 and 1816, the imperial administration thus also made provisions for the construction of new caged carts to safely transport this lion through the city to its new home.[136]

The existence of a lion house in Istanbul created the need for an animal administration to oversee all matters related to the presence of lions in the imperial capital. Perhaps the most basic concern of this administration was to ensure that these animals were always happily fed. An order from September 1574, for example, authorized the daily supply of ninety-three sheep heads (s. *kelle-i ghanem*) to feed an unspecified number of lions, tigers, wolves, and civets, as well as two monkeys, two lynx, and a sable that were kept separate from the cats in the lion house.[137] Each sheep head cost a quarter of an akçe; thus the total daily cost of feeding these animals in the *arslanhane* was 23.25 akçes. For comparative purposes, the average daily wage of a janissary in 1600 was around three akçes.[138] Thus at the turn of the seventeenth century, the sultan spent the equivalent of about eight soldiers' salaries feeding his cats. An order from 1743 specified that two lions and two tigers that arrived to Istanbul from western Tripoli that year were each to be fed three sheep heads every day, along with a suitable amount of mutton.[139]

The responsibility of overseeing the feeding and well-being of these animals fell to the empire's chief lion keeper (*arslancı*), who was a regular salaried member of the imperial bureaucracy.[140] Evliya Çelebi reports that the lion house employed over one hundred men, adding (perhaps editorially) that these men venerated Imam ʿAli, "the lion of God," as their patron.[141] Details survive about several appointments of lion keepers over the course of the eighteenth century. In October 1701, Ferraş Arapzade Abdullah replaced Külhani Mehmed as chief lion keeper.[142] At the accession of Mustafa III in 1757, a certain Ismail was appointed chief lion keeper, a position he had previously held under both Osman III and Mahmud I.[143] A few years later in 1773, a new chief lion keeper was appointed,[144] and in 1801 Ibrahim ibn Mustafa replaced Ibrahim ibn Khalil in the position.[145] In each of these appointments of a new lion keeper—except Ismail's in 1757—the daily assigned stipend was set at seventeen akçes.[146] The regular and timely reappointment of lion keepers and the consistency of their wages point to the lion house's institutional stability and evidence the state's continued interest in upholding it as a strong administrative unit of the imperial bureaucracy. This permanence of the lion house and the empire's investment in lion keepers stand in stark contrast to other administrative positions related to the care and maintenance of charismatic felines in Istanbul. At the end of the 1680s, for instance, the office of the chief leopard keeper (*parsçı*) was abolished as part of cost-cutting measures instituted as a result of war with Austria and the reforms associated with the deposition of the hunter-sultan Mehmed IV in 1687.[147] One of the critiques of Mehmed IV was that he allowed the empire to deteriorate because he spent too much time and energy on the hunt.[148] Abolishing the office of the leopard keeper was thus emblematic of the many attempts to create a new order after his reign.

The large-scale, intricate, and sophisticated administration built around the maintenance of lions in Istanbul and the priority given to lions over leopards and other animals were results of the important functions these animals played in projecting Ottoman imperial authority and articulating ideas of state sovereignty. There were, though, other more important roles for lions in the empire. They were effectively used in Ottoman miniature painting as symbols of the dynasty's authority over nature. An image of the founder of the empire, Osman I, composed at the end of the sixteenth century amply demonstrates this (figure 5.5).[149]

In the image, the proverbial king of the jungle prostrates himself at the foot of the sultan. Even more significantly, the lion licks Osman's boot. The cowering of nature's power at the literal feet of the progenitor of the Ottoman dynasty needs no further explanation. This image again points to the role of charismatic megafauna in early modern imperial rivalries. As the lion is often associated with the Shia imam ʿAli, this lion may possibly be symbolic of the Shia Safavid state licking Osman's boot in recognition of and deference to universal Ottoman authority.[150]

In another scene of sultan and lion, Selim II is depicted during a hunt in Üsküdar (figure 5.6).[151] Selim and the lion gaze unflinchingly, and perhaps even understandingly, at one another as befuddled onlookers stare in amazement. There are several possible readings of this image. The message is possibly one about the sheer bravery of Selim in the face of such a potentially deadly animal. The sultan does not move from his seat, and his frightened assistant's sword remains in its sheath. Perhaps the artist would even have viewers believe that the lion—not Selim—is the scared creature in this scene, that the powerful feline has finally met his match in the person of the Ottoman sovereign. Alternatively, perhaps the sultan's calm manner and peaceable ability to reconcile all of creation under his felicitous rule has tamed the beast, removing any reason for fright or panic on the part of the assembled humans. Through some combination of fear, harmonious imperial aura, and interspecies understanding, lion and sultan are at ease in this image. The sultan's commanding yet serene influence on the lion is meant to serve as an analogy for the Ottoman state's abilities to reconcile disparate parts of nature under the harmonious balance of its rule.[152]

Lions were also frequently theatrically employed in imperial processions through Istanbul as public displays of dynastic authority. One of the most detailed accounts of such a procession is given by Evliya Çelebi about a parade through Istanbul commemorating one of Murad IV's military campaigns in the second quarter of the seventeenth century.[153] Ten lions, five leopards, twelve tigers, and a group of hyenas, foxes, wolves, and jackals walked alongside artisans, soldiers, and imperial officials. About fifty-five handlers accompanied the animals. The leopard keepers wore leopard skins on their backs and used staffs to keep the felines in check. The leopards themselves were also adorned with fine fabrics. The lions were held by chains and accompanied by their keepers. In

Figure 5.5 Osman I and Lion. Seyyid Lokman, *Hünername I,* Topkapı Sarayı Müzesi Kütüphanesi, H. 1523, 57b. Used by permission.

the event that one of them broke free from his handler, the keepers had a load of gazelle meat laced with opium and other drugs ready to attract and—it was hoped—subdue the animal.

These precautions notwithstanding, this procession is notable for the fact that the lions were purposely walked through the city without cages. Ottoman authorities chose not to cage the animals in order to elicit an amount of fear from

Figure 5.6 Selim II and Lion. *Şehnâme-i Âli Osman*, Topkapı Sarayı Müzesi Kütüphanesi, A. 3592, 39b. Used by permission.

onlookers as the imperial parade passed. The sultan used lions to induce dread and wonder in his imperial subjects. As with one glance from the stately sovereign, a lion's sudden jolt could effortlessly kill anyone in the gathered crowd. The goal was not brute terror, but a more sophisticated message about the harmonious reconciliation and coexistence of opposing natural forces. Evoking a much older theme of sovereignty in the Muslim world—epitomized in the famous Mughal image of lion and lamb known as Jahangir's Dream—the lion walking peacefully through the city suggested the sultan's power to provide a social and political sanctuary in which all creatures, weak and strong, meek and violent, could live together in serenity and security.[154]

This dynamic of fright and fancy was central to many public Ottoman festivities and ceremonies. Alongside the real animals in Murad IV's parade, furriers dressed as lions, leopards, and tigers made their way through the crowd scaring young and old alike, in the words of Evliya Çelebi, "just for the fun of it."[155] These threats of bodily harm and pain were performed as theater on the streets of Istanbul to impress upon the empire's peoples the sultan's ability to provide refuge and protection for all his subjects—human and animal. Brought to the edge of terror and physical pain, spectators left these staged events unscathed but hopefully having gained some understanding of the nature of their sultan and his state.

Imperial Pursuits

In considering the roles of animals in Ottoman projections of imperial sovereignty, one aspect of the human-animal relationship looms larger than all others. Hunting by Ottoman sultans, Egyptian provincial governors, and, indeed, most other political figures in Islamic history was a key facet of the articulation and maintenance of notions of kingship, governance, and power.[156] At first glance, the hunt may seem to be the quintessential institution of human domination over other animals, but the hunt in the early modern world was primarily about the competition *between* animals—and thus the very real fear (and frequent fact) of human death—rather than the subjugation of species.[157] The dangers of the hunt were made all too clear to Ottoman sultans by the fact that the grandson (Süleyman Pasha) of the very founder of the empire died in 1357 during a hunting expedition when his horse tripped and threw him to the ground.[158] At least until the beginning of the seventeenth century, the hunt functioned as an arena of productive potential danger in which Ottoman sultans could hone and display their bravery, chivalry, cunning, equestrian skills, strength, military instincts, and sovereign ideals.[159] Indeed, the hunter-sultan emerged as a major topos of Ottoman rulers. The notion of the political sovereign as hunter (*avcı* in Turkish) is a very old one in Eurasia with roots in various traditions—Sassanian,

Figure 5.7 The Animal World at Ease from the Ottoman Translation of Qazvīnī's *ʿAcaʾib ül-Mahlukat*. Walters Art Museum, Ms. W.659, 267b. Used by permission of Images for Academic Publishing.

Vedic, Arab, and Turkic.[160] The use and efficacy of the hunt in portraying political power was most intense in the Ottoman Empire during its early centuries of expansion; it then waned as the polity moved toward a period of consolidation, sedentary power, and administrative refinement—the bureaucratic state—in the late fifteenth and sixteenth centuries.[161] In the sixteenth century, as actual hunting most likely decreased, texts about hunting increased. To chronicle a fading art, to recast earlier sultans in the glowing light of nostalgia, and to critique contemporary sultans for lacking their predecessors' skills, several new genres of imperial chronicle writing emerged in this period expounding on the role of the sultan as hunter. One of the most prominent of these new literary forms was the *Hünernāme*, or *Book of Talents*.[162]

As both patron and subject of these books, and as sponsor of much imperial reform in the sixteenth century, Süleyman I (Süleyman Kanuni, not to be confused with Süleyman Pasha who died in 1357) not surprisingly emerges as the grandest of hunter-sultans in these texts.[163] He purportedly went hunting nearly every day when in Istanbul.[164] These texts also recast the reigns of earlier sultans through a hunting lens to describe their royal attributes and ferocity on the horse.[165] Both of the late fourteenth-century sultans Murad I and Beyazid I, readers learn, created large hunting reserves near Edirne and in other parts of the Ottoman Balkans that were maintained by a corps of no less than five to six thousand men.[166] Murad I also introduced into the Ottoman military a series of animal fighting regiments: *samsuncular* (mastiff keepers), *segbanlar* (hound keepers), *zağarcılar* (a different group of houndsmen), and *turnacılar* (crane keepers). Dogs and to a lesser extent birds—most prominently falcons—were already in widespread use in Ottoman hunts by the time of Murad I's reign, and their adoption by the imperial military suggests both the familiarity of Ottoman sovereigns with hunting dogs and the importance Murad I placed on them.[167] Murad I's interests in hunting were fully evidenced and further strengthened by his construction of a grand hunting lodge in Çömlek near Edirne.[168]

Enjoying the institution Murad I founded, later sultans regularly frequented this and other hunting lodges around Edirne.[169] Murad II, whose reign (in two periods) roughly spanned the second quarter of the fifteenth century, was an avid hunter, famous for his collection of over one thousand hunting hounds and two hundred birds. He spent a great deal of time at the Çömlek hunting lodge and even took the opportunities of his army's battles at Vidin and Zlatica to enjoy a few hunting expeditions in the vicinity. His son Mehmed II and grandson Beyazid II likewise enjoyed the imperial hunting lodge at Çömlek, using it as a base of operations for military campaigns in the Balkans and as a kind of vacation mini-palace where they received envoys and conducted official imperial business. Even after the conquest of Constantinople in 1453 and the transfer of the capital from Edirne, Mehmed II and his son Beyazid II continued to return to areas around the former capital for elaborate hunting parties aimed at building camaraderie and esprit de corps among the palace and military elite.[170] Selim I was similarly well-regarded for his hunting prowess. He dedicated most of his winters to hunting expeditions and traveled as far as Trabzon for the thrill of the chase.[171]

From Osman to Süleyman, hunting was thus not only a favorite pastime of sultans, but also an activity with clear political implications for both the contemporary and subsequent historical images of Ottoman imperial sovereigns.[172]

How should this deep and abiding interest of Ottoman sultans in hunting be understood and interpreted? First, hunting was important because it helped construct notions of what a just, proper, and successful Ottoman ruler was to be. Hunting was a tool to forge Ottoman sovereignty. A hunting manual from the early seventeenth century, for example, clearly articulates that it was essential for Ottoman sultans to hunt because the experience instilled in them the courage they would need to rule effectively and face enemies of the state.[173] Hunting, moreover, helped dispel pride and laziness in the ruler and make his soul strong.[174] The indolent passive ruler accustomed to pleasure and luxury would allow enemies of the state, both internal and external, to infiltrate his realm. The activity demanded by hunting worked to overcome this lethargy and impress upon the sovereign the need for vigilance in the face of the Ottoman state's many imminent threats.[175] Reflecting a post-Süleymanic understanding of palace politics, hunting was thought to be useful in upholding moral principle and preventing the sovereign from falling into the commoners' world of intrigue, interests, bribery, and petty desire. If the sultan spent too much time in his palace with his cronies, concubines, and advisors—thereby lowering himself to the level of the household—he would become caught up in the customs and concerns of this world and would allow his advisors to think themselves his equals and to begin acting against his interests and those of the empire, a commonly cited trope of why an Ottoman ruler falls.[176]

A related benefit of hunting for the Ottoman sultan derived from another common motif of sovereignty—the juxtaposition between city and forest. This distinction tapped into a long tradition in the Ottoman Empire (and elsewhere of course) that contrasted imperial civilization, the city, the palace, and the protected imperial domain on the one hand, with the wild, the barbaric, the rural, and the uncouth on the other.[177] In opposition to much of the literature on this subject, from the perspective of the sultan-as-hunter, these dangerous elements are taken to be productive rather than destructive of imperial sovereignty. The sultan quickly gets bored in the city.[178] It is too crowded to ride a horse and too comfortable and well-fed for any sort of moral training. In the spartan forest, by contrast, the sultan can escape the worldly cares of his palace and the city—distractions that take him away from the reflective spirit required for the cultivation of sovereign ideals like justice, sagacity, and moral uprightness. Fresh air, open skies, and empty terrain help him to clear his mind and purify his soul and aid in the maintenance of good health. In ways that are difficult to achieve in the city, the open space of the countryside allows the sultan to test himself and his body in order to keep physically fit and mentally sharp-witted. Indeed, according to this prescriptive literature, hunting expeditions in the countryside allowed the sultan and his retinue to experience extremities of weather and terrain and to push their horses to full gallop in the chase, jump them over obstacles, and let them roughhouse a bit.[179]

Significantly, the countryside and hunting reserves were void of human onlookers. If one of the purposes of the hunt was to test the resolve and physical attributes of the sultan by pushing him to his limits, then failure was a real—and obviously undesirable—possibility. Thankfully, however, if the sultan fell off his horse or allowed his prey to escape, no one would see these imperial shortcomings in the largely uninhabited spaces of the hunt. It seems to have been the common understanding of Ottoman writers that sultans were not very accomplished equestrians. Osman II fell off his horse and injured his head, and Süleyman Pasha (never sultan himself but nonetheless grandson of the founder of the Ottoman dynasty) was killed when thrown from his horse.[180] If the sultan were to ride his horse in the city, there was a real risk that a commoner or some vulgar low-ranking soldier would think or—even worse—publicly say, "I am more stable in the saddle, I am more powerful, and I am a better rider than the padishah."[181] These derogatory remarks had the potential to make the sultan seem less powerful and dignified in his subjects' eyes. Such a humiliating possibility was naturally intolerable. Thus, hunting in the relative safety and seclusion of the open countryside afforded the sultan both corporeal and moral benefits as well as defense against potential embarrassment.

Above all, hunting served an important military purpose. A proactive sultan who regularly projected Ottoman power through his hunting prowess could more effectively defend and preserve the state against both external and internal enemies.[182] Chasing a deer or leopard in sport thus came to symbolize chasing the enemies of Islam. As such, many Ottoman hunting treatises (again, in line with a long literary tradition) cite horsemanship and equine veterinary knowledge as of paramount political and military significance for the state—for the person of the sultan, as well as his army.[183] Because most of the Council for Princes literature concurred that the sultan should not personally engage in active combat for fear of his injury or death, the hunt proved even more important as a symbol of warfare and as a controlled staging ground to display the sovereign's military acumen.[184] As a kind of early modern imperial publicity stunt, the hunt thus allowed for the recreation of battle scenarios in carefully orchestrated and far less dangerous environments in which the sultan could display his talents as a fierce warrior.

The skills and lessons the sovereign gained through hunting—bravery, strength, self-reliance, cunning, and resolve—helped protect his realms.[185] As one hunting treatise puts it:

> As the sultan pursues and hunts these predators [here specified as leopards, tigers, wolves, and hawks], he will gain courage and self-confidence; as he observes their many ways, he will note how fiercely they seize and grab, and how ferociously they rage. As he fights

> these beasts and overcomes some of them, he will observe how they seek to evade pursuit through all kinds of trickery and thus make their escape. After witnessing all sorts of situations in which predatory beasts hunt, the sultan's character will come to partake of their temper and nature. The sultan thus acquires characteristics such as strength, determination, focus, and greatness, as well as public spirit, a sense of protectiveness toward his realm and his subjects, and perseverance against his enemies. By watching and observing the behavior of predators and those they prey upon, the sultan learns how to wage war. Brave fighters and warriors who acquire and apply their martial skills in this way are able to defeat the enemy on the field of battle.[186]

Animals, both predator and prey, thus taught sultans how to be proper rulers, and it was they—not the shadow of God on earth—that represented the ideal toward which all creation should strive.[187] The sultan's character would indeed be improved by adopting the "temper and nature" of certain animals.[188] Yet again, the animal served a very productive role for the human world.

Hunting on the part of sultans was not without its critics, however. Certain elements of the palace elite and religious establishment strongly denounced Mehmed IV's and Ahmet I's intense interests in hunting. Mehmed IV was derided by later chroniclers for his "addiction to the hunt."[189] Ahmet I's attachment to the hunt was said to take him away from Friday prayers, to lead him to neglect his duties as a sovereign, and to walk him down the path of excess and pleasure.[190] The fact that hunting was very expensive only added to such critiques.[191] In 1685, Mehmed IV's hunting trips cost half of the total amount expended that year on war with the Habsburgs.[192] Ahmet I was criticized for his excessive spending on hunting preserves, horses, dogs, weaponry, carriages, and hunters; he hired attendants to keep his horses' faces clean at all times and to hold his falcon.[193] Moreover, land devoted to hunting parks and reserves could have been used more productively, it was thought, to cultivate agricultural goods or harvest wood.[194] To counter these critiques, defenders of Ahmet I wrote that the hunt allowed the sultan to inspect the conditions in which the general population lived, interact with them, and understand the best ways to serve their needs.[195] Similarly, recent scholarship has argued that Mehmed IV also effectively used the hunt as a means of interfacing with imperial subjects in the Ottoman Balkans in order to win over converts to Islam.[196] This image of both Ahmet I and Mehmed IV then is one of a public sultan who engaged in hunting as a way to interact with the general populace, a way of "bringing the state to the people."

This potentially interactive element notwithstanding, hunting was clearly the sole domain of the sultan and a small group of advisors around him—a "royal art."[197] It was legally prohibited to common subjects, and they were strictly barred from entering hunting reserves.[198] Similarly, soldiers were not allowed to hunt for fear it would take them away from their military duties and thereby endanger the empire. Hunting was principally a show of royal pageantry with strict rules that excluded all but the sultan and his retinue.[199] One of these rules, for example, stated that when addressing the sultan during a winter hunt, his attendants should stand so as to avoid blocking the sun's full warmth from reaching the padishah.[200] Above all, hunting was a guarded and cultivated domain, a tool the sultan used to produce, project, and protect his image as a just, brave, and engaged monarch. By imitating and then overtaking other animals, the sultan proved that there was only one true king of the jungle, only one true hunter in the forest of empire.

Elephants, lions, tigers, and even those animals Ottoman sultans hunted were all part of a globalized Indian Ocean economy of gift exchange and imperial one-upmanship in which fantastic creatures were strategically used as symbols of power and prestige. The circulation of animals in this early modern world moved horses from the Middle East to South Asia, and elephants and sometimes lions and tigers from South Asia or occasionally North Africa to Egypt and the Ottoman and Safavid Empires. In Egypt and throughout the Ottoman world, a whole infrastructure of care and upkeep therefore developed to support the presence of these animals in Istanbul, Cairo, and elsewhere in the empire. Sultans usefully deployed these creatures to impress upon foreign travelers and domestic subjects the ability of the Ottoman dynasty to command, control, and display the natural world. Likewise, the imperial use of the hunt as a technique to both stage and inculcate ideals of kingship was a cornerstone of Ottoman imperial life between the fourteenth and eighteenth centuries.

The institutions that came to govern the human relationship with charismatic megafauna in the early modern period—the menagerie, imperial processions, the lion house, and the hunt—were all built on a foundation of awe, mystery, spectacle, fear, danger, and imagination. Certain wondrous creatures were still rare and unknown enough to be deployed by imperial sovereigns in Ottoman Egypt, Istanbul, and throughout the Muslim world and beyond in their symbolic and military rivalries against one another. These early modern institutions of the human-animal relationship would give way in the nineteenth century to very different kinds of interspecies structures. The charismatic economy would soon become an economy of brutality.

6

Encagement

Four legs good, two legs bad.
—George Orwell

In 1826, Rifaʿa al-Ṭahṭāwī, a twenty-five-year-old Azhari imam, arrived in Marseille as part of the first educational delegation dispatched by Mehmet ʿAli to Europe. al-Ṭahṭāwī and his other Egyptian companions were sent to France to learn about European educational institutions and technical expertise in the hopes that they would return to Egypt to apply some of this know-how at home.[1] This generation of students would prove central to the organization of Egyptian bureaucratic, educational, and technical institutions over the course of the nineteenth century. al-Ṭahṭāwī in particular has been seen as one of the progenitors of modernity and liberal and enlightened thought in Egypt.[2] His time in Paris is credited, rightly or wrongly, with planting the seeds of this developmentalist attitude early on in his life.[3]

Six months later, another ship from Alexandria arrived in Marseille. The tallest of its passengers was a twelve-foot female giraffe sent by Mehmet ʿAli to Charles X.[4] By the time this creature arrived on October 31, 1826, she had earned the distinction of being the first giraffe to ever set foot on French soil.[5] She was promised to the Jardin des Plantes in Paris, but it was quickly determined that there was no easy way to transport her from Marseille to the capital. A train was out of the question since no car could accommodate a giraffe and because of the many low tunnels along the way. Some other sort of wheeled contraption seemed precarious as well. It was therefore determined that the best solution was to walk France's first giraffe the more than 450 miles between Marseille and Paris. The wonder, fascination, and delight occasioned by people seeing this strange creature for the very first time in their lives was repeated in each small village and town the giraffe passed through on her way to Paris.

The year 1826 thus saw two important beginnings for both Egypt and France: the first Egyptian students to be trained in Paris with the goal of importing European technical knowledge and educational skills to Egypt and the

Figure 6.1 An Earlier Giraffe Gift from Egypt, 1404. The Ambassadors of the Egyptian Sultan al-Nasir Faraj ibn Barquq Present their Gifts of Tribute, Including a Giraffe, to Timur (1370–1405). Sharaf al-Din ʿAli Yazdi, *Zafarnama* (1436). Worcester Art Museum, Worcester, Massachusetts, Jerome Wheelock Fund, 1935.26. Used by permission.

beginning of decades of imagination for French spectators who were afforded the opportunity to marvel at the sight of France's very first giraffe.[6] In this nineteenth-century economy of import and export, Egypt got supposed enlightenment and France got a giraffe. It was not a fair exchange.

The story of this giraffe moving from Egypt to France is but one example of the very different human-animal landscape that existed in Egypt after 1820.[7]

Chapter Two explains how decades of diminishing domestic animal populations and efforts by Egyptian elites to gain landed wealth resulted in a shift away from animal labor in the countryside. Chapter Four shows how changing notions of disease, urban health, and sanitation pushed dogs out of the city and out of their evolutionarily forged social niches. By focusing on charismatic megafauna (and a few other animals) and on some of the institutions that took hold of them in the nineteenth century, this chapter explores how the human-animal relationship in the middle decades of that century was of a very different conceptual form than what it had been in the early modern period or, indeed, ever before.

The most important reason for this newly emergent difference in the nineteenth century was the fact that various institutions of state came to mediate relationships between humans and animals. These institutions included the Egyptian School of Veterinary Medicine, the military, industrial farms, the silk industry, slaughterhouses, colonial hunting parks, and—above all—the zoo. Most of the animals targeted by these new zoological structures were charismatic megafauna. Some, however, were not. Animals like horses and silkworms, while of course technically not megafauna, are essential to consider because the establishments created to manage these animals significantly shaped how humans came to interact not only with them but also with the entirety of the animal world and with the human world as well. Collectively, these new sorts of institutions were the crucial force molding the character of the human-animal relationship in the nineteenth century.

All of these institutions were nineteenth-century attempts to deal with much older exigencies—labor, disease, transport, public health, diplomacy, sanitation, and war—that Egyptian human and nonhuman animals had quite successfully dealt with for centuries. The veterinary school, for example, served to medicalize the human-animal relationship in the nineteenth century. Animals came to be seen as disease carriers that had to be understood primarily through epidemiology and zoology rather than through experience, localized knowledge, or home remedies as had previously been the case. Moreover, similar to the example of meat production Chapter Two discusses, Mehmet ʿAli's cavalry undertook the centralization and monopolization of horses in the 1820s and 1830s to support various military campaigns. As evidenced by the giraffe sent to France in 1826, Egypt's human-animal relationship increasingly came to be shaped by capitalist modes of exchange, with market demands affecting Egypt and its animals in various significant ways. Egypt indeed became one of this period's primary suppliers of megafauna for zoological collections in Europe and later North America.[8]

The most famous and successful European animal supplier of the nineteenth century, the German Carl Hagenbeck, secured a large portion of his animals from Egypt.[9] Although the Hagenbeck family's attempts to import animals from

Figure 6.2 London's First Hippo with His Egyptian Handler, 1850. *The Illustrated London News*, May 25, 1850.

Egypt and other parts of Africa began in the late 1840s, their business rapidly accelerated and expanded in the late 1850s and 1860s and peaked in the 1870s after the opening of the Suez Canal.[10] In the spring of 1870, Carl Hagenbeck traveled to Suez on an animal purchasing and transport mission. When he arrived at the hotel where his Italian supplier was staying, he was floored by the collection waiting for him in the lobby. "I will never forget the unique scene with which I was greeted when we entered the court of the hotel. Had a painter seen this sight, he would perhaps have immortalized it under the title, 'chained wilderness.' Elephants and giraffes, antelopes and buffalo were bound to the palm trees. In the background sixteen full-grown ostriches ran about freely, and in sixty crates moved lions, leopards, cheetahs, thirty spotted hyenas, jackals, lynx, serval cats, monkeys, marabou storks, rhinoceroses, birds, and a large number of birds of prey."[11]

All of this "chained wilderness" was transported by train to Alexandria, where it joined another shipment of animals waiting for Hagenbeck.[12] The entire group was then loaded onto a ship bound for Trieste, constituting—according to Hagenbeck—"the largest animal transport that to that point had been brought to Europe."[13] The amazing assortment and number of animals Hagenbeck shipped from Egypt in 1870 to zoological collections in Vienna, Dresden,

Figure 6.3 European Depiction of Egypt's Animal Bounty. From author's collection.

Berlin, and Hamburg solidified his position as the continent's leading purveyor of charismatics.

As a major source of animals for European zoos, Egypt emerged as a leading player in the acquisition, purchase, sale, trade, supply, and distribution of animals all over the world in the nineteenth century.[14] This was only part of Egypt's new anthrozoological order. As it became more and more enmeshed in capitalist networks of trade and transport, Egypt grew increasingly susceptible to more frequent outbreaks of epidemic diseases and epizootics. Even an older institution like the hunt became transformed in the nineteenth century into a project of empire and capital. And as the silkworm industry shows, new interspecies economies often reduced the animal body to nothing more than a market commodity.

The exchange of charismatic megafauna in the nineteenth century was thus of a fundamentally different order than it had been in previous centuries. Whereas in the early modern period elephants and lions were unique wonders sovereigns strategically deployed as anomalous living forms meant to impress visitors and subjects alike, in the nineteenth century these and other animals came to feed an increasing hunger for the display and consumption of never-before-seen animal specimens.[15] This demand was driven by wholly different interests, desires, and forces, and it operated on a much wider scale.[16] The nineteenth-century animal economy was principally propelled by Mediterranean, and later Atlantic, demand rather than Indian Ocean commerce.[17] This was due in part to the expansion of European empire in parts of Africa, which allowed for the exploitation of animals formerly secured only in South Asia. Because of its quasi-colonial political rule in Sudan, Egypt in particular served as a kind of proxy for European empire in Africa.[18] France's first giraffe, for example, in all likelihood came from the Sudan.[19] This geographic shift in the movement of charismatic megafauna toward the Mediterranean was primarily a function of Europe's zoological gardens.

Egypt played a vital role in this new brutish animal economy not only as a supplier of animals to foreign zoos but also as a consumer of them for domestic purposes.[20] By the middle decades of the nineteenth century, Egypt was well on its way to assembling an impressive zoological collection of its own. Like all zoological endeavors, the Egyptian zoo was an imperial institution. It symbolized the nineteenth century's new human-animal relationship—one based on the fundamental separation of large numbers of animals from humans. The medicalization of the animal world, treating animals primarily as commodities to be exchanged on capitalist world markets, and putting them behind bars were all new phenomena in the nineteenth century that severed the historic ties between humans and animals in ways that both embodied and contributed to the wholesale redefinition of Egyptian society, economy, and politics.

The Animal Body

The first major new animal institution of the nineteenth century was the school of veterinary medicine. The founding of the school in 1827 was obviously an attempt to address one of the oldest issues affecting the human-animal relationship—disease—through several new techniques. The first was the employment of French expertise to help with the founding and operation of the new animal institution—an outgrowth of both the French expedition to Egypt between 1798 and 1801 and the sort of educational delegations that took al-Ṭahṭāwī and other Egyptians to France in the nineteenth century.[21] Second was the school's clear objective to instrumentalize and maintain healthy animals for warfare. The third was the new notion that medical understandings of the animal body were the best means to forge a more productive and sustainable society.[22] Disease in all its forms—human, animal, social—was a problem that had to be removed from Egypt, and the school of veterinary medicine was a major part of the effort to do that.

The Egyptian School of Veterinary Medicine was not the first opportunity for French scientists to study Egyptian animals. Indeed, Napoleon's expedition

Figure 6.4 Étienne Geoffroy Saint-Hilaire's Egyptian Crocodile Sketches. Commission des sciences et arts d'Egypte, *Histoire naturelle*, vol. 1, pt. 3 of *Description de l'Égypte*, Reptiles, pl. 2. Beinecke Rare Book and Manuscript Library, Yale University.

to Egypt at the turn of the nineteenth century led to gigantic leaps forward in French and wider European zoology, anatomy, and botany.[23] The person most immediately responsible for the zoological work that came out of the expedition was Étienne Geoffroy Saint-Hilaire.[24] He spent much of his years in Egypt examining fish, ostriches, and crocodiles, and collecting mummified animal bodies and other zoological specimens to take back with him to Paris.[25] His major contribution to nineteenth-century zoology and anatomy was his idea of the "unity of type," which for the first time put forth the notion that homologous structures existed across organisms, a concept that angered many of his contemporaries but would later prove to be a major influence on Charles Darwin.[26] Identifying homologous anatomical features shared by seemingly unrelated species quickly led to recognizing the roles of evolutionary relationships and environmental factors in natural selection. Darwin understood the importance of Geoffroy Saint-Hilaire's contributions to nineteenth-century science and to his own work.[27] He wrote:

> This is the most interesting department of natural history, and may be said to be its very soul. What can be more curious than that the hand of a man, formed for grasping, that of a mole for digging, the leg of the horse, the paddle of the porpoise, and the wing of the bat, should all be constructed on the same pattern, and should include the same bones, in the same relative positions? Geoffroy St. Hilaire has insisted strongly on the high importance of relative connexion in homologous organs: the parts may change to almost any extent in form and size, and yet they always remain connected together in the same order.[28]

Zoology was quickly becoming the primary means through which humans were engaging with and understanding animals.[29] More importantly, the human

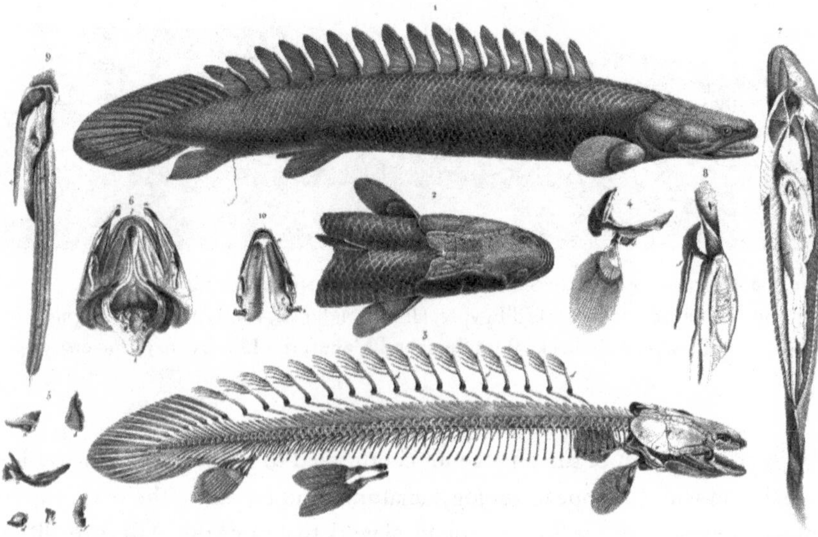

Figure 6.5 Étienne Geoffroy Saint-Hilaire's Anatomical Drawings of *Polypterus bichir*. Commission des sciences et arts d'Egypte, *Histoire naturelle*, vol. 1, pt. 3 of *Description de l'Égypte*, Zoologie, Poissons du Nil, pl. 3. Beinecke Rare Book and Manuscript Library, Yale University.

experience of animals in Egypt heavily shaped European and other globalizing scientific understandings of the animal world.

Epizootics increasingly came to be understood and dealt with through veterinary medicine and zoological conceptions of animals' bodies. In point of fact, the immediate impetus for the founding of the Egyptian School of Veterinary Medicine was an epizootic among livestock working in rice fields in Rosetta in 1827. Ever the believer in European expertise, Mehmet ʿAli hired two French veterinarians to deal with this crisis in 1827.[30] Pierre Nicolas Hamont and his associate Pretot were both distinguished graduates of the renowned French National Veterinary School of Alfort.[31] In dealing with the Rosetta epizootic, Hamont identified a combination of mostly climatic factors as the main cause of the animals' diseases—high rates of humidity, exposure to extreme winter rainfall, inadequate amounts of food, and too much work.[32] He and Pretot quickly set out to ameliorate these conditions by constructing better stables and by preventing animals from drinking from the Nile. Having proven themselves, Mehmet ʿAli quickly took advantage of the veterinarians' expertise. He sent ten students from the military academy to Rosetta to be trained in techniques useful in the prevention of animal diseases. Unlike almost all of the major educational institutions founded in Egypt in the first third of the nineteenth century, the school of veterinary medicine did not initially operate in Cairo.[33] A short time after the first ten students were sent to the coastal city of Rosetta in

1827, a few more were dispatched by Mehmet ʿAli to train with Hamont, and the school started to improve and extend its facilities, adding stables and drinking troughs for horses, dorms for students, kitchens, more doctors, and an administrative structure. Soon, however, the school began to face many problems. Hamont bitterly complained that the students lacked concentration and seriousness and that the translators employed by the school did not have the veterinary expertise needed to adequately express in Arabic the concepts Hamont related in French.[34] Moreover, Hamont grumbled that the chief translator was lazy and would often speak badly of him to the governor of Rosetta.

Discussions about moving the school to Abu Zaʿabel near Cairo began in 1829, but the move did not actually occur until two years later.[35] In addition to being near the capital and having better facilities, the new location was closer to the school of medicine. Since both schools taught their students anatomy and surgery, Mehmet ʿAli's government thought the institutions could usefully combine resources to share facilities and teaching responsibilities. Hamont strongly resisted merging the schools of human and animal medicine under one administration and pedagogical structure and sent a letter to Ibrahim Pasha making his case that a country as dependent on agriculture as Egypt had to maintain an independent and autonomous school of veterinary medicine. He added that each cavalry division should employ at least two of the school's graduates to oversee the well-being of its horses.[36]

These reasons aside, Hamont's primary opposition to combining the two schools came from his professional competition with his fellow countryman Clot-Bey, whom Mehmet ʿAli had also recently hired to establish a school of human medicine in Egypt.[37] One of Clot-Bey's assistants, a doctor named Bouzari who headed up the Medical Council in the 1830s, repeatedly told Hamont that it would be nearly impossible to establish a regime of human medicine in Egypt as long as he continued to demand a separate veterinary school.[38] Nevertheless, Hamont proved largely successful in maintaining his independent school in Abu Zaʿabel in these early years. A separate veterinary facility was built, complete with classrooms, lecture halls, surgical theaters, chemical labs, a pharmacy, and a recovery unit large enough for 140 horses. Student dormitories and housing for professors were also constructed. Between fifty and one hundred students were enrolled in the school in any given year in the early 1830s.[39]

In the early 1830s, Hamont also began overseeing the health of Egyptian cavalry horses in the Shubra stables. Much of his efforts went toward ensuring a sizeable animal fighting force for various missions of the Egyptian army in this period. In addition to training the army's horses, the cavalry school consistently had an enrollment of two to five hundred soldiers studying the various military uses of horses.[40] Not coincidently, the major push for veterinary medicine in Egypt came at the end of the 1820s and through the 1830s, when Mehmet ʿAli

Figure 6.6 Cavalry Horse Training from Nineteenth-Century Ottoman Cavalry Manual. Ahmed ʿAta Tayyarzade, *Tuhfet ül-Farisin fi Ahval-i Huyul ül-Mücahidin* (1854–55). Walters Art Museum, Ms. W.661, 70a. Used by permission of Images for Academic Publishing.

invaded Greater Syria and parts of Anatolia with the aim of marching to Istanbul to bring down the entire Ottoman Empire.[41]

When Hamont first arrived to the stables in Shubra, he found the cavalry horses inadequately fed and exercised.[42] Many were sick, and these sick horses were allowed to mix freely with the healthy animals.[43] Hamont's first task was to institute a new system for feeding and exercising the horses. This effort yielded immediate results, greatly improving the animals' overall health. Despite this early success, pressure to combine the veterinary school with the school of medicine remained strong. When the school of medicine's main hospital was transferred to Qasr al-ʿAini in 1837, Clot-Bey pushed hard to fold the veterinary school into the new facility. The animal institution was nevertheless able to remain essentially independent, thanks mostly to its indispensable military function. In 1837, the school was moved to Shubra and reached its highest enrollment—122 students.[44]

A compromise was eventually reached between the school of veterinary medicine and the medical school whereby veterinary students would complete their first year of preparatory training at the school of medicine before undertaking more specialized training with animals at Shubra.[45] In 1846, the system was slightly modified to allow students to spend all four years at the veterinary school, with a professor of medicine brought in during the students' first year to teach them physics, chemistry, and botany.[46] In addition to these introductory subjects, students studied muscular and skeletal structures and general anatomy in year one; surgery, pathology, specialized anatomy, and physiology in year two; and internal medicine, surgery, general and endemic disease pathologies, physiology, and general medicine in the third and fourth years.[47] French was studied in all years. The academic year was nine months long, split into three three-month segments.[48] The first of these was spent in book study and classroom lecture; the last two consisted of various practica.

Upon graduation, new veterinarians were dispersed throughout Egypt to oversee animal health in all the country's provinces. Given the Delta's higher concentration of agricultural land, the majority were sent there, though many were dispatched to Fayyum and the south as well. They maintained connections with Cairo once in the field, mostly to secure medicines, equipment, and other necessities that were doled out by the capital's central veterinary administration.[49] One of the first major reform efforts pushed by these veterinarians was to prevent animal labor during the hottest periods of the summer for fear of increasing pulse rates and certain inflammatory diseases.[50] Mehmet ʿAli's new veterinarians recommended that animals not be used during the afternoon, that they be housed in shaded quarters, that they be given copious amounts of drinking water, and that the ground underneath them be regularly watered to keep it cool. This burgeoning veterinary establishment also recommended how long and at what times of day each animal in the countryside was to work and rest. Livestock used to push waterwheels and threshers (*al-nawārij*) were only to perform these tasks for one and a half hours each day. Buffalo cows were not to labor in the afternoons, and camel labor was recommended only at harvest and then only at sunset. Furthermore, Mehmet ʿAli's government directed that trees be planted near waterwheels and other areas where livestock worked to provide them with adequate shade.

These efforts to reform the agricultural utilization of animals show that the relationship between Egyptian farmers and beasts of burden was becoming increasingly subsumed under the scientific gaze of the medical bureaucracy that purported to understand this relationship better than peasants who had been using animals their entire lives. The interspecies relationship was to be managed and controlled to ensure the most effective uses of animal productivity and to defend human and animal populations from disease. Veterinarians' recommendations were ultimately meant to fundamentally reconfigure various aspects of the countryside. It was no longer the intense intimacy between humans and animals that was the primary motor force of the human-animal relationship—and hence of the social world built around it—but rather a logic of order, separation, cleanliness, and disease prevention driven by objective science.

Figure 6.7 Heavily Grassed Area of Rice Cultivation. Norden, *Voyage d'Égypte et de Nubie*, 2, pl. 30. Beinecke Rare Book and Manuscript Library, Yale University.

Needless to say, Egypt's new veterinarians worked hard to prevent the incidence and spread of certain animal diseases. Especially in the northern Delta, where the majority of Egypt's rice was cultivated, large populations of flies and mosquitoes were sustained in the wide pools of standing water that covered rice fields. In 1847, the veterinarian assigned to Kafr al-Shaykh made provisions for large swaths of canvas to cover camels in the city in an effort to protect them from insects and heat.[51] The following year the veterinary council outlawed the importation of southern (*ṣaʿīdī*) buffalo cows to Sharqiyya and elsewhere in the Delta. Most animals brought from the south died quickly in their new homes of angina and liver failure resulting from—it was determined—elevated temperatures and increased humidity in the north and exposure to insects living in the standing water of rice fields and in heavily weeded areas.[52] Temperature fluctuations caused further problems.[53] The cooling of fall and the attendant increase in humidity and precipitation resulted in a spike in the incidence of typhoid, dropsy, fevers, tapeworms, respiratory diseases, and skin infections.[54] Fall, therefore, brought with it various regulations and rules the veterinary establishment believed would curb the severity of these outbreaks. Maximum effort was to be undertaken to ensure that animals were well-fed and especially that they consumed enough salts.[55] They were not to drink from the Nile after the flood or to

Figure 6.8 Drinking from the Nile in Asyūṭ. Commission des sciences et arts d'Egypte, *État moderne*, vol. 1, pt. 2 of *Description de l'Égypte*, Syout, pl. 3. Beinecke Rare Book and Manuscript Library, Yale University.

wade in canals. They were only to consume filtered well water, and their coats were to be washed and cleaned twice a day.

For those animals who nevertheless succumbed there were also new rules about the disposal of their bodies.[56] Whereas previously peasants might have left the bodies of their dead animals on roads or in fields, or thrown them into ditches or canals, early nineteenth-century veterinarians outlawed these practices, which they deemed unclean, unfit, uncivilized, and unhelpful in the prevention of disease. Veterinarians performed autopsies on dead animals in the countryside to determine their cause of death and would then quickly bury the carcasses after their investigations. Furthermore, living animals found to be stricken with contagion were immediately slaughtered and their bodies burned in the hopes of preventing an epizootic. The slaughter site would then be doused with lime or ammonia.[57] In cases in which an animal was clearly sick and on the brink of death, it was often buried alive (*wa'ada*) so that no human would be exposed to the sick body or diseased blood while killing the creature. If it was feared that a contagion might have already spread in a certain location, veterinarians were given the necessary resources to establish a quarantine in the suspected area and were required to send daily reports to Cairo about the situation.[58]

Other efforts to manage and harness the fertility, fecundity, and reproductive capacities of animals exemplify how the school of veterinary medicine contributed to the bureaucratic control of the biological lives of animals and their commodification in the service of a growing capitalist economy. Every Friday, veterinary students undertook controlled breeding operations aimed at increasing the number of desirable animals in Egypt.[59] There were always enough suitable females, but securing an adequate number of healthy and fit males as mates proved a consistent problem, due mostly to their slaughter for meat and, less so, to their use in the countryside. Thus in 1839, Mehmet ʿAli issued several decrees outlawing the slaughter of bulls and also limiting the duration and amount of labor they were allowed to perform on rural estates. Cows too were not to be killed before they reached a productive age. Anyone caught breaking these rules would be severely punished.

Mehmet ʿAli also sent some of his own veterinarians and hired others to secure premium animals from outside of Egypt to mate with domestic females and thereby "improve" Egypt's gene pool.[60] In the late 1820s and throughout the 1830s, oxen were regularly imported from Marseille and donkeys from Italy. In the early 1830s, Mehmet ʿAli sought new kinds of sheep because the wool of most Egyptian sheep varieties was not suitable for commercial manufacturing. He serendipitously acquired sheep in 1834 when the King of Sardinia, in recognition of the appointment of a new Egyptian consul to the island, sent Mehmet ʿAli a gift of 150 Spanish merinos—prized, of course, for their fine soft wool.[61] The king also sent along twelve shepherds to show their Egyptian counterparts how best to care for the animals. As part of the effort to maintain this flock and to improve local sheep stocks, the Egyptian veterinarians put in charge of the merinos demanded that Bedouin shepherds provide them with local (*baladī*) sheep of two or three years of age that, if found suitable after an examination, would be bred with the imported animals.[62] To keep his supply going, Mehmet ʿAli also regularly sent some of his officials to Sardinia and Spain to secure more male merinos. He also sent the king an elephant.[63]

With the veterinary school, Mehmet ʿAli's government instituted a very deliberate biopolitical program to control the bodies, muscular output, pathologies, and reproductive capacities of various animal species in Egypt. This was the state's first major attempt to systematically manage the biological lives of animals—nonhuman or otherwise—and interspecies interactions. These efforts resulted in a growing separation between the worlds of humans and animals as objective science, experimentation, veterinarians, surgical tools, capitalist desires for production, and market-driven reproductive demands became the most important factors determining how humans viewed, used, and related to animals. The intimacy an individual farmer had developed with his own ox or cow, with whom he had lived and worked for years, was no longer the dominant mode of interaction between human and animal. While this dynamic did not entirely disappear from Egypt, it was no longer how the vast majority of Egyptians engaged with animals. Lingering remnants of this older human-animal relationship were no match for the powerful march toward a very different anthrozoological world in the nineteenth century, a world embodied by the emergence of new institutions like the Egyptian School of Veterinary Medicine.

The school came to an end nearly as quickly as it began. Twenty-two years after its founding in 1827, an order by Mehmet ʿAli's grandson and successor ʿAbbas on March 24, 1849, closed it down. Many structural problems with the school's training and the implementation of its mission contributed to its closing.[64] Declining enrollments clearly evidenced some of the school's troubles.[65] Too few veterinarians were being produced to meet the large demand for their services. By the middle of the 1840s, each of the school's graduates was responsible for an average of twenty villages. The distances between these villages were often very great, and it proved impossible for one person to oversee so many animals in so many different places. It also became impossible to effect any unity of action among the country's veterinarians. Some implemented orders immediately, some only when it was in their personal interests, and some never. Moreover, the small number of suitable veterinary facilities in the countryside also meant that animals could often not be transported quickly enough to receive the treatments they needed. And as they did with their sick familial relations, many peasants continued to hide their sick animals from authorities for fear of losing them forever if taken to a government clinic.[66]

All of this contributed to a general sense among some government officials and the populace at large that the veterinary school did not serve an identifiable and useful purpose. It was ultimately unclear whether the school's graduates improved the lives, productivity, quality, or usability of Egypt's animals or increased their number. By the middle of the 1840s, it was determined to be much more cost effective and efficacious to send a few Egyptian students to Paris for veterinary training than to maintain a fledging domestic school.[67] For ʿAbbas, the ineffectiveness and futility of the school were made patently obvious when several of his favorite horses fell sick in 1849. He brought a few veterinarians from the school to try to save his animals, but they could not. Their failure, he wrote, was "as clear as the sun," and he ordered the school to be shut down immediately.[68]

A World of Disease

Although the approximately twenty-year experiment of the school of veterinary medicine was judged a failure, near-constant epizootics made clear that the original justification for the school remained. As Egypt became ever more deeply enmeshed in the global animal economy, the regularity and severity of disease increased dramatically. Unlike epizootics at the end of the eighteenth century or earlier, those in the middle of the nineteenth century did not result from solely domestic environmental phenomena like drought or famine; rather, these diseases most often came to Egypt through the various trading links that connected Egypt to the world. This intertwined history of epizootics and animal movement was not always simply the story of animals bringing diseases to Egypt. A year before the closing of the veterinary school, for example, so many animals were dying in Egypt that Ottoman authorities successfully moved healthy cows and oxen from Teke and Hamid in Anatolia to replace the Egyptian livestock.[69] Still, in most cases the movement of animals across the seas caused rather than solved the problem of widespread epizootic death.

Throughout the 1850s, 1860s, and 1870s, there were numerous outbreaks of various epizootics triggered by animals brought to Egypt through its increasingly global economic connections.[70] In the late 1860s, an outbreak of typhus and rinderpest in Ismailia, Port Said, and other cities in the canal zone caused the Egyptian government to momentarily regret its decision to close the school of veterinary medicine. The French Compagnie Universelle, which managed the canal's construction and later operation, had its own professional veterinarian—itself a sign of the frequency and severity of epizootics in Egypt—who wrote to the Egyptian government about an outbreak of animal disease in late 1865 and early 1866. In a letter dated February 15, 1866, this veterinarian reported on the autopsies he had performed on a dozen cows.[71] He had continually observed these animals from their first signs of disease until their quick deaths twenty-four to thirty-six hours later. Their autopsies revealed hemorrhaged intestines, traces of emphysema, weakened muscles, ulcerations of the rectum, large amounts of phlegm, and lacerations of the esophagus. These years' epizootics were indeed very deadly, with numerous reports of extremely high mortality rates. For example, of 347 infected animals in two villages in 1871, 223 of them died.[72] Other villages also frequently reported animal deaths numbering in the hundreds.

In the face of these serious epizootics in the canal zone, Khedive Ismail entertained the possibility of restarting government-sponsored veterinary instruction in Egypt. In October 1865 he therefore hired yet another French veterinarian to train a group of soldiers in the military academy.[73] These new attempts to raise a fresh crop of Egyptian veterinarians did not get very far. They were hampered by a shortage of equipment and books and, as before, a lack of qualified translators.[74] In the face of the Egyptian government's failures to stop disease, the French Canal Company's veterinarian suggested two possible strategies to break the cycle of epizootics ravaging Egypt in the late 1860s.

The first was to keep animals isolated from one another as much as possible. Animal owners were instructed to prevent their animals from mixing as best they could.[75] At Gabari near Alexandria, for example, cows were separated by variety and kept in different barns in an old cotton mill.[76] In addition to these rudimentary quarantine efforts, various bovine vaccines, such as coal tar, were

Figure 6.9 ADMINISTERING SOURED WINE AND OLIVE OIL TO TREAT HORSE WITH COLIC. Ahmed ʿAta Tayyarzade, *Tuhfet ül-Farisin fi Ahval-i Huyul ül-Mücahidin* (1854–55). Walters Art Museum, Ms. W.661, 74a. Used by permission of Images for Academic Publishing.

also used.[77] Coal tar was not available in large quantities in the canal zone and therefore had to be sent from Cairo, which added cost and complexity to the operation.[78] Other medicinal solutions to cattle plagues included ensuring that livestock stables were kept clean and that animals were adequately fed. The Canal Company's veterinarian also wrote that he had had some success in reducing the incidence of various epizootics by washing animals' mouths daily with vinegar, salt, and crushed garlic. He also cleaned their eyes with phenol diluted in water and made them consume twenty-five grams of phenol dissolved in a large bucket of water daily. He also regularly performed enemas on animals using coal tar and phenol. As all of this suggests, the strategy was to keep all the animals' orifices as clean as possible.

Of greatest concern was explaining the etiology of the epizootics of the 1860s and 1870s. Determining how and whence infections came to Egypt was part of a nineteenth-century competition among nations to measure and compare the relative epidemiological and civilizational health and disease of each against the others. Health became a marker of civilization. The globalization of disease etiology was also of course a consequence of the rapidly increasing trade and movement of animals in this period. It was thought, for instance, that the typhus and cattle plague that came to Egypt in the 1860s originated in the Russian steppe.[79] As a result, Ottoman authorities temporarily outlawed the importation of animals from Odessa to Egypt in 1863.[80] To make up for some of the animals lost in Egypt in this year, geldings (*bargir*) and mares (*kısrak*) were imported from unspecified areas of the Balkans.[81] There was such high demand and short supply in 1863 that even when a universal ban was imposed on exporting livestock from Adana due to a disease outbreak there, an official exception was made to send to Egypt those animals deemed healthy enough for travel.[82]

Cattle plague moving from the Russian steppe to the Ottoman Empire was thought to (and most likely did) spread from the Danube Basin through Hungary to Trieste, Istanbul, and beyond.[83] Many of these cities, and even ports further afield in England and France, took measures in the 1860s and 1870s to prevent the spread of epizootics from the Steppe and Ottoman Empire.[84] British officials in Egypt and elsewhere were particularly worried about the global spread of diseases like rinderpest and wrote a great deal about various diseases' origins, symptoms, and pathways.[85] They too placed most epizootics' origins in the Russian steppe.[86] Cattle imported to Egypt from the Danube basin usually traveled from Trieste to Alexandria, and from there would be distributed to various locations throughout Egypt. Those on their way to Cairo would first be transported to Tanta in the central Delta. Those destined for the canal zone would go east to Damietta and then on to Port Said. Other animals were sent to various smaller towns all across the Delta—al-Manṣūra, Kafr al-Shaykh, and Damanhūr, for example.

Alongside concerns about the importation of cattle plague from Europe, there was also a great deal of attention devoted to the threat of animal diseases coming from Egypt's southern Red Sea ports.[87] Arabian horses—renowned for their strength, size, speed, and beauty[88]—were often brought to Egypt through Red Sea shipping channels both for domestic purposes and to be transported to other parts of the Mediterranean world.[89] Many cattle were also often imported from the Hijaz to Egypt. During the epizootic scares of the late 1860s and 1870s, horses and cattle imported from Jidda and other Red Sea ports became subject to intense quarantine regimes and other strictures on their movement.[90] Before being allowed to dock, Red Sea ship captains were required to provide proper bills of health and other paperwork for the animals onboard their vessels.[91] From their offices in Jidda, British officials regularly dispatched reports about the state of cattle plague in this period.[92] Throughout the 1870s, there were numerous quarantines placed on cattle and horses arriving from the Arabian Peninsula to Egypt and parts of the Sudan.[93] In the spring of 1880, horses arriving in Egypt from the Hijaz were placed in quarantine for twenty-one days.[94] The application of these disease prevention regimes on animals, of course, came alongside those regularly imposed on humans arriving in Egypt from the Hijaz and other points south and east.[95]

The foreign importation of disease was not the only reason animals coming to Egypt from across the Mediterranean and Red Seas fell sick and died once in the country. Healthy arrivals regularly fell ill on their journeys from Alexandria or Suez after drinking from the Nile or its tributaries. On one such occasion in 1866, waterborne diseases hit a convoy of cattle being transported from Alexandria through Damietta to Port Said so severely that corpses began piling up in the river near Damietta to the point that dogs were able to cross from one bank of the Nile's tributary to the other (a distance measured at the time to be nearly two thousand feet) by walking on the animals' carcasses.[96] In an ironic—if rather revolting—twist of fate, many of these dead animals did eventually make it to their final destination, as a number of their corpses were found washed up on empty beaches around Port Said. Given the sources, it is impossible to determine the exact etiology of intense epizootics like this one or the many others that hit Egypt and elsewhere in the 1860s.

One thing the sources do make clear is that Egyptians understood epizootics to be a result of the regular movement of animals across the Mediterranean and Red Seas. While Egyptians had been dealing with animal diseases for millennia, there were two main novelties to the history of epizootics in the nineteenth century. More than at any other time in Egypt's recorded history, the health of its animals was deeply impacted by global commerce, exchange, and transport. Etiological and epizoological forces outside of Egypt could now kill its domestic animals. The second innovation was how the proposed institutional means for dealing with these animal diseases—chief among them veterinary medicine and quarantine—changed the human-animal relationship.[97] In a world in which animals were increasingly treated as potential disease vectors, new technologies of disease management became some of the primary means through which humans and animals interacted.

The State at Full Gallop

Nearly all animals in nineteenth-century Egypt were being progressively displaced from particular facets of the social and economic worlds of humans and given new roles to play in society as exports and imports, trophy possessions, disease carriers, and—above all—commodities. Many of these new roles for animals were consequences of various political and economic reforms that tied Egypt to global networks of capital and politics through processes like cotton cultivation, the linking of European colonial fortunes to the prospects of the Suez Canal, Egypt's growing foreign debt, and Mehmet ʿAli's desires to emulate European forms of knowledge and modes of government. The history of horses in nineteenth-century Egypt puts this shifting anthrozoological terrain in stark relief.

In the early modern period, horses were common signs of status and prestige in the Ottoman-Egyptian countryside. They were some of the most expensive animals an individual could own. Ottoman imperial officials and other elites rode them to project their elevated station within rural society.[98] Bedouin employed horses for raids on village communities and to carry out various military missions.[99] While the massive importation and centralization of horses in the middle of the nineteenth century bore some resemblance to these earlier processes, it radically departed from them in several important ways.[100] Most significantly, horse importations fed the rapid and unprecedented militarization of Egyptian society and the maintenance, cultivation, and extension of the royal family's prestige and public displays of power.[101]

Military efforts to collect domestic horses and import others from abroad went hand in hand with the founding of the school of veterinary medicine. After the Battle of Navarino at the end of 1827, when the Egyptian military observed French cavalry on land, it became clear to Mehmet ʿAli and his military commander and son Ibrahim that the Egyptian army needed a strong cavalry.[102] Therefore, in the first few months of 1829, Mehmet ʿAli set out to collect as many horses as he could. He imported many from Syria and from Dongola in Northern Sudan and also forcibly seized twenty-five hundred head from amirs and other officials in Egypt.[103] Soldiers began to be trained in equestrianism, cavalry warfare, and equine medicine. As part of efforts to co-opt the Bedouin, many of them were brought into military service for their experience and expertise in riding horses into battle. These men and horses were sent to Tura south of Cairo for training under the direction of another hired French hand, Paulin de Tarlé. Assisted by two Italian cavalry officers, he set out to train six hundred men. It soon became clear that this was a wholly insufficient number of teachers for so many soldiers, and thus at the end of 1829, Mehmet ʿAli hired four more French cavalry officers. These European military men bitterly complained about the Bedouins' lack of discipline in training properly for cavalry warfare. By the start of 1831, however, cavalry training, exercises, and pedagogy had become more regularized and established enough to be formally institutionalized in the Egyptian School of Cavalry, housed in the former palace of Murad Bey in Giza.[104]

The first real test of Egypt's new cavalry came less than a year later, during the military operations of the 1832 Syrian campaign. The Battle of Bilan south of the Taurus Mountains was fought almost entirely on horseback. When it ended on July 29, 1832, the Egyptian army found itself victorious but also suffering from the loss of 172 horses.[105] To make up for these casualties, Egyptian soldiers took to stealing horses and other animals from villages across Anatolia and Greater Syria. Thus retooled after Bilan, Egyptian forces prepared for the Battle of Konya that December. Unaccustomed to such violence, the mostly agricultural animals the Egyptian army rode into battle were said to be "trembling" in the face of heavy enemy gunfire.[106] Reeling once again from large numbers of horse fatalities after the battle, Mehmet ʿAli's nephew and military commander

Ibrahim Yeğen was sorely in need of more horses, reinforcements, and supplies from Egypt. While waiting for them, he took five to ten horses as a "tax" from local villages under his control in Anatolia.

Like the school of veterinary medicine, the cavalry school was both a result of and then an important shaper of the nineteenth century's changed human-animal world. While horses and other animals had been used in warfare for centuries in Egypt, never before had they been systematically acquired, trained, and reserved solely for that purpose.[107] Cavalry was thus part of the logic and processes of mechanization, bureaucratization, instrumentalization, and order that took over the human-animal relationship in Egypt in this period.[108] Another result of the further institutionalization and commodification of animals was the emergence of a set of bureaucratic technologies of rule surrounding the cavalry horse. Just as earlier labor charts had counted both humans and animals together as two varieties of the basic unit of "laborer," both horses and soldiers had to be dutifully enumerated at the end of each day of battle during the Syrian campaign.[109] These animals and humans were listed alongside equipment, ammunitions, and other nonliving things as military materiel. As with other animals in the early nineteenth century, the demands of a militarizing and bureaucratizing society gave horses new social roles in Egypt. No longer were they signs of status or objects of exchange. They largely became machines of war, mechanical objects to be thought of, enumerated, and treated like a gun or cannon.

Perhaps less violent, but no less important, than the military uses of horses was the Egyptian ruling elite's importation of the animal to pull carriages. Arrangements to import English horses to Egypt were part of the growing international competition between states to bring animals to their territories from as far away as possible and, in the case of Egypt, to mark similarity with Europe. Whereas previously simply riding a horse was the mark of a rural official's higher status, in the middle of the nineteenth century the Egyptian royal family strove to publicly display its prestige by importing horses from England to draw its European-style carriages.

A telling example of the use of horses to mark class difference in Egypt is the letterhead of the Shepheard's Hotel from the 1870s (figure 6.10). A favorite of European and American tourists to Egypt (Herman Melville stayed there in 1857, for example), the hotel catered to foreign tastes by offering guests an appealing "authentically eastern" experience with all the creature comforts of a European-style lodging. As the hotel's letterhead prominently displayed at its center, horse-drawn carriages were an important part of the self-image of the hotel both for its foreign guests and elite Egyptians. Typical of many tourists in the middle of the nineteenth century, American traveler George Freeman Noyes stayed at the Shepheard's during his time in Cairo in May 1854 and toured the city on horseback and donkey-back and in horse-drawn carriages.[110] In the context of

Figure 6.10 Shepheard's Hotel Letterhead, Late 1870s. TNA, FO 141/131 (1879). Used by permission.

the globalizing animal economy of the middle of the nineteenth century, horses demonstrated what some believed to be the new proper roles for animals in human society—leisure, comfort, class distinction, and bourgeois sociability.

The Egyptian royal family most certainly conceived of horses in this way. In the late 1860s, the importation of English horses to the Khedive's royal stables in Alexandria was managed by the Englishman George Smart.[111] By 1869, there were close to seventy carriage horses in the stables, with more on the way.[112] There were, in fact, far too many horses for the sixteen coachmen currently in the employ of the royal family. So Nubar Pasha, one of Ismail's most trusted advisors, wrote to Smart from Paris, where he was engaged in negotiations with the Suez Canal Company, to arrange for the hire of twelve additional coachmen.[113] The increasing use of carriages in Cairo by both foreigners and Egyptians, and also of more modest carts by Egyptian laborers, put additional pressure on the labor of horses, donkeys, and camels.[114] Not only did the widespread use of wheeled vehicles create more traffic on the streets, but the increasing popularity of human push and pull carts cut into the hauling work animals used to undertake almost exclusively on their own.

Thus, the importation of horses to Egypt for two very different purposes—one military and the other imperial and rather mundane—highlighted two important aspects of the new nineteenth-century human-animal relationship. Cavalry horses showed how living animals came to be objectified as weaponry and how institutions largely came to manage the human experience of these animals. Likewise, the use of horses to pull the elite's carriages represented a similar objectification of animals as part of ostentatious public shows of political and class power.

Threads of Capital

The story of the silkworm in Egypt in the third quarter of the nineteenth century further demonstrates how a market-driven capitalist ethos fueling the globalized movement of animals fundamentally changed relationships between humans and other animals.[115] Unlike early modern beasts of burden, the utility of silkworms derived not from their capacities to help humans perform useful work in the countryside. Rather, these animals were brought to Egypt for an incredibly specific economic goal—to extract and abstract their very physicality as a material commodity. The only reason silkworm importation to Egypt was deemed desirable was because it allowed Egyptians to participate in the global market for silk. For the very first time in Egypt's history, an animal was valued for no other reason than the desire to accumulate wealth. Silkworms in nineteenth-century Egypt were thus not only alienated from their labor but from their very bodies. Animals were literally becoming capital.[116]

Efforts to create and maintain a domestic Egyptian silk industry began in earnest in the 1870s.[117] Like horses, silkworms were imported to Egypt through networks of international capital and trade. The globalized context in which human Egyptians were coming to interact with animals is immediately clear from the record of conversations about the possibility and utility of a silk industry in Egypt. India, Japan, Cuba, and China were all cited as examples of countries in which the domestic silk industry was thriving and actively contributing to the economy, industrial production levels, and living standards.[118] A Persian official named Mirza Ibrahim Khan was consulted on the Iranian experience of sericulture;[119] mulberry trees to support silkworms were imported from Italy; and French experts were employed to help in the worms' proper rearing. In April 1871, 42,300 "grafted [mulberry] trees of the finest quality procured in Italy" were planted in a town in the Egyptian Delta identified as Desumis.[120] This planting was overseen by Maxwell Anketell, a British agriculturalist and entrepreneur in the employ of Khedive Ismail's government.[121] In May 1871, Anketell wrote to Riaz Pasha, one of Ismail's advisors, about what was needed in Egypt to create a silk industry and, more specifically, about the importance of the mulberry tree itself. "Amongst other advantages, the leaf of the

Mulberry tree is a valuable food for Cattle and the small shade which it throws upon the roads and water ways is an advantage to both man and beast in all civilized countries, and most of all in Egypt."[122] Planting trees to feed the worms that would produce the silk was, however, only the first step.[123] Once a sufficient number of trees were in the ground, a central magnanery had to be constructed to house the silkworms and a reeling establishment was also needed to get silk from their cocoons. M. Sauvadon, a Frenchman put in charge of the Khedive's zoological gardens in 1871;[124] the Iranian Mirza Ibrahim Khan; Jean-Pierre Barillet-Deschamps, another hired Frenchman and the Khedive's chief gardener from 1870 to 1873;[125] and several others were all part of the effort to further the Egyptian economy through the utilization and commercial maximization of silkworms and their products. Appertaining to the trend toward scientific agriculture, these men and their Egyptian bosses discussed the proper conditions needed to maintain healthy silkworms, how to maximize cocoon yields, how much revenue could be generated per worm, and future experiments to further maximize profits.

By referencing Japan, India, China, and Cuba, industrial sericulturalists in Egypt invoked the civilizing potential of the silkworm to justify the development of a domestic silk industry. In addition to his previous comments about the mulberry tree's civilizing potential, Anketell articulated the idea that Egyptian families would benefit from the presence of the silkworm in their midst. "The cultivation of the Silk Worm is essentially a civilising agriculture. It causes the peasant to feel that it would be better to keep his house a little cleaner; it causes his children to try to do better, and finally it produces in him a desire to improve his habitation. Thus it is that in all countries where silk cultivation is carried on, the habitations of the poor are clean and healthy."[126] The Khedive and his retinue of advisors clearly seem to have heard this message about the civilizing and economic benefits of silk and accepted it without reservation. Thus, by the spring of 1872, there were over one thousand acres in the Delta dedicated to industrial sericulture, with upward of 140,000 mulberry trees under cultivation.[127] A complex network of canals and roads serviced these trees. Anketell estimated that 60,000 of the 140,000 trees would be ready to produce a crop in the first year alone, leading to an overall yield of approximately four hundred pounds of silk.[128]

Sauvadon, head of the zoological garden, soon replaced Anketell in overseeing the raising of silkworms in the Delta. Assigning this industry to the chief administrator of the zoological collection made sense, since the cultivation of these worms for the products of their bodies was increasingly becoming conceptualized as similar to the management of other animals in the Khedive's garden. In December 1876, Sauvadon carried out further experiments in the zoological garden aimed at improving the yield and quality of silkworms.[129] These trials were clearly successful, as only a few years later in June 1878 Egyptian silk was honored with an award from the French Imperial Zoological Society of Acclimatization.[130] In May 1879, seeking further economic validation from the Lyon Chamber of Commerce, Sauvadon sent samples of Egyptian silk, some worms, and their cocoons to France.[131] As these attempts to gain foreign recognition attest, the cultivation of the silkworm in Egypt was both a colonial and a capitalist project of civilizational comparison and competition.

Animals like the silkworm reflected Egypt's developing position as a junior partner in global imperialism and its role as a producer of raw materials for global capitalist markets.[132] The animals at the heart of these dual processes—whether horses needed for imperial carriages, silkworms sought for their profit-generating potential, or various other creatures—largely lost out. They no longer stood as the productive and integral working members of society they had been in the early modern period. They were not supported materially as dogs had been in seventeenth-century Ottoman Egypt. And they were certainly not recognized, as elephants and lions had been in early modern Cairo and Istanbul, as wondrous anomalies of creation to be contemplated, emulated, and feared. Rather, they were alienated and abstracted objects exploited for raw wealth accumulation.

Egypt was clearly attempting to affirm its place at the table of what European modernity—as stated quite explicitly by Anketell—termed "civilization." This entailed asserting Egypt's difference from certain other regions of the world—that it was *not* in Africa or the Middle East but indeed part of Europe. Khedive Ismail said quite plainly in 1879, "My country is no longer in Africa; we are now part of Europe. It is therefore natural for us to abandon our former ways and to adopt a new system adapted to our social conditions."[133] The consequences for the human-animal relationship of this "new system"—indeed, of this entirely new reimagination of Egypt—could not have been greater. To more powerfully make the case for its own modernity, Egypt engaged in sericulture and various colonial missions to acquire animals in the late nineteenth century. In so doing, Egypt's ruling elite sought to prove that their society had the resources, knowledge, and expertise needed to develop its own domestic industries and menagerie, just like other "civilized countries."[134] More importantly, Egypt's rulers wanted to showcase their abilities to dominate the natural world of chaos, wilderness, disorder, and brutality. If Egypt could fully dominate the nonhuman nature of various sorts of beasts—be they silkworms, donkeys, eagles, or hippopotami—then surely it could produce and then control a successful and economically viable human social and political order.

A Garden of Cages

Cairo's zoo, like the zoos of most other cities, emerged in the nineteenth century during a period that saw—in socialist art critic John Berger's words—"the disappearance of animals from daily life."[135] The development of commercial agriculture and urban public health and sanitation regimes were part and parcel of this process of disappearance in Egypt. Zoological collections in menageries and zoos are obviously very different phenomena, but they nevertheless share some family resemblances with what happened earlier in the countryside and city and constitute an essential part of the new human-animal landscape that emerged in Egypt in the nineteenth century. This was a world in which humans and animals interacted far less frequently and in which the fate of Egypt's humans was beginning to be separated from that of its animals. As Berger continues, "The zoo to which people go to meet animals, to observe them, to see them, is, in fact, a monument to the impossibility of such encounters. Modern zoos are an epitaph to a relationship which was as old as man."[136]

The new world of human and animal in Egypt that replaced an older one of mutual interspecies reliance and intense cooperation is best thought of as a colonial relationship of capture and domination.[137] One of the outcomes of this "disappearance of animals" from the daily lives of most Egyptians was their progressive cordoning off from human society in separate spaces set aside for the display of the sovereign's domination over nature. In the early modern period, political sovereigns used charismatic megafauna to impart a message about their sultanic capacity to order, harmonize, and reconcile nature; by contrast, these animals were used in the nineteenth century as examples of the brute power of human political leaders to capture, control, and render nature at their will.[138] Indeed, politics in Egypt in the late nineteenth century was largely about the exercise of power over other parts of creation, whether human, animal, or otherwise. State institutions of education, discipline, enumeration, conquest, management, and control emerged as the most dominant forms of government under Mehmet ʿAli and continued to be refined over the course of the nineteenth century and beyond. The formation and operation of the zoo was cut from the same cloth as colonial attempts to conquer territory in the Sudan, the Hijaz, Greater Syria, and elsewhere; urban reform efforts that displaced poorer residents of Cairo; public works projects to reclaim land, manipulate waterways, and dig canals; and the establishment of prisons, hospitals, quarantines, and various other institutions of enclosure.

The animals that would eventually populate the Cairo zoo were first collected in Khedive Ismail's menagerie on the grounds of his palace garden on the island of Gezira. The first evidence of serious efforts to constitute the collection comes from 1868, a year after Ismail's visit to the Exposition Universelle in Paris, where he visited menageries and other institutions of ostentatious public animal display, including the exhibition of a whale carcass.[139] Ismail's attempts to create a collection of animals were part of the changes to Cairo's urban landscape that accompanied the massive celebration of the Suez Canal opening in 1869, an event meant to impress upon European visitors the grandeur of the city and—perhaps more importantly for Ismail—of Ismail's family.[140] Like the circulation of veterinary expertise, the use of horses for military purposes, and the importation of silkworms, Ismail's animal collection participated in the process of Egypt joining the nineteenth-century global trade in and competition for charismatic megafauna and other animals. Zoological collections across Europe, North America, and elsewhere were engaged in a heavy contest to secure animals from as many parts of the globe as possible.[141] A zoo's stature was in large measure built on having as many different species from as many faraway places as possible.[142]

For nineteenth-century Egypt, "faraway" very often meant the Sudan and other parts of east Africa. In late 1868, Ismail's government made arrangements to have large numbers of various kinds of animals and birds moved to Egypt from different parts of Sudan, Nubia, and Abyssinia.[143] He enlisted the support of various government ministers in this affair, including the Egyptian governor general of Sudan and members of the medical establishment.[144] Those moving these animals from the south to Egypt were instructed to carefully observe the climate, diet, and behavior of the creatures in their native habitats so that these conditions could be replicated as much as possible during the journey to Egypt and then once in Cairo. These officials were afforded the full financial backing of the government and were instructed to make every effort to bring both a male and female of each species. These were colonial ventures to capture the living nonhuman animal, made possible by the possession of a colonial territory that had historically served as a source of human slaves.[145] The Khedive's instructions to secure both sexes of each species further evidence the state's intentions to capture, exhibit, manipulate, and control the natural world for some time to come. Nature was to be so fully dominated that it would literally reproduce itself on display in the controlled space of the Khedive's garden.

By 1871, Ismail's hired French landscape gardener Barillet-Deschamps, who had previously worked in both the Bois de Boulogne and Champ de Mars gardens in Paris, had appointed M. Sauvadon to oversee the growing collection of animals in the Gezira palace garden.[146] Under Sauvadon's care and direction, the zoological gardens were greatly expanded and refurbished in the early 1870s.[147] Shaping the physical landscape and built environment of animal enclosures so as to effectively contain animals while still affording the illusions of a "natural" setting and the freedom of movement was one of the major innovations in zoological display in this period, and Sauvadon was well aware of some of

Table 6.1 **Animals in the Palace Garden, October 4, 1878**

Animal	*Total Number*	*Animal*	*Total Number*
Lion	4	Shoveler	1
Lion cub	2	Pintail	1
Panther	1	Blue peacock	6
African civet	2	Blue peachick	8
Bear	1	White peacock[1]	3
Hippopotamus	1	White peachick	3
Giraffe	1	Plumed peacock[2]	1
Sudanese buffalo	2	Barbel hocco	1
European deer	4	Prince Albert hocco	2
European fawn	1	Silver pheasant	21
Javan Rusa deer	3	Ordinary pheasant	7
White deer	5	Swinhoe's pheasant	1
Common deer	13	Eperonnier	1
Common fawn	10	Melanotte	3
Oryx leucoryx	1	Guinea fowl	50
Mixed oryx leucoryx and pasan[3]	1	Australian crowned goura	1
Springbok	1	Emerald dove	20
Brazilian agoutis	8	Collared dove	30
Gray-green cercopithecidae	6	Egyptian dove	40
Tortoise	3	Varied birds	300
Tortoise hatchling	1	Salmon crested cockatoo	1
Nile tortoise	1	Yellow crested cockatoo	1
Ostrich	1	Galah	1
Ostrich chick	4	Parakeet crick d'Owards (?)	4
Crowned crane	1	Caleopsite parakeet	4
Bustard	1	African green parakeet	4
Flamingo	1	Budgerigar	3
White swan[4]	1	Abyssinian owl	1
Egyptian goose	3	Golden eagle	1
Toulouse goose	10	Bald eagle	1
Common goose	6	Secretary	1

Continued

Animal	*Total Number*	*Animal*	*Total Number*
Cute (*mignon*) white duck	9	Kite	2
Cute white duckling	1	European magpie	1

From DWQ, 'AI, Promenades et Plantations, Jardin Zoologique, 62/4 (4 Oct. 1878). For the initial lists that came to inform this more complete one, see DWQ, 'AI, Promenades et Plantations, Jardin Zoologique, 62/4 (Oct. 1878); DWQ, 'AI, Promenades et Plantations, Jardin Zoologique, 62/4 C (n.d.). By late nineteenth-century global standards, this collection was rather small and unimpressive. I thank Harriet Ritvo for this observation.

[1]One of the males was very ill. DWQ, 'AI, Promenades et Plantations, Jardin Zoologique, 62/4 (4 Oct. 1878).

[2]Promised to His Highness Prince Hussein. Ibid.

[3]Died when only a few days old. Ibid.

[4]Very ill. Ibid.

the novelties that were being developed in other zoological collections at the time.[148] He launched several intense planting campaigns and constructed various earthenworks to add to the topography and visual experience of the garden. Animal pens were built so sick animals could be cleaned and cared for away from the healthy ones, as well as to provide shelter for caretakers wishing to sleep in them when they were not being used. Aviaries, storage sheds, and several cabins were also built in the garden. Sauvadon oversaw all these construction activities and was directed to give special attention to the well-being of the animals while undertaking these refurbishments.

Certain aspects of these building and renovation projects prompted a great deal of careful attention from Sauvadon and his assistants.[149] One was fence height. Various directives were issued in the 1870s to increase the height of the fences and walls of certain internal structures on the grounds and also the wall enclosing the garden as a whole. Of particular concern was the construction of a protected and reinforced service corridor from one of the storage cabins that allowed access to various areas of the garden without fear of attack. These and other measures obviously aimed to enclose animals in the garden and prevent their escape. They also represented an important conceptual innovation: mechanisms that allowed one to service animals without having to actually interact with them directly. Walls, fences, and protective corridors emerged as the primary means through which Egyptians on the palace grounds came to interact—or rather came to avoid interacting—with animals. While early modern menageries allowed humans and animals to commingle in open, undifferentiated spaces, the nineteenth-century versions of these institutions were spaces of strict division and separation. It was in the Khedive's zoological garden that this newer sort of encaged relationship between species first occurred in Egypt.

In October 1878, the Khedive's menagerie staff undertook the first full enumeration of the institution's animals, determining the total population to be 635 (see table 6.1).[150] This was the first bureaucratic listing of a collection of animals in Egypt owned solely for purposes of display, recreation, and exhibition rather than for any productive agricultural, laboring, or economic function.

Like timesheets, labor charts, travel passes, birth and death certificates, censuses, and other nineteenth-century technologies of enumeration, commodification, administration, objectification, and classification, this animal list was a taxonomical tool created by a bureaucracy to make society more legible and hence easier to govern.[151] Classifying nature's overwhelming abundance—human and nonhuman—into discrete and recognizable categories advanced in lockstep with the collection of animals in the menagerie.[152]

This enumeration of living creatures, moreover, makes clear the colonial venture that was the zoo in the nineteenth century. Because a zoological collection's prestige was measured in large part by the geographic range represented by its animals, great effort was put toward securing animals from as many different places outside of Egypt as possible: deer from Java and Europe, agoutis from Brazil, green parakeets from Africa, geese from Toulouse, birds from Australia, and buffalo from the Sudan. The Khedive was quite explicit about his desire to raise the profile of his living collection by importing from abroad as many animals as possible.[153]

To this end, he directed Sauvadon to undertake various missions to bring to Egypt animals from locations throughout the world. In late 1877 and early 1878, Sauvadon was instructed to get various species of bird and bovine from parts of France. The Khedive was absolutely clear in his directives that he desired these animals not for their capacities as producers of milk or meat or for any agricultural role they might play but for their beauty and ornamental function. Thus, even though these fowl and cattle were best known for their milk and meat, the Khedive wanted them as showpieces. Their potentially productive social and economic features and abilities were completely irrelevant to a logic that interpreted them principally as requisite living specimens to be checked off the list of what a modern zoological collection was to own.[154]

As part of this new engagement with living animals, Ismail also asked Sauvadon to get him the most beautiful hens he could find.[155] Sauvadon answered this call by importing a few pairs of wild American turkeys, esteemed for their flesh, and groups of Roman pigeons, half as large as ordinary varieties. The discussion around these birds is crucial since it makes clear that both Sauvadon and the Khedive fully understood the potentially productive capacities of these animals' bodies.[156] Yet they did not want these birds for their meat or eggs but for the message their possession imparted: the Egyptian Khedive could possess and control the world's living bounty in his garden.[157]

Elephants remained a prize for zoological collections throughout the world in the 1870s, and India was still the best place to get them.[158] In April 1876, Khedive Ismail coordinated with British imperial officials to organize the purchase of six trained elephants from India.[159] Unlike the elephants that used to arrive as gifts, these were bought on the global animal market and brought to Egypt through the resources and organization of European empire. The Khedive sent letters first to the British consul in Egypt, who then wrote to the secretary of the government of India, who then coordinated with his assistants to make contacts with local mahouts around Delhi to acquire the animals and later to secure payment from the Khedive in Egypt.[160] To ensure that the elephants arrived to Egypt without complication, an Indian mahout and four aides accompanied them on their journey.[161] These five men stayed on in Egypt for two years to make sure the elephants were properly settled in their new home and to train a few Egyptian elephant handlers.[162] Once back in India, these men complained to British colonial officials that they had not received their full promised payment, a complaint which precipitated a good deal of correspondence between British officials in Egypt and India.[163]

More important than birds or elephants for Khedive Ismail were bovine. Sauvadon was thus charged with weighing the possibilities of importing dairy cows and some other "continental" varieties from Normandy.[164] The Khedive assured Sauvadon that price was of no concern, and so arrangements were made to begin moving cows from Normandy through Paris to Marseille, where a fleet of Egyptian ships would meet them to bring them to Alexandria. The French suppliers from the Jardin Zoologique d'Acclimatation du Bois de Boulogne in Paris warned Sauvadon and his associates that these Norman cows were accustomed to a steady supply of rich grasslands, copious provisions of water, and a humid climate. The Jardin's scientists doubted the cows would acclimatize to Cairo's warm and dry air and suggested that the Khedive consider importing Bazadaise, Savoisienne, or Breton cattle.[165] Citing evidence garnered from their decades of experience with the global circulation of bovine, the French zoologists added that Breton cattle had been shown to thrive in Algeria and might therefore also do well in Cairo. Sauvadon agreed with the Jardin's zoologists' assessment that Breton and similar varieties would be the best suited to Egypt.[166] As evidence, he cited the fact that when he first arrived to the country in 1869 he knew of a Breton cow that was always in very good health. Unfortunately, this hearty and still-healthy animal was slaughtered as a precautionary measure during an outbreak of cattle plague. Sauvadon also referenced an earlier instance of the Khedive's importation of a group of Italian, Breton, Norman, and Béarnaise cows from Istanbul to Alexandria. Of these, the Breton and Béarnaise animals adapted best to Cairo, while half of the Italians died.[167]

In considering which animals were best suited to import to Egypt, disease and the proper care these creatures would require once in Cairo were of the utmost concern for Sauvadon and his zoological staff. A successful zoological venture had to bring as many different animals from as many faraway places as possible *and* keep them alive and thriving for as long as possible.[168] Needless to say, a dead American turkey or Javanese deer would do little to further the prestige of the Khedive's menagerie. The case of a group of three young hippopotami imported from an unspecified part of Africa makes the imperative and difficulty of animal maintenance patently clear.[169] On the evening of October 14, 1878, these animals arrived at the garden in Gezira bruised and battered, exhausted and weak from their long journey. One of them was in extremely critical condition. Just two days after arriving to Cairo, this most debilitated of the three began to refuse his food and steadily grew frailer.[170] At one in the morning, he began to stumble and then fainted from exhaustion. He died a few hours later. Sauvadon cursed those who had delivered the animal to Cairo in such an awful state. He felt that the young hippo was all but dead on arrival and that there was therefore nothing he could have done to save the poor creature. Sauvadon swore to the Khedive that he would spare no effort in caring for the two remaining hippos, both of whom were also in very fragile condition.

Like all zoos founded during the institution's heyday at the end of the nineteenth century, Cairo's collection was built through colonial ventures of animal capture; capitalist modes of competitive economic exchange on a global animal market; and technologies of statecraft and governance that sought to control and order society into recognizable and reproducible social, political, and economic units. The institution thus represented the end of an earlier relationship between humans and animals in Egypt. By the second half of the nineteenth century, commodification, empire, separation, disease, specialized knowledge, technical expertise, species competition, and fear had replaced intimacy, cooperation, utility, wonder, codependence, and productivity as the defining features of the human-animal relationship.[171]

Disappearance

The sad fate of the three hippopotami brought to Egypt in October 1878 clearly marks what changed in human-animal relations in Ottoman Egypt—from animals being constructive and active members of human communities to their becoming set apart from humans through various modes of governance and new economic arrangements in the nineteenth century. The histories of two rather charismatic animals—hippos and ostriches—make explicit what the "disappearance of animals from daily life" meant for the human-animal relationship and for Egyptian society and politics more generally. Animals like hippos and ostriches—but also cows, oxen, dogs, elephants, and other species—"disappeared" from Egypt in the nineteenth century both because their populations precipitously declined due to disease, hunting, market-driven export, and eradication campaigns, and—more significantly—because Egypt became a society in which animals were no longer the bases of economic, social, and political life.

Hippos were widespread throughout Egypt in the seventeenth and eighteenth centuries.[172] An account from 1631 about a group of approximately twenty-three hippos moving from the Nile's edge to fields near Damietta shows how the city's residents attempted to prevent the animals from destroying their planted fields by building fences and setting traps to deter them.[173] Suggesting a long history of human-hippo interactions, this account makes clear that these sorts of protective measures were regularly undertaken to prevent hippos from eating from and stomping on humans' planted fields. Further evidence from Rosetta in 1697 confirms that Egyptians in the early modern period greatly disdained the harming or forcible possession of hippopotami.[174] A French merchant visiting Rosetta wanted to catch one of the hippopotami he saw near the Nile to take it home with him. When he tried to put his plan of capture into action, however, he could find no locals to help him. According to local custom, anyone who forcibly captured a hippo would be cursed with bad luck for life. By denying the Frenchman any assistance, Egyptians therefore prevented him from laying hands on the animal. This kind of local lore associated with the hippopotamus impeded its capture and captivity by humans and thereby helped preserve hippopotami populations throughout the Delta.

One reason hippopotami were protected in Ottoman Egypt was because they served a constructive medicinal function for human communities. By closely observing hippos, Egyptian medical practitioners learned, for example, how bloodletting techniques could be used to relieve the human body of numerous maladies.[175] After gorging themselves on food in fields along the Nile, sick but newly fattened hippopotami were often observed seeking out a sharp reed or stick near the river. Knowing the correct location of a particular vein, the hippopotami would suitably position their bodies on the stick and then jerk with a swift and forceful motion to puncture the vein. After an appropriate period of bloodletting, they would then lie on mud or clay to close the wound, allowing it to heal correctly. Egyptian doctors who observed this phenomenon were able to successfully transfer these bloodletting techniques to their human patients. Hippopotami were thus observed frequently enough in the Egyptian countryside to be subsumed within medical thinking about the treatment of the human body and to even improve these treatments.

Underlining the fundamental difference between the seventeenth century and the beginning of the nineteenth was an extreme reduction in the number of hippopotami in Egypt over this period. Unlike livestock and urban dogs, hunting and habitat destruction were the primary reasons for declines in hippo populations. At the start of the nineteenth century, hippopotami were therefore no longer a regular component of rural Egyptians' worldviews, and when they were seen in the countryside, they were treated more as threats and unknown dangers than expected and accepted, instructive and constructive parts of the Egyptian environment. Take, for example, the following description from al-Jabartī of a hippopotamus found in Damietta in October 1818:

> The appearance in the rice fields on the eastern bank near Damietta of an animal the size and color of a water buffalo. Since its first appearance the previous year, it would emerge from the Nile and eat an acre of crops, most of which it regurgitated. Many people of the district would gather and attack it with stones and rifle fire, but its hide could not be penetrated and it would flee into the Nile. Once a man was devoured by this beast. Finally, it was felled with a wound in its eye and killed by a throng that gathered round it. They stripped the skin, stuffed it with straw, and brought it to Būlāq so that the pasha [Mehmet ʿAli] and the public could view it. More than one person who saw it told me that it was larger than a big water buffalo. 30 feet long, it had the same color as a water buffalo but its skin was smoother. It had a large, weasel-like head, with eyes set high and a wide mouth. Its tail was like that of a fish, and its feet, bulkier than an elephant's, had four long cloven hooves at the rear but smooth on the bottom, like a camel's.[176]

The detail given by al-Jabartī in this passage is essential since it seems it was unclear both to the writer and those who witnessed the animal firsthand what exactly the creature was. As if describing a never-before-seen mythical beast, al-Jabartī carefully details each of the animal's parts so that should one see it again, one would immediately recognize the animal.

Also significant in this passage is the description of the sheer violence the humans perpetrated on this hippopotamus. As in other such tales of mythical creatures, the hippo evoked terror, awe, and rumor in the residents of Damietta and far beyond as well.[177] The size of a water buffalo and possessing the head of a weasel, the tail of a fish, and feet like a camel, the monster was both of enormous stature and heterogeneous constitution. It ate fields and men and was virtually immune to attack. Finding one of its few weaknesses, the resourceful villagers were able to finally conquer the beast and display their trophy kill in the capital. al-Jabartī's reference to "many people of the district" suggests that the reactions and actions of these rural cultivators were indicative of more general Egyptian attitudes toward hippos in this period. And the mention of rifle fire indicates that the felling of animals was becoming more common with the widespread diffusion of new weaponry and the growing popularity of sport hunting.

The celebration of the hippo's carcass, effected by stuffing it, transporting it from Damietta, and then displaying it in Cairo, also suggests something of the growing separation between the human and animal worlds.[178] The animal here was something monstrous, anomalous, and extra-social, an almost singularly freakish creation that had to be seen by all to fathom its remarkable terror and unimaginable uniqueness. As the Khedive attempted to do in his menagerie and as Egypt's sericulturalists tried to do on their silk farms, so the people of Damietta killed this hippo, stuffed it, and staged it to demonstrate their domination over the nonhuman world.

As landed elites at the turn of the nineteenth century invested more heavily in increasingly commercialized agriculture on ever larger estates, they had more of an incentive to destroy animals like hippopotami who ate and stepped on their planted assets. The formation of these large landed estates also radically altered the Nile's ecology as Egypt's irrigation network was greatly expanded to feed water to more land. Again, during Mehmet ʿAli's reign the total length of Egypt's canals more than doubled.[179] Such massive environmental manipulation resulted in the destruction of sections of hippopotami habitats, further hurting their populations.[180] Sport hunting by both Europeans and Egyptian elites helped seal the fate of the hippopotamus in Egypt. In sum, the animal went from being a regularly expected and accepted part of the rural world in the seventeenth century to an anomalously threatening wonder in the nineteenth. Indeed, by the last third of the nineteenth century, hippos were so seldom found in Egypt that they had to be imported from other parts of Africa when the Khedive wanted some for his garden in 1878.[181] And even then, it was not at all clear they would survive in their new home.

The hippopotamus was, of course, not the only animal in Egypt to experience such a drastic and rapid decine over this period. The ostrich, also on the 1878 list of creatures in the Khedive's zoological collection, traces a similar story of disappearance from its former vital social, economic, and laboring roles in Egyptian society.[182] Beginning as early as the late Palaeolithic era, evidence based mostly on fragments of ostrich eggshells and eggshell beads suggests the continual presence and social utility of ostriches in Egypt from antiquity until the late eighteenth century.[183] Some of these data suggest that ostriches were not much hunted in this early period and that consuming the bird was perhaps even considered taboo in parts of North Africa. As late as the beginning of the Pharaonic period, there were still very few scenes of ostrich hunts. Only in the Middle and New Kingdom periods do depictions of hunts begin to emerge. Later during the

Byzantine period, large wild ostrich populations existed throughout Egypt and some birds were even raised on specialized farms. Concentrated evidence for ostrich life picks up again near the beginning of the Ottoman period. The largest populations were in the north and west of the Sinai Peninsula, in southeastern Egypt, in certain pockets along the Mediterranean coast, and in the desert between Cairo and Suez. Many observers noted robust numbers of ostrich herds in the sixteenth and seventeenth centuries, despite the fact that they increasingly became targets of hunts in this period. One account of an ostrich hunt from the early 1610s, for example, describes a group of fifty or sixty Sinai Bedouin armed with javelins setting off on horseback to chase ostrich herds.[184]

Despite the increase in ostrich hunting in the seventeenth and eighteenth centuries, the bird was still regularly seen in large numbers throughout Egypt. Several reports from 1631 indicate that there were many ostriches in the desert between Cairo and the Red Sea.[185] In January 1658, several large herds were sighted in Sinai, and later in the seventeenth century, many herds were seen in the western Delta, near Suez, and in areas of the desert between Cairo and the Red Sea.[186] Various Egyptian amirs and other notables kept flocks of ostriches to evidence their power, property, and status.[187] In 1738, Richard Pococke wrote that the ostrich was the most common bird in Egypt, although it was starting to be intensely hunted in this period. In the spring of 1792, British traveler W. G. Browne continually discovered fresh ostrich tracks as he journeyed west from Alexandria to the Siwa Oasis.[188] Numerous other European travelers noted seeing both ostriches and their tracks in various parts of Egypt during the eighteenth century.

By as early as 1800 and certainly after 1850, the number of ostrich sightings in Egypt had decreased precipitously.[189] In 1821, a herd of just ten to fifteen birds was seen on the road between Alexandria and Siwa. Perhaps the last European to see wild ostriches in Sinai, Swiss traveler J. L. Burckhardt reported spotting two birds in the north of the peninsula just east of Suez on April 22, 1816.[190] Beginning in the early nineteenth century, Egypt's ostriches were confined mostly to areas of very sparse human population. But even in relatively remote regions such as the southeast or extreme northwest, ostrich populations steadily declined over the course of the nineteenth century.[191] In 1891, French traveler E. A. Floyer guessed that no more of the bird existed in the southeast.[192] A few years later in 1905, John Ball, director of the Desert Survey Department of the Geological Survey of Egypt, also conjectured that the southeast no longer had any ostriches. In other parts of Egypt, ostriches seem to have been eliminated even earlier. Although far from certain given its vast size, in large portions of Egypt's western desert, the bird seems to have disappeared sometime around 1850. In sum, the western desert and parts of Sinai were most likely the first regions in Egypt to see the widespread elimination of the ostrich; the southeast

Figure 6.11 Typical Ostrich Territory in the Southeastern Desert between Asyūṭ and the Red Sea. Also important are the guns depicted here. Commission des sciences et arts d'Egypte, *État moderne*, vol. 2, pt. 2 of *Description de l'Égypte*, Déserts arabiques, pl. 100. Beinecke Rare Book and Manuscript Library, Yale University.

was likely next, and then areas of the Mediterranean coast and the desert between Cairo and Suez.

What explains the steady decline of ostrich populations in Ottoman Egypt? One factor was climate. Generally warm and dry summers and very cold winters between 1780 and 1820 led to drought. This climatic stress eventually resulted in less overall vegetation coverage in Egypt's deserts and elsewhere and hence less food for ostrich and other animal populations—less, for example, of the ostrich's preferred plant, *Trichodesma africana* or, tellingly, *shūk al-naʿām* (ostrich thorn). In the later decades of the first half of the nineteenth century, rainfall totals were higher in Egypt and vegetation levels increased.[193] Ostrich populations likely saw an uptick in this period before another period of drought and elevated temperatures at the turn of the twentieth century.[194]

The more significant reason for the decline in ostrich populations in the nineteenth century was human hunting. Mehmet ʿAli's government established a number of hunting reserves throughout Egypt and generally supported sport hunting as an elite, but no longer exclusively sovereign, activity.[195] As hunting by Egyptians and foreigners alike became more widespread in Egypt, ostrich populations were dramatically reduced; ostriches in particular became a favorite target since killing the fast and agile bird was an easy and quick way for a hunter to prove his skills.[196] At the same time, firearms technology was becoming cheaper and more readily available in Egypt, which allowed the average hunter to increase his efficiency and kill rates.[197]

Beyond sport and status, another impetus behind killing ostriches was the demand for their meat and eggs, both of which were highly valued delicacies. Various commentators in the nineteenth century noted the tenderness and taste of both ostrich brain and the flesh of ostrich chicks.[198] The bird's meat contains twice as much protein as beef and far fewer calories and less cholesterol.[199] Additionally, especially in the seventeenth and eighteenth centuries, ostrich fat was thought to have useful medicinal properties as a treatment for scalp disorders, paralysis, and rheumatism.[200] Ostrich eggs were coveted mostly for decorative functions—especially in Greek and Coptic Orthodox churches, as well as in many mosques.[201]

Ostrich feathers were also highly valued economic commodities, and the global feather trade was heavily impacted by ostrich population reductions in Egypt in the latter half of the nineteenth century.[202] From before the Renaissance through the nineteenth century, ostrich feathers were all the rage in Europe, Egypt, the Ottoman Empire, and elsewhere for their use in hats, fans, and clothing.[203] In Egypt, high demand for ostrich feathers meant there was a thriving import-export market for feathers coming mostly from the Sudan and Yemen, but also from Ethiopia and Somalia, that were then traded across the Mediterranean and beyond.[204] Throughout the early modern period and into the first half of the nineteenth century, Egyptian merchants, especially Egyptian Jews, in Cairo and Alexandria made impressive profits from exporting ostrich feathers to Europe. However, at the end of the nineteenth century, this trade began to peter out in Egypt as Tripoli emerged as North Africa's primary feather-exporting center.[205] Egypt's declining share of the Mediterranean feather trade was primarily a function of Egypt's—and more generally east Africa's—dwindling ostrich populations, as well as several changes in Ottoman protectionist policies.[206]

What the histories of hippos and ostriches in Ottoman Egypt show is that their reduced populations in the nineteenth century made human interactions with them less common and therefore less integral to Egyptian social and economic life. With hippos, ostriches, and countless other animals removed from the productive roles they used to play in Egypt, new institutions of governance, social and economic structures, and logics of control came to mediate how, when, where, and why humans experienced and interacted with the other animals around them.

The human-animal world France's first giraffe left behind in Egypt in 1826 was of a fundamentally different order than what it had been for the roughly three centuries before 1826. Animals were rapidly becoming instrumentalized in new ways in the nineteenth century—as patients in a school of veterinary medicine, as commodities governed by a globalized capitalist market economy, and as objects of competition among zoological collections around the world. New animal institutions and the integration of Egypt into the world economy radically reshaped the human-animal relationship. This new anthrozoological order resulted in multiple kinds of animal disappearances. While some animals lost their vital and productive social and economic functions, others lost their very lives. Slaughter, overhunting, climate change, and the rigors of long-distance transport all contributed to dwindling animal populations in the middle of the nineteenth century. Animals such as the ostrich may indeed have been driven near the point of domestic extinction in this period.

The end result of these new realities for humans and animals in Egypt was the rapid, increasing, and unprecedented separation of humans from other species. Instead of the mixed human-animal rural labor pools of the early modern period, work in nineteenth-century Egypt was almost exclusively the domain

Figure 6.12 "Obeying Security's Directives and those of the Animal Handler will Keep You Away from a Multitude of Harms. With the Administration's Regards." A Sign of Separation in the Contemporary Zoo, 2010. Photograph by author.

of humans and, later, of their machines. Dogs were removed from urban spaces. Charismatic megafauna and other animals became monopolized by just a few royals, industrialists, and entrepreneurs. By the end of the nineteenth century, many animals, even some dogs, were only to be found in the cages of the zoo. Itself a product of European empire, capitalist aspirations, and Egypt's deleterious imperial ambitions in the nineteenth century, Cairo's zoo officially opened to the general public in 1891.[207] As Egyptians walked through its gates for the first time that year, few knew they were stepping into a human-animal world never before seen in Egypt; even fewer knew of all that had transpired to make that world possible; and none knew what this history meant for them.

Conclusion

The Human Ends

> What might be called a society's "threshold of modernity" has been reached when the life of the species is wagered on its own political strategies. For millennia, man remained what he was for Aristotle: a living animal with the additional capacity for a political existence; modern man is an animal whose politics places his existence as a living being in question.
>
> —Michel Foucault

What happened to animals in Ottoman Egypt would soon happen to humans. Just as domestic animals were removed from Egypt's labor regime, dogs were cast outside of Cairo and Alexandria, and charismatic megafauna were excluded from changing notions of Ottoman imperial pageantry and reciprocal gift exchange networks, certain classes of humans were largely removed from definitions of the productive social and economic realms of nineteenth-century Egypt. From the perspective of Mehmet ʿAli's increasingly centralized state bureaucracy, the sick, the uneducated, the insane, and the poor, as well as criminals and often peasants, were seen as useless—and hence removable—parts of the social and economic sphere. All these classes of society were thus subjected to some form of coercive sequestering, exclusion, or enclosing institutionalization in the nineteenth century. Such disciplining forms of governance thereafter served as the primary mediator of these people's relationships to the rest of society. "The people of Egypt," in Timothy Mitchell's words, "were made inmates of their own villages."[1] Mehmet ʿAli's state and those of his successors built prisons, conscription camps, quarantine regimes, hospitals, schools, asylums, and police stations in ways unprecedented in Egyptian history.[2] All of these institutions—which, of course, regularly faced resistance to their illusory efforts at control—were charged with managing living creatures deemed socially useless or economically unproductive. This process was first worked out in the field of human-animal relations.[3] As countless animals came to be defined as no longer socially, economically, or politically productive, they were killed, caged, or physically removed from society. The nineteenth-century Egyptian state did much the same to its human population.[4] To adequately understand this human story, therefore, means coming to terms with the human-animal histories of the transition from early modernity to the nineteenth century that made these enclosures, exclusions, and eradications possible and productive.

Modern psychologists and sociologists of violence have long observed—as did Thomas Aquinas centuries before them—that human violence against animals is often the first step toward human violence against other humans.[5] Taking the long view of this idea, known as the progression thesis, shows that the historical processes that made possible mass violence against humans in the nineteenth century, the cordoning off of spaces throughout Ottoman Egypt, and the strict definitions of productive social and economic realms were part of deeper, slower, and more complex trajectories in the early modern human-animal relationship.[6] These changes included active manipulation of the labor regime to maximize productivity on large landed estates, the sanitation and management of urban spaces to open them up to as much human habitation and policing as possible, and the transition to capitalist modes of exchange that facilitated competition in global markets.

Thus, changes to the human-animal relationship were detrimental to the lives of both humans and animals. This may seem rather surprising and counterintuitive given the stories of progress, freedom, liberalism, and individualism that have usually been told about the beginnings of the nineteenth century in Egypt and elsewhere.[7] How could all of this "improvement" result in the degraded lives of so many living creatures? The answer to this question is that the history of modernity is usually told as a story of the state from inside of the state.[8] If, however, the story of modernity is told from a different set of perspectives—such as the human-animal one—very different stories come into focus. Modernity emerges as a project of violence, species separation, cages, structural inequality, exclusion, and ecological degradation. The social was built in a particular way at a particular time. In Egypt, as in many other places throughout the world, this construction was largely carried out by a nineteenth-century state through coercion, forcible relocation, and violence.[9]

The ferocious nineteenth century did nevertheless produce some new human-animal engagements. Petkeeping, for example, emerged in this period as an antidote to the loneliness of bourgeois existence.[10] The symbolic uses of animals in political cartoons, social iconography, film, and advertising exploded in this period as well.[11] Even as industrial society in many ways separated humans and animals, species were brought closer together through new kinds of interactions—affective relationships and the widespread consumption of meat, to name the most obvious examples. Indeed, the writing of human-animal histories is itself a consequence of the nineteenth century's cleavages and reconstitutions of interspecies relationships. Perhaps one of the most important outgrowths of changes to these relationships was the articulation of a liberal discourse of both human and animal rights at the end of the nineteenth century.[12] Humans and animals only came to need rights in this period precisely because their lives had slowly been subjected to violence and injury.[13] "Modern man," Foucault writes, "is an animal whose politics places his existence as a living being in question."[14] At the same time that the brute politics of modernity were threatening biological life, other kinds of politics were developing to attempt to save it.

In the case of Egypt, concerns over the humane treatment of nonhuman animals emerged at the end of the nineteenth century in the context of British colonialism.[15] Lord Cromer, the highest British colonial official in Egypt and staunch upholder of the white man's burden—a man who clearly believed in the inferiority of non-European humans—was a strong advocate for the rights of animals in Egypt. As he wrote from Cairo to one of his associates in December 1892, "The prevention of cruelty to animals in Egypt is a subject which has for a long time past interested me greatly."[16] Cromer was actively involved in the founding, operation, and financing of a society for the prevention of cruelty to animals in Cairo that ran an infirmary for sick and maltreated animals and that advocated for police action to punish those who committed violence against animals.[17] During his subsequent retirement in England, he was president of the Research Defence Society, an anti-animal-vivisection organization; chairman of the Entomological Research Committee; and vice president of the British Society for the Prevention of Cruelty to Animals.[18] In 1906, the number of cases the Egyptian anti-cruelty society successfully brought to British courts in Egypt peaked, with over five thousand convictions in animal cruelty and mistreatment cases.[19] Many British anti-cruelty advocates in colonial Egypt expressed their animal welfare politics through violence toward human animals: "thrashing a camel-driver who had ill-treated his animal," for example.[20] At the turn of the twentieth century, there were numerous other instances of Britons beating Egyptians for mistreating animals—examples of human violence against other humans to protect nonhumans.[21] Not only do Cromer's ties to anti-cruelty efforts and more general British attitudes toward both Egyptian animals and Egyptian humans show the emergence of a new discourse about animal rights at the end of the nineteenth century—one in which animals became objects of bourgeois European human sympathy, rescue, and action, and Egyptian humans became objects of violence—but they also put in stark relief the contradictory ways colonial ideas about the hierarchy of humanity stood unproblematically alongside imperial support for animal welfare.[22] In these colonial hierarchies, nonhumans often stood above certain kinds of (nearly always non-European) humans.

Having now walked, galloped, slithered, swam, crawled, ridden, crept, and hopped through these pages, what insights derive from a history of human-animal relations in Ottoman Egypt between 1517 and 1882? This book proposes four answers that deserve summation as potential areas for future research.

First, animal histories matter. They are fundamental to understanding any society at any time. Animals were everywhere in the past and help us to understand historical processes and phenomena as disparate as public health, commercial agriculture, trade, bourgeois cultural sensibilities, labor history, religious conceptions of humanity, and violence. An interspecies perspective allows one to analyze seemingly unconnected processes, institutions, and histories within a conceptual frame that illuminates their connections, resonances, and interdependence. As with the conceptual tools of gender, class, and race, historians can productively and innovatively interpret nearly any aspect of the past through the lens of human-animal relations. Historical writing about animals also raises important methodological questions about how to write histories of subalterns (nonhuman and otherwise) who do not generally author texts. Questions about historical agency, the use of sources, and the integration of evolutionary and other scientific data into historical analyses also emerge from considering human-animal histories. Ignoring these interspecies histories thus misses a significant part of the past.

Second, animal histories offer new insights into the global transition of early modern rural societies into state bureaucracies—a development that occurred throughout the world between approximately 1750 and 1850. In the specific case of the Ottoman Empire, historians are now focusing more squarely on this period as a distinct historical epoch in its own right—not simply one that sits uncomfortably between early modernity and the nineteenth century—to understand how various parts of the empire achieved and experienced these historical transformations. A concentration on the decades straddling either side of the turn of the nineteenth century highlights the many continuities and multiple ruptures between the two periods, as well as the complex connections and intricate inner workings of the imperial ecosystem across time and space. Among other things, these transformations were violent processes that stripped animals of their longstanding productive social, economic, and political roles. This made possible their encagement, their castigation outside city walls, and their slaughter in centralized meat production facilities. Thus, one of the most important features of the move from early modernity to the nineteenth century was the revamping of human relationships with animals. Because nonhumans were so enmeshed in all facets of daily social and economic life in the early modern period, changes to multispecies relationships affected all aspects of humans' social, economic, political, and ecological worlds.

Third, the technologies of species separation embedded in modernity often resulted in the animalization of certain humans. Processes like the commodification and capitalization of living creatures in global commodity exchange

Figure C.1 Encaged Interspecies Interaction in Cairo Zoo, 2010. Photograph by author.

networks and the relegation and replacement of certain kinds of labor occurred first in the field of human-animal relations. "Up to the end of the eighteenth century," Foucault writes, "life does not exist: only living beings."[23] If we accept Foucault's general concept and chronology, then biological life—whether human, animal, or plant—emerged as an identifiable and discrete object of state power for the first time at the turn of the nineteenth century. It was thereafter to be manipulated, harnessed, managed, and understood through new fields of knowledge such as veterinary and human medicine, capitalist economics, and urban sanitation. Therefore, replacing animals with humans in the economy of rural labor in Ottoman Egypt made possible, both conceptually and materially, the later replacement of humans with machines. In the same way that humans came to supplant animals as better working technologies at the turn of the nineteenth century, so too did machines later replace humans as mechanisms deemed more efficient, economical, and effective. This transformation in rural Egypt's labor regime thus tracks the same chronological and conceptual domain as the transition Carolyn Merchant identifies "from the organism to the machine as the dominant metaphor binding together the cosmos, society, and the self into a single cultural reality."[24] Moreover, the increasingly strict definitions of social spaces in Egypt that began with the removal of garbage and dogs from Cairo and Alexandria would be further refined in the nineteenth century to enclose criminals, the infirm, and the conscripted in various state institutions.

Thus through parallel technologies of rule, conceptions of life, and state practices, certain humans came to be treated in the same ways nonhuman animals had before them.

Ultimately then, animal histories show some of the particular, historically specific ways that coercive state projects built certain social, political, and economic realms. "Control over animals," philosopher Reviel Netz reminds us, "is rather like control over humans: you can either make them do what you like them to do or else get them out of the way. This is how societies are made: human societies as well as the larger, multispecies societies that humans have created."[25] In Ottoman Egypt at the turn of the nineteenth century, the social world was one that came to be defined and produced by excluding animals and, through various engineered technologies of power, replacing their former constructive social and economic functions predominantly with human action and agency. Part and parcel of the definition of the social was which humans were allowed to inhabit it. How, why, where, and when the social was built in Ottoman Egypt and elsewhere makes obvious that there is nothing preordained, complete, scripted, or static about the way a society functions and changes. As the arc of Egyptian history from 1611 to 1811 to 2011 indeed affirms, other futures are constantly being built.

NOTES

Abbreviations

ACM	Archives of the Cathedral of Mdina, Mdina
LC	Lettere Consolari
ASV	Archivio di Stato, Venice
Algeri, Egitto, Marocco	Miscellanea Documenti Algeri, Egitto, Marocco, Persia, Tripoli, Tunisi
Turchi	Miscellanea Documenti Turchi
Beinecke	Beinecke Rare Book and Manuscript Library, Yale University, New Haven
BOA	Başbakanlık Osmanlı Arşivi, Istanbul
BEO	Bâb-ı Âlî Evrak Odası
CS	Cevdet Saray
DMK	Dahiliye Mektubi Kalemi
HAT	Hatt-ı Hümayun
MM	Mühimme-i Mısır
DKM	Dār al-Kutub al-Miṣriyya, Cairo
DWQ	Dār al-Wathā'iq al-Qawmiyya, Cairo
'AI	'Aṣr Ismā'īl (Documents concernant Le regime de Khédive Ismail)
KLA	Kaireios Library of the Island of Andros, Chora
SK	Süleymaniye Kütüphanesi, Istanbul
TNA	The National Archives of the United Kingdom, Kew
BJ	Records of the Meteorological Office
CO	Colonial Office
FO	Foreign Office
MT	Ministry of Transport
PC	Privy Council

T	Treasury
TSMA	Topkapı Sarayı Müzesi Arşivi, Istanbul
E.	Evrak

Ottoman Turkish Islamic Month Abbreviations (Arabic in Parentheses)

M	Muharrem (Muḥarram)
S	Safer (Ṣafar)
Ra	Rebiülevvel (Rabīʿ al-Awwal)
R	Rebiülahir (Rabīʿ al-Thānī)
Ca	Cemazilevvel (Jumādā al-Ūlā)
C	Cemaziyel'ahır (Jumādā al-Ākhira)
B	Receb (Rajab)
Ş	Şaʿban (Shaʿbān)
N	Ramazan (Ramaḍān)
L	Şevval (Shawwāl)
Za	Zilkade (Dhū al-Qaʿda)
Z	Zilhicce (Dhū al-Ḥijja)

Introduction: Cephalopods in the Nile

1. For an illuminating description and discussion of the biological and ecological importance of cephalopods, see Ellen Prager, *Sex, Drugs, and Sea Slime: The Oceans' Oddest Creatures and Why They Matter* (Chicago: University of Chicago Press, 2011), 80–95.

   For a fifteenth-century Egyptian account of various aquatic animals, see Jalāl al-Dīn ʿAbd al-Raḥmān al-Suyūṭī, *Kawkab al-Rawḍa*, Beinecke, Landberg MSS 202. Some of the animals covered in this text are (with translations and page numbers in parentheses): the electric ray (*al-raʿʿād*, 49r–49v), the crocodile (*al-timsāḥ*, 47v–49r), the hippopotamus (*faras al-baḥr*, 49v–50v), the sea calf (*shaykh al-baḥr*, 50v), the skink (*al-saqinqūr*, 49r) and the otter (*kalb al-māʾ*, 50v). For more on many of these same animals, see Nūḥ ibn Muṣṭafā Rūmī, *Kitāb Taʾrīkh Miṣr wa al-Nīl wa Khabar man Malakahā min Ibtidāʾ al-Zamān*, Beinecke, Landberg MSS 301, 98v–100v. See also the following versions of al-Suyūṭī's text: Jalāl al-Dīn ʿAbd al-Raḥmān al-Suyūṭī, *Kawkab al-Rawḍa*, Beinecke, Landberg MSS 566; idem., *Kawkab al-Rawḍa*, ed. Muḥammad al-Shishtāwī (Cairo: Dār al-Āfāq al-ʿArabiyya, 2002). For another work by al-Suyūṭī dedicated solely to animals, see idem., *Kitāb Dīwān al-Ḥayawān*, SK, Fatih 4170.

   al-Suyūṭī's *Kawkab al-Rawḍa* is a history of the island of al-Rawḍa in the Nile near Cairo. This slender piece of land is perhaps most famous for its Nilometer (*Miqyās al-Nīl*), a device used to measure the annual flood. For an imaginary competitive "conversation" between the island and the city about each of their relative merits, see idem., *Mufākhara bayna al-Rawḍa wa al-Miqyās wa Miṣr al-Qāhira*, Beinecke, Landberg MSS 525, 28r–32v. For another of al-Suyūṭī's texts about the island of al-Rawḍa, see idem., *Bulbul al-Rawḍa fī Waṣf Nīl Miṣr*, SK, Reşid Efendi 865/7, 75r–82v. For a published version of this text, see idem., *Bulbul al-Rawḍa, maʿa Dirāsa ʿan Jazīrat al-Rawḍa*, ed. Nabīl Muḥammad ʿAbd al-Azīz Aḥmad (Cairo: Maktabat al-Anjilū al-Miṣriyya, 1981). On the history of the island's Nilometer, see Amīn Sāmī, *Taqwīm al-Nīl*, 5 vols. in 3 pts. (Cairo: Dār al-Kutub wa al-Wathāʾiq al-Qawmiyya, 2003), pt. 1, 65–95; William Popper, *The Cairo Nilometer: Studies in Ibn Taghrî Birdî's Chronicles of Egypt, I* (Berkeley: University of California Press, 1951); Nicholas Warner, *The True Description of Cairo: A Sixteenth-Century Venetian View*, 3 vols. (Oxford: Arcadian Library, in association with Oxford University Press, 2006), 2: 123–25.
2. I am well aware of the inadequacy of the terms *human* and *animal*. While *human* refers to just one species, *animal* is far too vague since it obliterates the complex distinctions between millions of species. Moreover, imprecise use of these words risks reifying the human-animal

divide as infinitely separate and impossible of transgression. That said, I nevertheless follow the scholarly convention of using both words in their general meaning. I also avoid repeated use of the qualifier *nonhuman* before *animal*.

3. For studies of various aspects of the history of animals in ancient Egypt, see Patrick F. Houlihan, "Animals in Egyptian Art and Hieroglyphs," in *A History of the Animal World in the Ancient Near East*, ed. Billie Jean Collins (Leiden: Brill, 2002), 97–143; Emily Teeter, "Animals in Egyptian Literature," in *A History of the Animal World in the Ancient Near East*, ed. Billie Jean Collins (Leiden: Brill, 2002), 251–70; idem., "Animals in Egyptian Religion," in *A History of the Animal World in the Ancient Near East*, ed. Billie Jean Collins (Leiden: Brill, 2002), 335–60; Douglas Brewer, "Hunting, Animal Husbandry and Diet in Ancient Egypt," in *A History of the Animal World in the Ancient Near East*, ed. Billie Jean Collins (Leiden: Brill, 2002), 427–56; Dorothea Arnold, "An Egyptian Bestiary," *Metropolitan Museum of Art Bulletin* 52 (1995): 1 and 7–64.

4. As will be discussed later, most histories of Ottoman Egypt concentrate on either the period before or after 1800. Very few treat the centuries of Ottoman rule as a whole. For studies of the period before 1800 see, for example, André Raymond, *Artisans et commerçants au Caire au XVIIIe siècle*, 2 vols. (Damascus: Institut français de Damas, 1973–74); Michael Winter, *Egyptian Society under Ottoman Rule, 1517–1798* (London: Routledge, 1992); Daniel Crecelius, *The Roots of Modern Egypt: A Study of the Regimes of ʿAli Bey al-Kabir and Muhammad Bey Abu al-Dhahab, 1760–1775* (Minneapolis: Bibliotheca Islamica, 1981); Stanford J. Shaw, *The Financial and Administrative Organization and Development of Ottoman Egypt, 1517–1798* (Princeton, NJ: Princeton University Press, 1962); Kammāl Ḥāmid Mughayth, *Miṣr fī al-ʿAṣr al-ʿUthmānī 1517–1798: al-Mujtamaʿ…wa al-Taʿlīm* (Cairo: Markaz al-Dirāsāt wa al-Maʿlūmāt al-Qānūniyya li-Ḥuqūq al-Insān, 1997); Laylā ʿAbd al-Laṭīf Aḥmad, *al-Idāra fī Miṣr fī al-ʿAṣr al-ʿUthmānī* (Cairo: Maṭbaʿat Jāmiʿat ʿAyn Shams, 1978); idem., *al-Mujtamaʿ al-Miṣrī fī al-ʿAṣr al-ʿUthmānī* (Cairo: Dār al-Kitāb al-Jāmiʿī, 1987); idem., *Tārīkh wa Muʾarrikhī Miṣr wa al-Shām ibbāna al-ʿAṣr al-ʿUthmānī* (Cairo: Maktabat al-Khānjī, 1980); Jane Hathaway, *A Tale of Two Factions: Myth, Memory, and Identity in Ottoman Egypt and Yemen* (Albany: State University of New York Press, 2003); idem., *The Politics of Households in Ottoman Egypt: The Rise of the Qazdağlıs* (Cambridge: Cambridge University Press, 1997); ʿAbd al-Raḥīm ʿAbd al-Raḥman ʿAbd al-Raḥīm, *al-Rīf al-Miṣrī fī al-Qarn al-Thāmin ʿAshar* (Cairo: Maktabat Madbūlī, 1986); Stanford J. Shaw, ed. and trans., *Ottoman Egypt in the Eighteenth Century: The Niẓâmznâme-i Mıṣır of Cezzâr Aḥmed Pasha* (Cambridge, MA: Center for Middle Eastern Studies of Harvard University, 1964); Galal H. El-Nahal, *The Judicial Administration of Ottoman Egypt in the Seventeenth Century* (Minneapolis: Bibliotheca Islamica, 1979); ʿIrāqī Yūsuf Muḥammad, *al-Wujūd al-ʿUthmānī fī Miṣr fī al-Qarnayn al-Sādis ʿAshar wa al-Sābiʿ ʿAshar (Dirāsa Wathāʾiqiyya)* (Cairo: Markaz Kliyūbātrā lil-Kumbiyūtar, 1996).

For some of the work that treats the period after 1800, see Khaled Fahmy, *All the Pasha's Men: Mehmed Ali, His Army and the Making of Modern Egypt* (Cambridge: Cambridge University Press, 1997); Afaf Lutfi al-Sayyid Marsot, *Egypt in the Reign of Muhammad Ali* (Cambridge: Cambridge University Press, 1984); Judith E. Tucker, *Women in Nineteenth-Century Egypt* (Cambridge: Cambridge University Press, 1985); ʿAbd al-Raḥman al-Rāfʿī, *ʿAṣr Muḥammad ʿAlī* (Cairo: Dār al-Maʿārif, 1989); F. Robert Hunter, *Egypt under the Khedives, 1805–1879: From Household Government to Modern Bureaucracy* (Pittsburgh: University of Pittsburgh Press, 1984); Ehud R. Toledano, *State and Society in Mid-Nineteenth-Century Egypt* (Cambridge: Cambridge University Press, 1990).

For studies that bridge the 1800 divide, see Peter Gran, *Islamic Roots of Capitalism: Egypt, 1760–1840* (Austin: University of Texas Press, 1979); Ibrahim el-Mouelhy, *Organisation et fonctionnement des institutions ottomanes en Egypte (1517-1917): étude documentaire, d'après les sources archivistiques égyptiennes* (Ankara?: Imprimerie de la Société turque d'histoire, 1989); P. M. Holt, *Egypt and the Fertile Crescent, 1516–1922: A Political History* (London: Longmans Green, 1966); idem., ed., *Political and Social Change in Modern Egypt: Historical Studies from the Ottoman Conquest to the United Arab Republic* (London: Oxford University Press, 1968); ʿIrāqī Yūsuf Muḥammad, *al-Wujūd al-ʿUthmānī al-Mamlūkī fī Miṣr fī al-Qarn al-Thāmin ʿAshar wa Awāʾil al-Qarn al-Tāsiʿ ʿAshar* (Cairo: Dār al-Maʿārif, 1985); Kenneth M. Cuno, *The Pasha's Peasants: Land, Society, and Economy in*

Lower Egypt, 1740–1858 (Cambridge: Cambridge University Press, 1992); M. W. Daly, ed., *Modern Egypt, from 1517 to the End of the Twentieth Century*, vol. 2 of *The Cambridge History of Egypt* (Cambridge: Cambridge University Press, 1998); Alan Mikhail, *Nature and Empire in Ottoman Egypt: An Environmental History* (Cambridge: Cambridge University Press, 2011).

5. The history of animals in Ottoman Egypt has not received much attention. Some of the work currently available includes: Michel Tuchscherer, "Some Reflections on the Place of the Camel in the Economy and Society of Ottoman Egypt," trans. Suraiya Faroqhi, in *Animals and People in the Ottoman Empire*, ed. Suraiya Faroqhi (Istanbul: Eren, 2010), 171–85; Catherine Mayeur-Jaouen, "Badawi and His Camel: An Animal as the Attribute of a Muslim Saint in Mamluk and Ottoman Egypt," trans. Suraiya Faroqhi, in *Animals and People in the Ottoman Empire*, ed. Suraiya Faroqhi (Istanbul: Eren, 2010), 113–28; Jean-Louis Bacqué-Grammont, Joséphine Lesur-Gebremariam, and Catherine Mayeur-Jaouen, "Quelques aspects de la faune nilotique dans la relation d'Evliyâ Çelebî, voyageur ottoman," *Journal Asiatique* 296 (2008): 331–74; Alan Mikhail, "Animals as Property in Early Modern Ottoman Egypt," *Journal of the Economic and Social History of the Orient* 53 (2010): 621–52.

   For works on the general cultural, symbolic, and religious history of animals in various Islamic traditions of the Middle East, see Annemarie Schimmel, *Islam and the Wonders of Creation: The Animal Kingdom* (London: al-Furqān Islamic Heritage Foundation, 2003); Sarra Tlili, *Animals in the Qur'an* (Cambridge: Cambridge University Press, 2012); Mohamed Hocine Benkheira, Catherine Mayeur-Jaouen, and Jacqueline Sublet, *L'animal en islam* (Paris: Indes savantes, 2005); Basheer Ahmad Masri, *Animals in Islam* (Petersfield: Athene Trust, 1989); idem., *Animal Welfare in Islam* (Markfield: Islamic Foundation, 2007); Richard C. Foltz, *Animals in Islamic Tradition and Muslim Cultures* (Oxford: Oneworld, 2006); Katherine Wills Perlo, *Kinship and Killing: The Animal in World Religions* (New York: Columbia University Press, 2009), 95–114; Thomas T. Allsen, *The Royal Hunt in Eurasian History* (Philadelphia: University of Pennsylvania Press, 2006). For studies of zoology in the Muslim world, see Herbert Eisenstein, *Einführung in die arabische Zoographie: Das tierkundliche Wissen in der arabisch-islamischen Literatur* (Berlin: Dietrich Reimer, 1990); Manfred Ullmann, *Die Natur- und Geheimwissenschaften im Islam* (Leiden: Brill, 1972), 5–61.

6. Reşat Kasaba, *The Ottoman Empire and the World Economy: The Nineteenth Century* (Albany: State University of New York Press, 1988), 23–27; Dina Rizk Khoury, "The Introduction of Commercial Agriculture in the Province of Mosul and its Effects on the Peasantry, 1750–1850," in *Landholding and Commercial Agriculture in the Middle East*, ed. Çağlar Keyder and Faruk Tabak (Albany: State University of New York Press, 1991), 155–71.

7. Much of this is recounted in the following: al-Rāfʿī, *ʿAṣr Muḥammad ʿAlī*; Marsot, *Egypt in the Reign of Muhammad Ali*. On Egypt's imperial military ventures in this period, see Fahmy, *All the Pasha's Men*; M. Abir, "Modernisation, Reaction and Muhammad Ali's 'Empire,'" *Middle Eastern Studies* 13 (1977): 295–313. Specifically on educational reforms, see Aḥmad ʿIzzat ʿAbd al-Karīm, *Tārīkh al-Taʿlīm fī ʿAṣr Muḥammad ʿAlī* (Cairo: Maktabat al-Nahḍa al-Miṣriyya, 1938).

8. André Raymond, "La population du Caire et de l'Égypte à l'époque ottomane et sous Muḥammad ʿAlî," in *Mémorial Ömer Lûtfi Barkan* (Paris: Librairie d'Amérique et d'Orient Adrien Maisonneuve, 1980), 169–78; Daniel Panzac, "Alexandrie: évolution d'une ville cosmopolite au XIXe siècle," in *Population et santé dans l'Empire ottoman (XVIIIe–XXe siècles)* (Istanbul: Isis, 1996), 141–59.

9. T. H. Aston and C. H. E. Philpin, eds., *The Brenner Debate: Agrarian Class Structure and Economic Development in Pre-Industrial Europe* (Cambridge: Cambridge University Press, 1985); Kenneth Pomeranz, *The Great Divergence: China, Europe, and the Making of the Modern World Economy* (Princeton, NJ: Princeton University Press, 2000); John F. Richards, "Toward a Global System of Property Rights in Land," in *The Environment and World History*, ed. Edmund Burke III and Kenneth Pomeranz (Berkeley: University of California Press, 2009), 54–78; idem., *The Unending Frontier: An Environmental History of the Early Modern World* (Berkeley: University of California Press, 2003); Robert B. Marks, *The Origins of the Modern World: A Global and Ecological Narrative* (Lanham: Rowman and Littlefield, 2002); Peter C. Perdue, *Exhausting the Earth: State and Peasant in Hunan, 1500–1850* (Cambridge, MA: Harvard University Press, 1987); Jean-Laurent Rosenthal and R. Bin Wong, *Before*

and Beyond Divergence: The Politics of Economic Change in China and Europe (Cambridge, MA: Harvard University Press, 2011); Richard P. Tucker, *Insatiable Appetite: The United States and the Ecological Degradation of the Tropical World* (Berkeley: University of California Press, 2000); Richard P. Tucker and J. F. Richards, eds., *Global Deforestation and the Nineteenth-Century World Economy* (Durham, NC: Duke University Press, 1983); E. A. Wrigley, *Continuity, Chance, and Change: The Character of the Industrial Revolution in England* (Cambridge: Cambridge University Press, 1988).

10. For general accounts of the French in Egypt, see André Raymond, *Égyptiens et Français au Caire (1798–1801)* (Cairo: Institut français d'archéologie orientale, 2004); Juan Cole, *Napoleon's Egypt: Invading the Middle East* (New York: Palgrave Macmillan, 2007).
11. Marshall G. S. Hodgson, *The Gunpowder Empires and Modern Times*, vol. 3 of *The Venture of Islam: Conscience and History in a World Civilization* (Chicago: University of Chicago Press, 1974), 216.
12. For example, in his pathbreaking work of Arab intellectual history, Albert Hourani draws a straight line from Napoleon, through the period of Mehmet ʿAli's reign, to what he terms the "first generation" of "officials, officers and teachers, alive to the importance of reforming the structure of the [Ottoman] empire, and convinced this could not be done unless some at least of the forms of European society were borrowed." Albert Hourani, *Arabic Thought in the Liberal Age, 1798–1939* (Cambridge: Cambridge University Press, 1983), 67. More generally in this regard see ibid., 49–71.
13. In this vein see, for example, Henry Dodwell, *The Founder of Modern Egypt: A Study of Muhammad ʿAli* (Cambridge: The University Press, 1931); al-Rāfʿī, *ʿAṣr Muḥammad ʿAlī*; Marsot, *Egypt in the Reign of Muhammad Ali*. For critiques of this school of history, see: Fahmy, *All the Pasha's Men*, 1–37; idem., *Mehmed Ali: From Ottoman Governor to Ruler of Egypt* (Oxford: Oneworld, 2009), 112–27; Ehud R. Toledano, "Mehmet Ali Paşa or Muhammad Ali Basha? An Historiographic Appraisal in the Wake of a Recent Book," *Middle Eastern Studies* 21 (1985): 141–59. Recent arguments for Mehmet ʿAli's "Ottomanness" include: M. Baha Tanman, ed., *Nil Kıyısından Boğaziçi'ne: Kavalalı Mehmed Ali Paşa Hanedanı'nın İstanbul'daki İzleri* (Istanbul: İstanbul Araştırmaları Enstitüsü, 2011); Heath W. Lowry and İsmail E. Erünsal, *Remembering One's Roots: Mehmed Ali Paşa of Egypt's Links to the Macedonian Town of Kavala: Architectural Monuments, Inscriptions, and Documents* (Istanbul: Bahçeşehir University Press, 2011). For a more general study of Ottoman cultural influence in nineteenth-century Egypt, see Ekmeleddin İhsanoğlu, *Mısır'da Türkler ve Kültürel Mirasları: Mehmed Ali Paşa'dan Günümüze Basılı Türk Kültürü Bibliyografyası ve Bir Değerlendirme* (Istanbul: İslam Tarih, Sanat, ve Kültür Araştırma Merkezi, 2006).
14. On the domestic production of capitalist economic relations in Egypt, see Gran, *Islamic Roots of Capitalism*; Cuno, *Pasha's Peasants*; Nelly Hanna, *Artisan Entrepreneurs in Cairo and Early Modern Capitalism (1600–1800)* (Syracuse, NY: Syracuse University Press, 2011); idem., *Making Big Money in 1600: The Life and Times of Ismaʿil Abu Taqiyya, Egyptian Merchant* (Syracuse, NY: Syracuse University Press, 1998). On Egyptian intellectuals and the emergence of secularity, see Gran, *Islamic Roots of Capitalism*; Khaled el-Rouayheb, "Was There a Revival of Logical Studies in Eighteenth-Century Egypt?" *Die Welt des Islams* 45 (2005): 1–19; Nelly Hanna, *In Praise of Books: A Cultural History of Cairo's Middle Class, Sixteenth to the Eighteenth Century* (Syracuse, NY: Syracuse University Press, 2003).
15. For a discussion of some of this work, see Mikhail, *Nature and Empire*, 38–81.
16. Throughout this book I will generally use the terms *caloric energy*, *caloric power*, and *caloric motor* to refer to the labor power and work capacities of animals and humans fueled by the expenditure of food calories stored in their bodies.
17. Michel Foucault's analysis of this same period seeks to understand how notions of overlapping and multi-sighted taxonomies in what he termed the Classical Period came to be disaggregated and fixed in the modern age as biology, philology, and political economy. Inspired by Foucault's work, this book similarly hopes to understand how an early modern world of humans and animals came to be decoupled into two separate realms in the nineteenth century. Michel Foucault, *The Order of Things: An Archaeology of the Human Sciences* (New York: Vintage Books, 1994).

18. About this period, Christine M. Philliou notes that there is only one detailed English-language study of the reign of Selim III from 1789 to 1807 and none about the reign of Mahmud II from 1808 to 1839. Christine M. Philliou, *Biography of an Empire: Governing Ottomans in an Age of Revolution* (Berkeley: University of California Press, 2011), xviii and 187, n. 2. The one study of the reign of Selim III is Stanford J. Shaw, *Between Old and New: The Ottoman Empire under Sultan Selim III, 1789–1807* (Cambridge, MA: Harvard University Press, 1971). This lack of scholarship on the periods of Selim III and Mahmud II seems set to change. See, for example, the following recent works: Betül Başaran, "Remaking the Gate of Felicity: Policing, Social Control, and Migration in Istanbul at the End of the Eighteenth Century, 1789–1793" (Ph.D. diss., University of Chicago, 2006); Ali Yaycıoğlu, "The Provincial Challenge: Regionalism, Crisis, and Integration in the Late Ottoman Empire (1792–1812)" (Ph.D. diss., Harvard University, 2008).
19. See note 4 of this chapter for studies of Ottoman Egypt that do treat this middle period.
20. On this point, see the following comparative study of the Ottoman, Safavid, and Mughal Empires: Halil Berktay, "Three Empires and the Societies They Governed: Iran, India, and the Ottoman Empire," *Journal of Peasant Studies* 18 (1991): 242–63.
21. On the idea of the circle of justice in the Ottoman Empire and how it operated on the ground, see Linda T. Darling, "'Do Justice, Do Justice, For That is Paradise': Middle Eastern Advice for Indian Muslim Rulers," *Comparative Studies of South Asia, Africa and the Middle East* 22 (2002): 3–19; Heather Lynn Ferguson, "The Circle of Justice as Genre, Practice, and Objectification: A Discursive Re-Mapping of the Early Modern Ottoman Empire" (Ph.D. diss., University of California, Berkeley, 2009); Karen Barkey, *Bandits and Bureaucrats: The Ottoman Route to State Centralization* (Ithaca, NY: Cornell University Press, 1994), 27–8; Halil İnalcık, *The Ottoman Empire: The Classical Age, 1300–1600*, trans. Norman Itzkowitz and Colin Imber (New York: Praeger, 1973), 65–9; Cornell Fleischer, "Royal Authority, Dynastic Cyclism, and 'Ibn Khaldûnism' in Sixteenth-Century Ottoman Letters," *Journal of Asian and African Studies* 18 (1983): 198–220. For the case of Mamluk Egypt, see also Linda T. Darling, "Medieval Egyptian Society and the Concept of the Circle of Justice," *Mamlūk Studies Review* 10 (2006): 1–17. And more generally, idem., *A History of Social Justice and Political Power in the Middle East: The Circle of Justice from Mesopotamia to Globalization* (New York: Routledge, 2013).
22. For conceptualizations of how to study this interconnected early modern world, see Sanjay Subrahmanyam, "Connected Histories: Notes towards a Reconfiguration of Early Modern Eurasia," *Modern Asian Studies* 31 (1997): 735–62; idem., *Explorations in Connected History: From the Tagus to the Ganges* (Delhi: Oxford University Press, 2005); idem., *Explorations in Connected History: Mughals and Franks* (Delhi: Oxford University Press, 2005).
23. On the history of gunpowder and other military technologies in the Ottoman Empire, see Gábor Ágoston, *Guns for the Sultan: Military Power and the Weapons Industry in the Ottoman Empire* (Cambridge: Cambridge University Press, 2005). For Ottoman imperial orders to Egypt concerning gunpowder, firearms, and related matters, see TSMA, E. 510 (7 B 1056/19 Aug. 1646); TSMA, E. 664/59 (n.d.); TSMA, E. 664/60 (n.d.); TSMA, E. 5207/49 (Evahir Ca 1056/5–14 July 1646); TSMA, E. 9320/1 (n.d.); BOA, MM, 11: 141 (Evahir S 1220/21–30 May 1805); BOA, MM, 12: 349 (Evahir S 1230/2–11 Feb. 1815); BOA, MM, 12: 440 (Evasıt B 1233/17–26 May 1818); BOA, MM, 13: 229 (Evasıt S 1255/26 Apr.–5 May 1839).
24. For a discussion of the history of the printing press in the Ottoman Empire, see Orlin Sabev, *İbrahim Müteferrika ya da İlk Osmanlı Matbaa Serüveni, 1726-1746: Yeniden Değerlendirme* (Istanbul: Yeditepe, 2006).
25. On the impact of, for example, maize in the Ottoman Empire, see J. R. McNeill, *The Mountains of the Mediterranean World: An Environmental History* (Cambridge: Cambridge University Press, 1992), 89–90; Faruk Tabak, *The Waning of the Mediterranean, 1550–1870: A Geohistorical Approach* (Baltimore: Johns Hopkins University Press, 2008), 255–69.
26. John F. Richards, "Early Modern India and World History," *Journal of World History* 8 (1997): 197–209.
27. Sanjay Subrahmanyam, "On World Historians in the Sixteenth Century," *Representations* 91 (2005): 26–57.

28. Huri İslamoğlu-İnan, *State and Peasant in the Ottoman Empire: Agrarian Power Relations and Regional Economic Development in Ottoman Anatolia during the Sixteenth Century* (Leiden: Brill, 1994); ʿAbd al-Raḥīm, *al-Rīf al-Miṣrī*.
29. On the sustainable uses of natural resources in the early modern world, see Conrad Totman, *The Green Archipelago: Forestry in Preindustrial Japan* (Berkeley: University of California Press, 1989); Paul Warde, *Ecology, Economy and State Formation in Early Modern Germany* (Cambridge: Cambridge University Press, 2006); Karl Appuhn, *A Forest on the Sea: Environmental Expertise in Renaissance Venice* (Baltimore: Johns Hopkins University Press, 2009); Mikhail, *Nature and Empire*; John Thomas Wing, "Roots of Empire: State Formation and the Politics of Timber Access in Early Modern Spain, 1556–1759" (Ph.D. diss., University of Minnesota, 2009); John T. Wing, "Keeping Spain Afloat: State Forestry and Imperial Defense in the Sixteenth Century," *Environmental History* 17 (2012): 116–45.
30. For a critical reading of the literature on early modern economic exchange and trust, see Francesca Trivellato, *The Familiarity of Strangers: The Sephardic Diaspora, Livorno, and Cross-Cultural Trade in the Early Modern Period* (New Haven, CT: Yale University Press, 2009), 10–16.
31. For studies that emphasize the differences between early modern and modern state structures, see Jack A. Goldstone, "The Problem of the 'Early Modern' World," *Journal of the Economic and Social History of the Orient* 41 (1998): 249–84; Steve Pincus, *1688: The First Modern Revolution* (New Haven, CT: Yale University Press, 2009).
32. Max Weber, "Politics as a Vocation," in *From Max Weber: Essays in Sociology*, trans. and ed. H. H. Gerth and C. Wright Mills (New York: Oxford University Press, 1958), 78. Emphasis in original.
33. One could cite Michel Foucault's entire oeuvre in explaining these phenomena. For a sampling, see the following: Michel Foucault, *Discipline and Punish: The Birth of the Prison*, trans. Alan Sheridan (New York: Pantheon Books, 1978); idem., *An Introduction*, vol. 1 of *The History of Sexuality*, trans. Robert Hurley (New York: Vintage Books, 1978); idem., *Madness and Civilization: A History of Insanity in the Age of Reason*, trans. Richard Howard (New York: Vintage Books, 1988). For the specific case of Egypt, see Timothy Mitchell, *Colonising Egypt* (Berkeley: University of California Press, 1991).
34. On these deleterious phenomena, in addition to the works of Foucault and Weber already cited, see Carolyn Merchant, *The Death of Nature: Women, Ecology, and the Scientific Revolution* (San Francisco: Harper & Row, 1980); James C. Scott, *Seeing Like a State: How Certain Schemes to Improve the Human Condition Have Failed* (New Haven, CT: Yale University Press, 1998).
35. Both Kant and Foucault, two hundred years apart, wrote essays entitled "What is Enlightenment?" Immanuel Kant, "An Answer to the Question: What is Enlightenment?" in *Perpetual Peace and Other Essays on Politics, History, and Morals*, trans. Ted Humphrey (Indianapolis: Hackett Publishing, 1983), 41–8; Michel Foucault, "What is Enlightenment?" in *Ethics: Subjectivity and Truth*, vol. 1 of *The Essential Works of Michel Foucault: 1954–1984*, ed. Paul Rabinow (New York: New Press, 1997), 303–19. As will no doubt become clear as one reads this book, my answer to the question of what is enlightenment is much closer to Foucault's than Kant's. The quote in the text comes from p. 318 of Foucault's essay.
36. In Randolph Starn's words, "The idea of early modernity vindicates the categorical separateness of the modern by conjuring up the binary of an archaic, conservative, traditional, or repressed opposition." Randolph Starn, "The Early Modern Muddle," *Journal of Early Modern History* 6 (2002), 304. Useful also in thinking about the connections between the early modern and the modern in this context is Shmuel N. Eisenstadt and Wolfgang Schluchter, "Introduction: Paths to Early Modernities—A Comparative View," *Daedalus* 127 (1998): 1–18; Goldstone, "Problem of the 'Early Modern' World"; Sanjay Subrahmanyam, "Hearing Voices: Vignettes of Early Modernity in South Asia, 1400–1750," *Daedalus* 127 (1998): 75–104; Björn Wittrock, "Early Modernities: Varieties and Transitions," *Daedalus* 127 (1998): 19–40. See also Dipesh Chakrabarty, "The Muddle of Modernity," *American Historical Review* 116 (2011): 663–75.
37. Environmental historians have identified the notion of "pristine nature" as both a fiction and a trap. Pristine nature is the idea that somehow nature was in a perfect state of harmony,

balance, and sustainability before humans came to destroy it. For articulations of this idea, see William M. Denevan, "The Pristine Myth: The Landscape of the Americas in 1492," *Annals of the Association of American Geographers* 82 (1992): 369–85; William Cronon, "The Trouble with Wilderness; or, Getting Back to the Wrong Nature," *Environmental History* 1 (1996): 7–28. For further discussion, see also the several essays on this topic and William Cronon's response in the same issue of *Environmental History*.

Embedded in the idea of pristine nature is the specter of ecological decline—an environmental narrative positing that the overwhelming majority of human interactions with nature have been detrimental. For an analysis of some of the political and ecological uses of a declensionist environmental narrative in colonial North Africa, see Diana K. Davis, *Resurrecting the Granary of Rome: Environmental History and French Colonial Expansion in North Africa* (Athens: Ohio University Press, 2007), 131–76; idem., "Potential Forests: Degradation Narratives, Science, and Environmental Policy in Protectorate Morocco, 1912–1956," *Environmental History* 10 (2005): 211–38.

38. For further discussion of this point, see Alan Mikhail and Christine M. Philliou, "The Ottoman Empire and the Imperial Turn," *Comparative Studies in Society and History* 54 (2012), 734–40.
39. For an argument for the democratization of the early modern Ottoman Empire, see Baki Tezcan, *The Second Ottoman Empire: Political and Social Transformation in the Early Modern World* (Cambridge: Cambridge University Press, 2010), 233.
40. In William Cronon's words, "environment may initially shape the range of choices available to a people at a given moment, but then culture reshapes environment in responding to those choices. The reshaped environment presents a new set of possibilities for cultural reproduction, thus setting up a new cycle of mutual determination." William Cronon, *Changes in the Land: Indians, Colonists, and the Ecology of New England*, rev. ed. (New York: Hill and Wang, 2003), 13.
41. Mikhail, *Nature and Empire*, 38–81; idem., "An Irrigated Empire: The View from Ottoman Fayyum," *International Journal of Middle East Studies* 42 (2010): 569–90. For examples of the Ottoman imperial administration's reliance on the knowledge, experience, and authority of Egyptian rural cultivators to manage irrigation, see BOA, Cevdet Nafia, 120 (Evasıt Ca 1125/5–14 June 1713); BOA, MM, 8: 469 (Evasıt L 1180/12–21 Mar. 1767); DWQ, Maḥkamat al-Manṣūra 18, p. 266, no case no. (B) (6 B 1121/10 Sept. 1709); BOA, MM, 9: 424 (Evail C 1194/4–13 June 1780); DWQ, Maḥkamat Asyūṭ 1, p. 201, case 583 (12 Za 1067/22 Aug. 1657); DWQ, Maḥkamat Asyūṭ 4, p. 206, case 645 (11 C 1156/2 Aug. 1743); DWQ, Maḥkamat Asyūṭ 2, p. 238, case 566 (13 M 1108/11 Aug. 1696); DWQ, Maḥkamat Asyūṭ 5, p. 179, case 343 (20 C 1189/17 Aug. 1775); DWQ, Maḥkamat Asyūṭ 8, p. 260, case 563 (14 S 1211/18 Aug. 1796); DWQ, Maḍābiṭ al-Daqahliyya 19, p. 299, case 878 (1185/1771 and 1772); DWQ, Maḍābiṭ al-Daqahliyya 19, p. 299, case 875 (1185/1771 and 1772); DWQ, Maḍābiṭ al-Daqahliyya 19, p. 299, case 874 (1185/1771 and 1772); DWQ, Maḍābiṭ al-Daqahliyya 19, p. 299, case 876 (1185/1771 and 1772); DWQ, Maḍābiṭ al-Daqahliyya 19, p. 299, case 877 (1185/1771 and 1772); DWQ, Maḍābiṭ al-Daqahliyya 19, p. 299, case 872 (1185/1771 and 1772); DWQ, Maḍābiṭ al-Daqahliyya 19, p. 299, case 873 (1185/1771 and 1772); DWQ, Maḍābiṭ al-Daqahliyya 19, p. 300, case 880 (1185/1771 and 1772); DWQ, Maḥkamat al-Manṣūra 4, p. 108, case 281 (1 M 1075/24 July 1664); DWQ, Maḥkamat al-Manṣūra 12, p. 448, no case no. (11 C 1104/17 Feb. 1693); DWQ, Maḥkamat al-Manṣūra 16, p. 402, no case no. (B) (30 Za 1115/5 Apr. 1704); DWQ, Maḥkamat al-Manṣūra 14, p. 206, case 482 (17 Za 1111/7 May 1700); DWQ, Maḥkamat al-Manṣūra 19, p. 33, case 88 (11 C 1122/7 Aug. 1710); DWQ, Maḥkamat al-Manṣūra 24, p. 146, case 333 (3 M 1136/3 Oct. 1723).
42. These were connections of push and pull, or what other historians—in a very different context—have usefully termed "tensions of empire." Frederick Cooper and Ann Laura Stoler, eds., *Tensions of Empire: Colonial Cultures in a Bourgeois World* (Berkeley: University of California Press, 1997).
43. "Economy," as William Cronon usefully reminds us, and to which we might also attach politics, society, and imperial sovereignty, is "a subset of ecology." Cronon, *Changes in the Land*, xv–xvi.
44. To be clear, I am not suggesting that everything in Egypt between 1517 and 1882 can only be studied or understood as an outcome of Ottoman rule. Clearly not. As is true in all

polities, whether ancient or modern, imperial or not, sources from the Ottoman Empire show that multiple histories of Egypt from this period bear no recognizable trace of the empire or any of its laws or institutions. For example, whole portions of the following seventeenth-century satirical chronicle of the Egyptian countryside seemingly have nothing to do with anything connected to Ottoman political presence in Egypt at the time: Yūsuf ibn Muḥammad al-Shirbīnī, *Kitāb Hazz al-Quḥūf bi-Sharḥ Qaṣīd Abī Shādūf*, ed. and trans. Humphrey Davies, 2 vols. (Leuven: Peeters, 2005–07). I also consulted the following version of this text: idem., *Hazz al-Quḥūf fī Sharḥ Qaṣīd Abī Shādūf*, Beinecke, Hartford Seminary Arabic MSS 56.

45. See, for example, Hathaway, *Politics of Households*; idem., "Eunuch Households in Istanbul, Medina, and Cairo during the Ottoman Era," *Turcica* 41 (2009): 291–303; Tezcan, *Second Ottoman Empire*; Ali Yaycıoğlu, "Provincial Power-Holders and the Empire in the Late Ottoman World: Conflict or Partnership?" in *The Ottoman World*, ed. Christine Woodhead (New York: Routledge, 2012), 436–52.

46. See, for example, Beshara Doumani, ed., *Family History in the Middle East: Household, Property, and Gender* (Albany: State University of New York Press, 2003); Philliou, *Biography of an Empire*; Margaret L. Meriwether, *The Kin Who Count: Family and Society in Ottoman Aleppo, 1770–1840* (Austin: University of Texas Press, 1999); Christopher H. Johnson, David Warren Sabean, Simon Teuscher, and Francesca Trivellato, eds., *Transregional and Transnational Families in Europe and Beyond: Experiences since the Middle Ages* (New York: Berghahn Books, 2011).

47. For accounts of the Ottoman conquest of Egypt, see *Sulṭān Selim'in İran ve Mısır Seferine dair Muḥāberāt*, Topkapı Sarayı Müzesi Kütüphanesi, Istanbul, R. 1955, 65r–106v; Muḥammad ibn Yūsuf al-Ḥallāq, *Tuḥfat al-Aḥbāb bi-man Malaka Miṣr min al-Mulūk wa al-Nūwāb*, Beinecke, Landberg MSS 229, 73r–78r; Rūmī, *Kitāb Ta'rīkh Miṣr wa al-Nīl*, 75r; Khalīl ibn Aḥmad al-Madābighī, *Ta'rīkh*, Beinecke, Landberg MSS 630, 34r-34v; Muḥammad ibn Abī al-Surūr al-Bakrī, *al-Rawḍa al-Ma'nūsa fī Akhbār Miṣr al-Maḥrūsa*, ed. ʿAbd al-Rāziq ʿAbd al-Rāziq ʿĪsā (Cairo: Maktabat al-Thaqāfa al-Dīniyya, 1997), 65–79; Muḥammad ibn Abī al-Surūr al-Ṣiddīqī al-Bakrī, *al-Tuḥfa al-Bahiyya fī Tamalluk Āl ʿUthmān al-Diyār al-Miṣriyya*, ed. ʿAbd al-Raḥīm ʿAbd al-Raḥmān ʿAbd al-Raḥīm (Cairo: Dār al-Kutub wa al-Wathā'iq al-Qawmiyya, 2005), 32–103; Muḥammad ibn Abī al-Surūr al-Bakrī al-Ṣiddīqī, *al-Minaḥ al-Raḥmāniyya fī al-Dawla al-ʿUthmāniyya, wa Dhayluhu al-Laṭā'if al-Rabbāniyya ʿalā al-Minaḥ al-Raḥmāniyya*, ed. Laylā al-Ṣabbāgh (Damascus: Dār al-Bashā'ir, 1995), 71–103; Celâl-zâde Sâlih Çelebi, *Tarih-i Mısr-ı Cedîd: İnceleme—Metin*, ed. Tuncay Bülbül (Ankara: Grafiker Yayınları, 2011), 236–39; Abdülkerim ibn Abdurrahman, *Tarih-i Mısır*, SK, Hekimoğlu 705, 2v–3v; *Hikaye-i Feth-i Mısır*, SK, Kemankeş 489/2, 42v–50v. See also the two versions of the following text: Çerkesler Katibi Yusuf Efendi, *Tarih-i Mısır*, SK, Esad Efendi 2148, 18v–53v; idem., *Tarih-i Fetih Mısır ve Hukmu*, SK, Esad Efendi 2146, 22r–59v. And the three different versions of this text: ʿAbdullah Çelebi Ridvan Paşazade, *Mısır Tarihi*, SK, Esad Efendi 2177, 211–23; idem., *Tarih-i Mısır*, SK, Fatih 4362, 101r–126v; idem., *Tarih-i Mısır*, SK, Reşid Efendi 624, 194v–238r. And the two slightly different versions of this one: Ahmed ibn Hemdem Süheyli, *Tarih-i Mısır*, SK, Mehmed Zeki Pakalın 99, 1v–37r; idem., *Tarih-i Mısır ül-Cedid*, SK, Reşid Efendi 631/3, 58r–86r. See also the following slightly different versions of this text: Marʿī ibn Yūsuf al-Karmī, *Nuzhat al-Nāẓirīn fī Ta'rīkh man Waliya Miṣr min al-Khulafā' wa al-Salāṭīn*, Beinecke, Arabic MSS suppl. 397, 187v–88r; idem., *Nuzhat al-Nāẓirīn fī Ta'rīkh man Waliya Miṣr min al-Khulafā' wa al-Salāṭīn*, Beinecke, Salisbury MSS 67, 65v–66v; idem., *Nuzhat al-Nāẓirīn fī Ta'rīkh man Waliya Miṣr min al-Khulafā' wa al-Salāṭīn*, Beinecke, Landberg MSS 11a, 59r–61v; idem., *Nuzhat al-Nāẓirīn fī Ta'rīkh man Waliya Miṣr min al-Khulafā' wa al-Salāṭīn*, Beinecke, Landberg MSS 232, 60v–63r; Marʿī al-Ḥanbalī al-Maqdisī, *Nuzhat al-Nāẓirīn fī Tārīkh man Waliya Miṣr min al-Khulafā' wa al-Salāṭīn*, Maʿhad al-Makhṭūṭāt al-ʿArabiyya, Cairo, al-Tārīkh 1283, 134–39. This text has been published as Marʿī al-Karmī al-Ḥanbalī, *Nuzhat al-Nāẓirīn fī Tārīkh man Waliya Miṣr min al-Khulafā' wa al-Salāṭīn*, ed. ʿAbd Allāh Muḥammad al-Kandarī (Beirut: Dār al-Nawādir, 2012). See also the two versions of the following text: Aḥmad ibn ʿAlī ibn Zunbul, *Ghazwat al-Sulṭān Salīm Khān maʿa al-Sulṭān al-Ghūrī*, Beinecke, Landberg MSS 461; idem., *Ākhirat al-Mamālīk: Wāqiʿat al-Sulṭān al-Ghūrī maʿa Salīm al-ʿUthmānī*,

ed. ʿAbd al-Munʿim ʿĀmir (Cairo: al-Dār al-Qawmiyya lil-Ṭibāʿa wa al-Nashr, 1962). And also the two versions of this text: Muḥammad ibn Abī al-Surūr al-Bakrī al-Ṣiddīqī, *al-Nuzha al-Zahiyya fī Dhikr Wulāt Miṣr wa al-Qāhira al-Muʿizziyya*, Beinecke, Landberg MSS 231, 20r–23v; Muḥammad ibn al-Surūr al-Bakrī al-Ṣiddīqī, *al-Nuzha al-Zahiyya fī Dhikr Wulāt Miṣr wa al-Qāhira al-Muʿizziyya*, ed. ʿAbd al-Rāziq ʿAbd al-Rāziq ʿĪsā (Cairo: al-ʿArabī lil-Nashr wa al-Tawzīʿ, 1998), 124–35.

For Ottoman guarantees to maintain peaceful trading relations between Egypt and Venice after the conquest of 1517, see ASV, Turchi, 167 (Ş 923/19 Aug.–17 Sept. 1517); ASV, Turchi, 169 (8 Sept. 1517). These and later examples of peace treaties between Venice and the Ottomans (and between the latter and other political powers) are known collectively as the capitulations. For mostly eighteenth-century examples of these treaties between Venice and North Africa, see ASV, Algeri, Egitto, Marocco, 1, Documenti Algeri 1 (7 M 1177/18 July 1763); ASV, Algeri, Egitto, Marocco, 1, Documenti Algeri 5 (1 S 1182/17 June 1768); ASV, Algeri, Egitto, Marocco, 1, Documenti Marocco 2 (9 N 913/22 Jan. 1508); ASV, Algeri, Egitto, Marocco, 1, Documenti Marocco 3 (24 Z 1178/14 June 1765); ASV, Algeri, Egitto, Marocco, 1, Documenti Marocco 4 (24 Z 1178/14 June 1765); ASV, Algeri, Egitto, Marocco, 1, Documenti Marocco 7 (12 R 1179/28 Sept. 1765); ASV, Algeri, Egitto, Marocco, 1, Documenti Marocco 9–10 (26 S 1210/11 Sept. 1795). On the capitulations, see Maurits H. van den Boogert, *The Capitulations and the Ottoman Legal System: Qadis, Consuls and Beratlıs in the 18th Century* (Leiden: Brill, 2005); Maurits H. van den Boogert and Kate Fleet, eds., *The Ottoman Capitulations: Text and Context* (Rome: Istituto per l'Oriente C. A. Nallino, 2004); Molly Greene, *Catholic Pirates and Greek Merchants: A Maritime History of the Mediterranean* (Princeton, NJ: Princeton University Press, 2010).

48. Shaw, *Financial and Administrative Organization and Development*. The annual remittance of Egypt to the empire was known as the *irsāliyye-i ḫazīne*. For examples of imperial orders concerning the preparation, payment, and use of these monies, see TSMA, E. 4675/2 (20 N 1061/6 Sept. 1651); TSMA, E. 664/4 (n.d.); TSMA, E. 664/7 (n.d.); TSMA, E. 664/10 (n.d.); TSMA, E. 664/40 (n.d.); TSMA, E. 664/51 (n.d.); TSMA, E. 664/64 (1 C 1059/12 June 1649); TSMA, E. 5207/57 (Evail B 1056/12–21 Aug. 1646); TSMA, E. 5207/58 (Evasıt B 1056/22–31 Aug. 1646); TSMA, E. 7016/95 (n.d.); TSMA, E. 5207/49 (Evahir Ca 1056/5–14 July 1646); TSMA, E. 664/52 (n.d.); TSMA, E. 664/63 (Evail Ra 1159/24 Mar.–2 Apr. 1746); TSMA, E. 664/66 (n.d.); TSMA, E. 3522 (24 Ş 1148/8 Jan. 1736); TSMA, E. 4830 (17 C 1194/20 June 1780); BOA, Cevdet Maliye, 15566 (15 C 1191/20 July 1777); BOA, MM, 1: 123 (Evahir R 1122/19–28 June 1710); BOA, MM, 1: 166 (Evasıt S 1123/31 Mar.–9 Apr. 1711); BOA, MM, 1: 285 (Evasıt N 1125/1–10 Oct. 1713); BOA, MM, 1: 334 (Evahir C 1126/4–13 July 1714); BOA, MM, 1: 429 (Evahir Ra 1128/15–24 Mar. 1716); BOA, MM, 3: 178 (Evasıt Ca 1133/10–19 Mar. 1721); BOA, MM, 4: 278 (Evasıt C 1142/1–10 Jan. 1730); BOA, MM, 4: 292 (Evahir L 1142/9–18 May 1730); BOA, MM, 8: 120 (Evahir C 1176/7–16 Jan. 1763); BOA, MM, 10: 364 (Evahir C 1207/3–12 Feb. 1793); BOA, HAT 95/3856A (27 Z 1215/11 May 1801); BOA, HAT 16/716A (29 Z 1189/19 Feb. 1776). There is no internal evidence for the date of this last case. The date given is the one assigned by the BOA. For a detailed accounting of each component of the *irsāliyye-i ḫazīne* from 1596 to 1597, see: Stanford J. Shaw, *The Budget of Ottoman Egypt, 1005–1006/1596–1597* (The Hague: Mouton, 1968).

49. Mikhail, *Nature and Empire*, 82–123. For examples of the shipment of food from Rosetta to Istanbul in the eighteenth century, see DWQ, Maḥkamat Rashīd 122, p. 67, case 113 (21 Ca 1131/11 Apr. 1719); DWQ, Maḥkamat Rashīd 123, p. 142, case 241 (25 B 1131/14 June 1719); DWQ, Maḥkamat Rashīd 124, p. 253, case 352 (1 Ca 1132/10 Mar. 1720); DWQ, Maḥkamat Rashīd 125, p. 318, case 516 (26 Ra 1133/25 Jan. 1721); DWQ, Maḥkamat Rashīd 125, p. 319, case 517 (28 Ra 1133/27 Jan. 1721); DWQ, Maḥkamat Rashīd 130, p. 404, no case no. (3 Ş 1136/27 Apr. 1724); DWQ, Maḥkamat Rashīd 132, p. 199, case 310 (2 C 1137/15 Feb. 1725); DWQ, Maḥkamat Rashīd 134, p. 344, case 462 (28 Ş 1140/8 Apr. 1728); DWQ, Maḥkamat Rashīd 137, p. 6, case 13 (19 Za 1143/26 May 1731); DWQ, Maḥkamat Rashīd 139, p. 60, case 95 (23 M 1146/5 July 1733); DWQ, Maḥkamat Rashīd 142, p. 51, case 45 (1 B 1149/4 Nov. 1736); DWQ, Maḥkamat Rashīd 146, p. 139, case 116 (1 C 1153/24 Aug. 1740); DWQ, Maḥkamat Rashīd 148, p. 176, case 219 (21 Z 1154/27

Feb. 1742); DWQ, Maḥkamat Rashīd 151, p. 315, case 357 (12 M 1160/23 Jan. 1747); DWQ, Maḥkamat Rashīd 154, p. 182, case 203 (25 Z 1162/6 Dec. 1749); DWQ, Maḥkamat Rashīd 157, p. 324, case 319 (15 R 1166/19 Feb. 1753).

50. Alexandria was especially important to Ottoman interests in the eastern Mediterranean. The port city was used as an Ottoman naval base for various operations around the sea. Ships from Alexandria, for example, supported Ottoman military expeditions to Chios in 1566, Malta in 1575, and Crete in 1666 and 1715. Moreover, the port was also crucial as a controlling hinge of trade between the Mediterranean and Red Sea and Indian Ocean. Indeed, in the first half of the eighteenth century the amount of trade and traffic moving through Alexandria was so great that there was even a floating crane in use in the port to aid in the loading and unloading of cargo and the repair of ships. Thus, because of its strategic military and economic importance, the Ottomans paid particularly close attention to the administration of Alexandria through a customs regime, legal and economic regulations, and a military presence. On these various roles of Ottoman Alexandria in the eastern Mediterranean, see İdris Bostan, "An Ottoman Base in Eastern Mediterranean: Alexandria of Egypt in the 18th Century," in *Proceedings of the International Conference on Egypt during the Ottoman Era: 26–30 November 2007, Cairo, Egypt*, ed. Research Centre for Islamic History, Art and Culture (Istanbul: IRCICA, 2010), 63–77; Michael J. Reimer, "Ottoman Alexandria: The Paradox of Decline and the Reconfiguration of Power in Eighteenth-Century Arab Provinces," *Journal of the Economic and Social History of the Orient* 37 (1994): 107–46; Daniel Panzac, "International and Domestic Maritime Trade in the Ottoman Empire during the 18th Century," *International Journal of Middle East Studies* 24 (1992): 189–206.

51. On connections between Egypt and the Hijaz in the Ottoman period, see Suraiya Faroqhi, "Trade Controls, Provisioning Policies, and Donations: The Egypt-Hijaz Connection during the Second Half of the Sixteenth Century," in *Süleymân the Second and His Time*, ed. Halil İnalcık and Cemal Kafadar (Istanbul: Isis Press, 1993), 131–43; idem., "Red Sea Trade and Communications as Observed by Evliya Çelebi (1671–72)," *New Perspectives on Turkey* 5–6 (1991): 87–105; idem., "Coffee and Spices: Official Ottoman Reactions to Egyptian Trade in the Later Sixteenth Century," *Wiener Zeitschrift für die Kunde des Morgenlandes* 76 (1986): 87–93; Michel Tuchscherer, "Commerce et production du café en mer Rouge au XVIe siècle," in *Le commerce du café avant l'ère des plantations coloniales: espaces, réseaux, sociétés (XVe–XIXe siècle)*, ed. Michel Tuchscherer (Cairo: Institut français d'archéologie orientale, 2001), 69–90; Ḥusām Muḥammad ʿAbd al-Muʿṭī, *al-ʿAlāqāt al-Miṣriyya al-Ḥijāziyya fī al-Qarn al-Thāmin ʿAshar* (Cairo: al-Hayʾa al-Miṣriyya al-ʿĀmma lil-Kitāb, 1999); Colin Heywood, "A Red Sea Shipping Register of the 1670s for the Supply of Foodstuffs from Egyptian *Wakf* Sources to Mecca and Medina (Turkish Documents from the Archive of ʿAbdurrahman 'ʿAbdi' Pasha of Buda, I)," *Anatolia Moderna* 6 (1996): 111–74; M. Abir, "The 'Arab Rebellion' of Amir Ghālib of Mecca (1788–1813)," *Middle Eastern Studies* 7 (1971): 185–200. For a discussion of earlier connections, see John L. Meloy, *Imperial Power and Maritime Trade: Mecca and Cairo in the Later Middle Ages* (Chicago: Middle East Documentation Center, 2010). For examples of the shipment of food from the southern Egyptian cities of Manfalūṭ and Asyūṭ to the Hijaz, see DWQ, Maḥkamat Manfalūṭ 2, p. 189, case 631 (24 Ca 1179/8 Nov. 1765); DWQ, Maḥkamat Manfalūṭ 2, p. 190, case 632 (20 C 1179/4 Dec. 1765); DWQ, Maḥkamat Manfalūṭ 2, p. 190, case 633 (3 Z 1180/2 May 1767); DWQ, Maḥkamat Asyūṭ 2, p. 235, no case no. (23 Z 1107/23 July 1696).

52. Generally on the Ottomans in the Red Sea and Indian Ocean, see Salih Özbaran, *Ottoman Expansion towards the Indian Ocean in the 16th Century* (Istanbul: Bilgi University Press, 2009); idem., *The Ottoman Response to European Expansion: Studies on Ottoman-Portuguese Relations in the Indian Ocean and Ottoman Administration in the Arab Lands during the Sixteenth Century* (Istanbul: Isis Press, 1994); idem., "A Turkish Report on the Red Sea and the Portuguese in the Indian Ocean (1525)," *Arabian Studies* 4 (1978): 81–88; idem., "Ottoman Naval Power in the Indian Ocean in the 16th Century," in *The Kapudan Pasha, His Office and His Domain: Halcyon Days in Crete IV*, ed. Elizabeth Zachariadou (Rethymnon: Crete University Press, 2002), 109–17; Giancarlo Casale, *The Ottoman Age of Exploration* (New York: Oxford University Press, 2010); idem., "The Ottoman Administration of the Spice Trade in the Sixteenth-Century Red Sea and Persian Gulf," *Journal of the Economic and Social History of*

the Orient 49 (2006): 170–98; Jean-Louis Bacqué-Grammont and Anne Kroell, *Mamlouks, Ottomans et Portugais en Mer Rouge: L'affaire de Djedda en 1517* (Cairo: Institut français d'archéologie orientale, 1988); Anthony Reid, "Sixteenth-Century Turkish Influence in Western Indonesia," *Journal of South East Asian History* 10 (1969): 395–414; Michel Tuchscherer, "La flotte impériale de Suez de 1694 à 1719," *Turcica* 29 (1997): 47–69.

53. Crecelius, *Roots of Modern Egypt*; Fahmy, *All the Pasha's Men*; Fred H. Lawson, *The Social Origins of Egyptian Expansionism during the Muhammad ʿAli Period* (New York: Columbia University Press, 1992); Muhammad H. Kutluoğlu, *The Egyptian Question (1831–1841): The Expansionist Policy of Mehmed Ali Paşa in Syria and Asia Minor and the Reaction of the Sublime Porte* (Istanbul: Eren, 1998); Şinasi Altundağ, *Kavalalı Mehmet Ali Paşa Isyanı: Mısır Meselesi, 1831–1841* (Ankara: Türk Tarih Kurumu, 1988); Asad J. Rustom, *The Royal Archives of Egypt and the Origins of the Egyptian Expedition to Syria, 1831–1841* (Beirut: American Press, 1936).

54. See, for example, the following about palace eunuchs' impressive economic investments and interests in Egypt: Jane Hathaway, *Beshir Agha: Chief Eunuch of the Ottoman Imperial Harem* (Oxford: Oneworld, 2005); idem., "The Wealth and Influence of an Exiled Ottoman Eunuch in Egypt: The *Waqf* Inventory of ʿAbbās Agha," *Journal of the Economic and Social History of the Orient* 37 (1994): 293–317; idem., "The Role of the Kızlar Ağası in 17th-18th Century Ottoman Egypt," *Studia Islamica* 75 (1992): 141–58.

55. For various examples of Venetian commercial interests in Alexandria, see ASV, Turchi, 253 (Evahir Z 936/16–25 Aug. 1530); ASV, Turchi, 254 (Evahir Z 936/16–25 Aug. 1530); ASV, Turchi, 554 (Evahir N 952/26 Nov.–5 Dec.1545); ASV, Turchi, 933 (7 Z 992/10 Dec. 1584); ASV, Lettere e Scritture Turchesche, 3: 187 (Evahir L 985/1–10 Jan. 1578); ASV, Lettere e Scritture Turchesche, 4: 159 (Evahir M 999/19–28 Nov. 1590); ASV, Algeri, Egitto, Marocco, 1, Documenti Egitto 1 (10 Ş 877/10 Jan. 1473); ASV, Algeri, Egitto, Marocco, 1, Documenti Egitto 4 (18 M 913/30 May 1507). For a statement about the very favorable treatment Venetian merchants received in Egypt, see ASV, Lettere e Scritture Turchesche, 2: 65 (Evasıt S 961/26 Jan.–4 Feb. 1554). For an example both of the intense interests that Venice and the Ottoman Porte maintained in Egypt in the sixteenth century and of their cooperation in the province, consider the following Ottoman request that the Doge dispatch Venetian ships to deliver needed steel to Egypt: ASV, Lettere e Scritture Turchesche, 2: 283 (Evasıt Ca 977/22–31 Oct. 1569). On Venetian merchants in Mamluk Egypt, see Georg Christ, *Trading Conflicts: Venetian Merchants and Mamluk Officials in Late Medieval Alexandria* (Leiden: Brill, 2012); Benjamin Arbel, "The Last Decades of Venice's Trade with the Mamluks: Importations into Egypt and Syria," *Mamlūk Studies Review* 8 (2004): 37–86.

56. Eve M. Troutt Powell, *A Different Shade of Colonialism: Egypt, Great Britain, and the Mastery of the Sudan* (Berkeley: University of California Press, 2003); Heather J. Sharkey, *Living with Colonialism: Nationalism and Culture in the Anglo-Egyptian Sudan* (Berkeley: University of California Press, 2003); Terence Walz, *Trade between Egypt and Bilād as-Sūdān, 1700–1820* (Cairo: Institut français d'archéologie orientale du Caire, 1978); M. F. Shukry, *The Khedive Ismail and Slavery in the Sudan (1863–1879)* (Cairo: Librairie La Renaissance d'Égypte, 1937).

57. While animals are, of course, as Claude Lévi-Strauss famously put it and as is constantly repeated, "good to think" with, they are clearly equally—if not much—better to plow, transport, and haul with. The full quotation is: "We can understand, too, that natural species are chosen not because they are 'good to eat' but because they are 'good to think.'" Claude Lévi-Strauss, *Totemism*, trans. Rodney Needham (Boston: Beacon Press, 1963), 89. In the vein of moving beyond the Lévi-Straussian dictum, the anthropologist Marvin Harris writes, "I believe the irrational, non-economic, and exotic aspects of the Indian cattle complex are greatly overemphasized at the expense of rational, economic, and mundane interpretations." Marvin Harris, "The Cultural Ecology of India's Sacred Cattle," *Current Anthropology* 7 (1966), 51. In seeking to get past culturalist and religious explanations of the role of cattle in Indian society that relied almost exclusively on interpretations of Hinduism, Harris thus shifted toward an economic and materialist understanding of animals in society. My goal is similar in this book.

Animal historians have focused on many different aspects of human-animal relations. For example, cultural historians interested in subjects such as the history of petkeeping have made important contributions to the historiography of human affect and the study of class and domesticity (in all its meanings). Kathleen Kete, *The Beast in the Boudoir: Petkeeping in Nineteenth-Century Paris* (Berkeley: University of California Press, 1994); Erica Fudge, *Pets* (Stocksfield: Acumen, 2008); Katherine C. Grier, *Pets in America: A History* (Chapel Hill: University of North Carolina Press, 2006); Yi-Fu Tuan, *Dominance and Affection: The Making of Pets* (New Haven, CT: Yale University Press, 1984); Kathleen Walker-Meikle, *Medieval Pets* (Woodbridge, UK: Boydell Press, 2012); Donna J. Haraway, *When Species Meet* (Minneapolis: University of Minnesota Press, 2008); Sarah Hand Meacham, "Pets, Status, and Slavery in the Late-Eighteenth-Century Chesapeake," *Journal of Southern History* 77 (2011): 521–54; Russell W. Belk, "Metaphoric Relationships with Pets," *Society and Animals* 4 (1996): 121–45.

For their part, political, intellectual, and environmental historians have considered the history of animals in the service of fleshing out genealogies of concepts such as brutality, animality, wildness, the exotic, tameness, humanity, morality, and human and animal rights. Harriet Ritvo, *The Animal Estate: The English and Other Creatures in the Victorian Age* (Cambridge, MA: Harvard University Press, 1987); idem., *The Platypus and the Mermaid and Other Figments of the Classifying Imagination* (Cambridge, MA: Harvard University Press, 1997); Dominick LaCapra, *History and Its Limits: Human, Animal, Violence* (Ithaca, NY: Cornell University Press, 2009); Erica Fudge, *Brutal Reasoning: Animals, Rationality, and Humanity in Early Modern England* (Ithaca, NY: Cornell University Press, 2006); idem., *Perceiving Animals: Humans and Beasts in Early Modern English Culture* (New York: St. Martin's Press, 1999); Nancy J. Jacobs, "The Great Bophuthatswana Donkey Massacre: Discourse on the Ass and the Politics of Class and Grass," *American Historical Review* 106 (2001): 485–507; Londa Schiebinger, "Why Mammals are Called Mammals: Gender Politics in Eighteenth-Century Natural History," *American Historical Review* 98 (1993): 382–411; Cary Wolfe, *What is Posthumanism?* (Minneapolis: University of Minnesota Press, 2010); Keith Tester, *Animals and Society: The Humanity of Animal Rights* (London: Routledge, 1991); Jacques Derrida, *The Animal That Therefore I Am*, ed. Marie-Louise Mallet, trans. David Wills (New York: Fordham University Press, 2008); Stephen R. L. Clark, *The Nature of the Beast: Are Animals Moral?* (Oxford: Oxford University Press, 1982); S. F. Sapontzis, *Morals, Reason, and Animals* (Philadelphia: Temple University Press, 1987); Hilda Kean, "Imagining Rabbits and Squirrels in the English Countryside," *Society and Animals* 9 (2001): 163–75; Evelyn B. Pluhar, *Beyond Prejudice: The Moral Significance of Human and Nonhuman Animals* (Durham, NC: Duke University Press, 1995); Eric Lambin, *An Ecology of Happiness*, trans. Teresa Lavender Fagan (Chicago: University of Chicago Press, 2012), 29–49; Lori Gruen, *Ethics and Animals: An Introduction* (Cambridge: Cambridge University Press, 2011); Joyce E. Salisbury, *The Beast Within: Animals in the Middle Ages* (New York: Routledge, 1994); Richard W. Bulliet, *Hunters, Herders, and Hamburgers: The Past and Future of Human-Animal Relationships* (New York: Columbia University Press, 2005); Thomas G. Andrews, "Contemplating Animal Histories: Pedagogy and Politics across Borders," *Radical History Review* 107 (2010): 139–65; Rod Preece, "Thoughts out of Season on the History of Animal Ethics," *Society and Animals* 15 (2007): 365–78; Nathaniel Wolloch, *Subjugated Animals: Animals and Anthropocentrism in Early Modern European Culture* (Amherst, NY: Humanity Books, 2006); S. Eben Kirksey and Stefan Helmreich, "The Emergence of Multispecies Ethnography," *Cultural Anthropology* 25 (2010): 545–76; Anita Guerrini, *Experimenting with Humans and Animals: From Galen to Animal Rights* (Baltimore: Johns Hopkins University Press, 2003); Deborah Rudacille, *The Scalpel and the Butterfly: The War between Animal Research and Animal Protection* (New York: Farrar, Straus, and Giroux, 2000); Craig Buettinger, "Women and Antivivisection in Late Nineteenth-Century America," *Journal of Social History* 30 (1997): 857–72; Hayley Rose Glaholt, "Vivisection as War: The 'Moral Diseases' of Animal Experimentation and Slavery in British Victorian Quaker Pacifist Ethics," *Society and Animals* 20 (2012): 154–72; Al-Hafiz B.A. Masri, "Animal Experimentation: The Muslim Viewpoint," in *Animal Sacrifices: Religious Perspectives on the Use of Animals in Science*, ed. Tom Regan (Philadelphia: Temple University Press, 1986), 171–98; Jake Kosek, "Ecologies of Empire: On the New Uses of the Honeybee," *Cultural Anthropology* 25 (2010): 650–78; E. P. Evans, *The Criminal Prosecution and Capital Punishment of Animals*

(New York: E. P. Dutton, 1906); Roel Sterchx, *The Animal and the Daemon in Early China* (Albany: State University of New York Press, 2002); Louise E. Robbins, *Elephant Slaves and Pampered Parrots: Exotic Animals in Eighteenth-Century Paris* (Baltimore: Johns Hopkins University Press, 2002). See also the relevant chapters in the following important volumes: Gregory M. Pflugfelder and Brett L. Walker, eds., *JAPANimals: History and Culture in Japan's Animal Life* (Ann Arbor: Center for Japanese Studies, University of Michigan, 2005); Nigel Rothfels, ed., *Representing Animals* (Bloomington: Indiana University Press, 2002); Erica Fudge, ed., *Renaissance Beasts: Of Animals, Humans, and Other Wonderful Creatures* (Urbana: University of Illinois Press, 2004); H. Peter Steeves, ed., *Animal Others: On Ethics, Ontology, and Animal Life* (Albany: State University of New York Press, 1999); Angela N. H. Creager and William Chester Jordan, eds., *The Animal/Human Boundary: Historical Perspectives* (Rochester, NY: University of Rochester Press, 2002); Mary J. Henninger-Voss, ed., *Animals in Human Histories: The Mirror of Nature and Culture* (Rochester, NY: University of Rochester Press, 2002). For an effort to identify a Marxist ethics of animals, see Katherine Perlo, "Marxism and the Underdog," *Society and Animals* 10 (2002): 303–18. For responses to Perlo's arguments, see Peter Dickens, "The Labor Process: How the Underdog is Kept Under," *Society and Animals* 11 (2003): 69–72; Ted Benton, "Marxism and the Moral Status of Animals," *Society and Animals* 11 (2003): 73–79.

Some environmental and agricultural historians have focused on the implications of animal husbandry, migration, hunting, and species depletion for landscapes and human communities alike. Brett L. Walker, *The Lost Wolves of Japan* (Seattle: University of Washington Press, 2005); Elinor G. K. Melville, *A Plague of Sheep: Environmental Consequences of the Conquest of Mexico* (Cambridge: Cambridge University Press, 1997); Alfred W. Crosby, *Ecological Imperialism: The Biological Expansion of Europe, 900–1900* (Cambridge: Cambridge University Press, 2004), 171–94; Joseph E. Taylor III, *Making Salmon: An Environmental History of the Northwest Fisheries Crisis* (Seattle: University of Washington Press, 1999); Virginia DeJohn Anderson, *Creatures of Empire: How Domestic Animals Transformed Early America* (New York: Oxford University Press, 2004); Jon T. Coleman, *Vicious: Wolves and Men in America* (New Haven, CT: Yale University Press, 2004); Matt Cartmill, *A View to a Death in the Morning: Hunting and Nature through History* (Cambridge, MA: Harvard University Press, 1993); Andrew C. Isenberg, *The Destruction of the Bison: An Environmental History, 1750–1920* (Cambridge: Cambridge University Press, 2000).

Historians of medicine make clear that most human diseases, from HIV/AIDS to the common flu, originated in the nonhuman animal world and made the species leap through intimate human interactions with other animals. David Quammen, *Spillover: Animal Infections and the Next Human Pandemic* (New York: W. W. Norton, 2012); Lambin, *Ecology of Happiness*, 80–91; Nathan Wolfe, *The Viral Storm: The Dawn of a New Pandemic Age* (New York: Times Books, 2011); Barbara Natterson-Horowitz and Kathryn Bowers, *Zoobiquity: The Astonishing Connection Between Human and Animal Health* (New York: Vintage Books, 2013).

And institutions like the zoo, its antecedent the zoological park, and schools of veterinary medicine have also received some attention from historians. On zoos and zoological parks, see Nigel Rothfels, *Savages and Beasts: The Birth of the Modern Zoo* (Baltimore: Johns Hopkins University Press, 2002); R. J. Hoage and William A. Deiss, eds., *New Worlds, New Animals: From Menagerie to Zoological Park in the Nineteenth Century* (Baltimore: Johns Hopkins University Press, 1996); Donna Haraway, "Teddy Bear Patriarchy: Taxidermy in the Garden of Eden, New York City, 1908–1936," *Social Text* 11 (1984): 19–64; Ian Jared Miller, "The Nature of the Beast: The Ueno Zoological Gardens and Imperial Modernity in Japan, 1882–1945" (Ph.D. diss., Columbia University, 2004); Randy Malamud, *Reading Zoos: Representations of Animals and Captivity* (New York: New York University Press, 1998). On veterinary medicine, see Joanna Swabe, *Animals, Disease, and Human Society: Human-Animal Relations and the Rise of Veterinary Medicine* (London: Routledge, 1999); Karen Brown and Daniel Gilfoyle, eds., *Healing the Herds: Disease, Livestock Economies, and the Globalization of Veterinary Medicine* (Athens: Ohio University Press, 2010); Susan D. Jones, *Valuing Animals: Veterinarians and Their Patients in Modern America* (Baltimore: Johns Hopkins University Press, 2002); Housni Alkhateeb Shehada, *Mamluks and Animals: Veterinary Medicine in Medieval Islam* (Leiden: Brill, 2013).

For a recent holistic take on the historiography of human-animal relations, see Harriet Ritvo, "Animal Planet," *Environmental History* 9 (2004): 204–20.

58. On the role of the car in shaping American cities, see Clay McShane, *Down the Asphalt Path: The Automobile and the American City* (New York: Columbia University Press, 1994).
59. One answer to this question is offered by Temple Grandin, a scholar of animal science who is also autistic. She has argued passionately that her autism allows her greater access than people without autism to understanding the ways animals think. Temple Grandin and Catherine Johnson, *Animals in Translation: Using the Mysteries of Autism to Decode Animal Behavior* (New York: Scribner, 2005); idem., *Thinking in Pictures: And Other Reports from My Life with Autism* (New York: Doubleday, 1995). Grandin has used her stated elevated understanding of the animal mind to design supposedly more humane animal-handling facilities for industrial meat production. Temple Grandin, with Mark Deesing, *Humane Livestock Handling: Understanding Livestock Behavior and Building Facilities for Healthier Animals* (North Adams, MA: Storey Publishing, 2008); Temple Grandin, ed., *Improving Animal Welfare: A Practical Approach* (Cambridge: CAB International, 2010). For a useful discussion of the role of empathy in Temple Grandin's writings, see Dennis A. Lynch, "Rhetorics of Proximity: Empathy in Temple Grandin and Cornel West," *Rhetoric Society Quarterly* 28 (1998): 5–23. For a searing ethnographic study of the industrial meat industry in the United States, see Timothy Pachirat, *Every Twelve Seconds: Industrialized Slaughter and the Politics of Sight* (New Haven, CT: Yale University Press, 2011). See also Natalie Purcell, "Cruel Intimacies and Risky Relationships: Accounting for Suffering in Industrial Livestock Production," *Society and Animals* 19 (2011): 59–81.
60. On anthropocentric interspecies hubris, see LaCapra, *History and Its Limits*.
61. The literature on animal emotions is currently one of the most fruitful bodies of work attempting to understand the subjective worlds and experiences of animals. Most useful for my thinking on the subject have been the following: Charles Darwin, *The Expression of the Emotions in Man and Animals*, ed. Joe Cain and Sharon Messenger (London: Penguin, 2009); Jeffrey Moussaieff Masson and Susan McCarthy, *When Elephants Weep: The Emotional Lives of Animals* (New York: Delacorte Press, 1995); Jeffrey Moussaieff Masson, *The Pig who Sang to the Moon: The Emotional World of Farm Animals* (New York: Ballantine Books, 2003); idem., *Dogs Never Lie about Love: Reflections on the Emotional World of Dogs* (London: J. Cape, 1997); Marc Bekoff, *The Smile of a Dolphin: Remarkable Accounts of Animal Emotions* (New York: Discovery Books, 2000); idem., *Minding Animals: Awareness, Emotions, and Heart* (Oxford: Oxford University Press, 2002); Vicki Hearne, *Adam's Task: Calling Animals by Name* (New York: Knopf, 1986); B. A. Dixon, *Animals, Emotion, and Morality: Marking the Boundary* (Amherst, NY: Prometheus Books, 2008). See also the following very useful collection on animal subjectivities: Julie A. Smith and Robert W. Mitchell, eds., *Experiencing Animal Minds: An Anthology of Animal-Human Encounters* (New York: Columbia University Press, 2012). For an attempt to measure the accuracy of human interpretations of horses' emotions, see Leslie A. Russell, "Decoding Equine Emotions," *Society and Animals* 11 (2003): 265–66. For a related argument about the language humans and horses share, see Keri Brandt, "A Language of Their Own: An Interactionist Approach to Human-Horse Communication," *Society and Animals* 12 (2004): 299–316.
62. For important histories of climate change in the Middle East, see Sam White, *The Climate of Rebellion in the Early Modern Ottoman Empire* (Cambridge: Cambridge University Press, 2011); Richard W. Bulliet, *Cotton, Climate, and Camels in Early Islamic Iran: A Moment in World History* (New York: Columbia University Press, 2009).
63. On animal agency, see Jason C. Hribal, "Animals, Agency, and Class: Writing the History of Animals from Below," *Human Ecology Review* 14 (2007): 101–12; Jennifer Adams Martin, "When Sharks (Don't) Attack: Wild Animal Agency in Historical Narratives," *Environmental History* 16 (2011): 451–55. For a broader perspective from environmental history, see Ted Steinberg, "Down to Earth: Nature, Agency, and Power in History," *American Historical Review* 107 (2002): 798–820.
64. Arnold Toynbee attempted just such an exercise in writing nonhuman histories with his examination of plants' experiences of ancient Rome. Arnold J. Toynbee, "The Roman Revolution from the Flora's Point of View," in *Rome and Her Neighbours after Hannibal's Exit*,

vol. 2 of *Hannibal's Legacy: The Hannibalic War's Effects on Roman Life* (London: Oxford University Press, 1965), 585–99.

65. Timothy Mitchell, "Can the Mosquito Speak?" in *Rule of Experts: Egypt, Techno-Politics, Modernity* (Berkeley: University of California Press, 2002), 19–53.

66. Aaron Skabelund, "Can the Subaltern Bark? Imperialism, Civilization, and Canine Cultures in Nineteenth-Century Japan," in *JAPANimals: History and Culture in Japan's Animal Life*, ed. Gregory M. Pflugfelder and Brett L. Walker (Ann Arbor: Center for Japanese Studies, University of Michigan, 2005), 195–243.

67. On this point, animal psychologist Alexandra Horowitz usefully writes, "I do not know the particulars of another person's experience, but I know enough about the feeling of being human myself that I can draw an analogy from my own experience to someone else's. I can imagine what the world is like to him by extrapolating from my own perception and transplanting it with him at its center. The more information I have about that person—physically, his life history, his behavior—the better my drawn analogy will be. So can we do this with dogs. The more information we have, the better the drawing will be. To this point, we have physical information (about their nervous systems, their sensory systems), historical knowledge (their evolutionary heritage, their developmental path from birth to adults), and a growing corpus of work about their behavior. In sum, we have a sketch of the dog umwelt." Alexandra Horowitz, *Inside of a Dog: What Dogs See, Smell, and Know* (New York: Scribner, 2009), 242–43. On animal umwelt—a particular animal's subjective experience and understanding of the world *as an animal*—see Jacob von Uexhüll, *A Foray into the Worlds of Animals and Humans; with A Theory of Meaning*, trans. Joseph D. O'Neil (Minneapolis: University of Minnesota Press, 2010). For more on the human ability (or inability) to understand animal subjectivities, see Thomas Nagel, "What Is It Like to Be a Bat?" *Philosophical Review* 83 (1974): 435–50.

68. Evliyâ Çelebi bin Derviş Mehemmed Zıllî, *Evliyâ Çelebi Seyahatnâmesi*, ed. Seyit Ali Kahraman, Yücel Dağlı, and Robert Dankoff, 10 vols. (Istanbul: Yapı Kredi Yayınları, 2011), 10: 186–89. A partial English translation of this section is provided in the following: Evliya Çelebi, *An Ottoman Traveller: Selections from the* Book of Travels *of Evliya Çelebi*, trans. Robert Dankoff and Sooyong Kim (London: Eland, 2010), 388–93. For more about the animals in Evliya Çelebi's writings on Egypt, see Bacqué-Grammont, Lesur-Gebremariam, and Mayeur-Jaouen, "La faune nilotique dans la relation d'Evliyâ Çelebî." For other early modern accounts of crocodiles in Egypt, see al-Suyūṭī, *Kawkab al-Rawḍa*, 47v–49r; Rūmī, *Kitāb Ta'rīkh Miṣr wa al-Nīl*, 98v–99v. For a sixteenth-century Ottoman account of bestiality, see Selim Sirri Kuru, "A Sixteenth-Century Ottoman Scholar, Deli Birader, and His Dāfiʿü'l-ġumūm ve Rāfiʿü'l-humūm" (Ph.D. diss., Harvard University, 2000), 252–57.

69. Evliyâ Çelebi, *Evliyâ Çelebi Seyahatnâmesi*, 10: 260. For a partial English translation of this section, see Evliya Çelebi, *Ottoman Traveller*, 400–02. For another seventeenth-century account of chicken egg incubation in Egypt, see Rūmī, *Kitāb Ta'rīkh Miṣr wa al-Nīl*, 83v–84r. On the use of ovens to incubate eggs in late eighteenth-century al-Manṣūra, see Mary Kerr, "Savary's Letters on Egypt. 1786," Notes on Visits to Various Country Houses and Towns in Great Britain, 1789–1826, Rare Books and Manuscripts, Yale Center for British Art, Yale University, New Haven, 10–11.

70. I consulted the following versions of the Kanunname-i Mısır: *Ḳānūn-nāme-i Mıṣr*, Topkapı Sarayı Müzesi Kütüphanesi, Istanbul, E. H. 2063; Aḥmad Fu'ād Mutawallī, trans. and intro., *Qānūn Nāmah Miṣr, alladhī Aṣdarahu al-Sulṭān al-Qānūnī li-Ḥukm Miṣr* (Cairo: Maktabat al-Anjlū al-Miṣriyya, 1986); Ömer Lûtfi Barkan, ed., *Kanunlar*, vol. 1 of *XV ve XVIinci asırlarda Osmanlı İmparatorluğunda Ziraî Ekonominin Hukukî ve Malî Esasları*, İstanbul Üniversitesi Yayınlarından 256 (Istanbul: Bürhaneddin Matbaası, 1943), 355–87. Mutawallī's Arabic translation also contains a copy of Barkan's Turkish text. For the section on Egypt's animal wealth, see Mutawallī, *Qānūn Nāmah Miṣr*, 31–32.

71. For a study that takes seriously this ubiquity of animals in the Ottoman world by using animal husbandry as a framing device to understand various iterations of Ottoman governance, see Molly Greene, "The Ottoman Experience," *Daedalus* 134 (2005): 88–99.

72. On the utility of the central Ottoman archives for the history of Egypt, see Stanford J. Shaw, "The Ottoman Archives as a Source for Egyptian History," *Journal of the American Oriental Society* 83 (1963): 447–52.
73. About the potential of some of these sources for narrating Ottoman-Egyptian history, see idem., "Cairo's Archives and the History of Ottoman Egypt," *Middle East Institute Report on Current Research* (1956): 59–72.
74. For a general introduction to some of the Ottoman Turkish and Italian materials available in Venice for the study of the Ottoman Empire, see Maria Pia Pedani Fabris, with Alessio Bombaci, *I "Documenti Turchi" dell'Archivio di Stato di Venezia* (Rome: Ministero per i Beni Culturali e Ambientali, Ufficio Centrale per i Beni Archivistici, 1994); Maria Pia Pedani, ed., based on the materials compiled by Alessio Bombaci, *Inventory of the* Lettere e Scritture Turchesche *in the Venetian State Archives* (Leiden: Brill, 2010). For recent studies that make good use of Venetian sources about the Ottomans, see Eric R. Dursteler, *Venetians in Constantinople: Nation, Identity, and Coexistence in the Early Modern Mediterranean* (Baltimore: Johns Hopkins University Press, 2006); E. Natalie Rothman, *Brokering Empire: Trans-Imperial Subjects between Venice and Istanbul* (Ithaca, NY: Cornell University Press, 2012).
75. For Maltese evidence of Egypt's role in efforts to address piracy in the eighteenth century, see ACM, LC 12, 1r–2r (27 Oct. 1756); ACM, LC 12, 3r–5v (15 Sept. 1758); ACM, LC 12, 60r–60v (20 Aug. 1779); ACM, LC 12, 133r–33v (21 Apr. 1790); ACM, LC 12, 201r–02r (19 Oct. 1778). For Venetian evidence concerning earlier instances of Egypt's part in Mediterranean piracy (both as perpetrator and victim), see ASV, Turchi, 1193 (Evail Ş 1023/6–15 Sept. 1614); ASV, Turchi, 1329 (1625?); ASV, Lettere e Scritture Turchesche, 2: 50 (Evail Ş 965/19–28 May 1558). On slavery, see ACM, LC 12, 148r–49r (30 Apr. 1791); ACM, LC 12, 180r–83r (22 Apr. 1763); ACM, LC 12, 199r–200r (21 June 1774). For examples of Egypt's importance to the trade in Yemeni coffee throughout the eastern Mediterranean, see ACM, LC 12, 98r–99v (21 June 1785); ACM, LC 12, 125r–25v (31 Mar. 1789); ACM, LC 12, 211r–12v (30 June 1787). On Egypt's role in the coffee trade throughout the Ottoman Empire, see Faroqhi, "Coffee and Spices"; Tuchscherer, "Commerce et production du café en mer Rouge"; Muhammad Husām al-Din Ismāʿil, "Le café dans la ville de Rosette à l'époque ottomane XVIe—XVIIe siècle," in *Le commerce du café avant l'ère des plantations coloniales: espaces, réseaux, sociétés (XVe–XIXe siècle)*, ed. Michel Tuchscherer (Cairo: Institut français d'archéologie orientale, 2001), 103–09; André Raymond, "Une famille de grands négociants en café au Caire dans la première moitié du XVIIIe siècle: les Sharāybī," in *Le commerce du café avant l'ère des plantations coloniales: espaces, réseaux, sociétés (XVe–XIXe siècle)*, ed. Michel Tuchscherer (Cairo: Institut français d'archéologie orientale, 2001), 111–24; idem., "A Divided Sea: The Cairo Coffee Trade in the Red Sea Area during the Seventeenth and Eighteenth Centuries," in *Modernity and Culture: From the Mediterranean to the Indian Ocean*, ed. Leila Tarazi Fawaz and C. A. Bayly (New York: Columbia University Press, 2002), 46–57; Nancy Um, *The Merchant Houses of Mocha: Trade and Architecture in an Indian Ocean Port* (Seattle: University of Washington Press, 2009), 34–36, 43, 46; Mehmet Genç, "Contrôle et taxation du commerce du café dans l'Empire ottoman fin XVIIe — première moitié du XVIIIe siècle," in *Le commerce du café avant l'ère des plantations coloniales: espaces, réseaux, sociétés (XVe–XIXe siècle)*, ed. Michel Tuchscherer (Cairo: Institut français d'archéologie orientale, 2001), 161–79.
76. John Berger, *About Looking* (London: Writers and Readers Publishing Cooperative, 1980), 19.

1. Early Modern Human and Animal

1. DWQ, Maḥkamat al-Baḥayra 21, p. 177, case 343 (24 N 1206/16 May 1792). For a description of the village of Surunbāy, see Muḥammad Ramzī, *al-Qāmūs al-Jughrāfī lil-Bilād al-Miṣriyya min ʿAhd Qudamāʾ al-Miṣriyyīn ilā Sanat 1945*, 6 vols. in 2 pts. (Cairo: al-Hayʾa al-Miṣriyya al-ʿĀmma lil-Kitāb, 1994), pt. 2, vol. 2: 270. On the courts and wider legal system of Ottoman Egypt, see El-Nahal, *Judicial Administration of Ottoman Egypt*. For the biographies of various judges from the period, see Aḥmad ibn Aḥmad al-Damīrī, *Quḍāt Miṣr fī*

al-Qarn al-ʿĀshir wa al-Rubʿ al-Awwal min al-Qarn al-Ḥādī ʿAshar al-Hijrī, ed. ʿAbd al-Rāziq ʿAbd al-Rāziq ʿĪsā and Yūsuf Muṣṭafā al-Maḥmūdī (Cairo: al-ʿArabī lil-Nashr wa al-Tawzīʿ, 2000); Ibn Abī al-Surūr, *al-Tuḥfa al-Bahiyya*, 163–74.

2. Another possible explanation for al-Ḥājj Muṣṭafā's seemingly futile action in bringing this case to court was that it represented some sort of mid-feud snapshot of a tumultuous relationship that had been going on for quite some time. In other words, perhaps al-Ḥājj Muṣṭafā brought this case to court in an attempt to get revenge on al-Shābb ʿAlī; or perhaps al-Ḥājj Muṣṭafā sought to raise in the minds of others in their community the possibility that al-Shābb ʿAlī was a trickster who acted dishonestly; or perhaps al-Ḥājj Muṣṭafā even meant to suggest that al-Shābb ʿAlī came in the middle of the night and killed the ox himself. There are several plausible scenarios. Whatever the case may have been, the important point to take away is that the decision to go to court was one that had many possible motivations with various consequences. It is not outside the realm of possibility—indeed, it is quite likely—that plaintiffs brought cases to court knowing full well they would lose them. The goal of bringing a case to court, in other words, was not always (or only) to win it. On these and other strategic uses of Ottoman courts, see Boğaç A. Ergene, *Local Court, Provincial Society and Justice in the Ottoman Empire: Legal Practice and Dispute Resolution in Çankırı and Kastamonu (1652–1744)* (Leiden: Brill, 2003); Leslie Peirce, *Morality Tales: Law and Gender in the Ottoman Court of Aintab* (Berkeley: University of California Press, 2003).
3. In 1731, for example, 500 mules were used to move supplies and troops to the Ottoman border with Iran. These animals were one of the few ways to transport an enormous load over vast distances and rugged terrain. BOA, Cevdet Askeriye 16011 (3 M 1144/8 July 1731).
4. Some writers even went so far as to say that the impressive size and power of Egypt's mules, donkeys, and horses were unmatched anywhere in the world. See the two slightly different versions of the following text: Muḥammad ibn Muḥammad ibn Ẓahīra, *al-Faḍāʾil al-Bāhira fī Maḥāsin Miṣr wa al-Qāhira*, Beinecke, Landberg MSS 105, 80v–83r; idem., *al-Faḍāʾil al-Bāhira fī Maḥāsin Miṣr wa al-Qāhira*, Beinecke, Arabic MSS suppl. 395, 71v–73v. For an eleventh-century description of the physical attributes of horses, see ʿUbayd Allah ibn Jibrīl ibn Bakhtīshūʿ, *ʿIqd al-Jumān fī Ṭabāʾiʿ al-Ḥayawān wa al-Insān*, Maʿhad al-Makhṭūṭāt al-ʿArabiyya, Cairo, al-Kīmiyāʾ wa al-Ṭabīʿiyāt 66, 15v–17r. For a description of mules, see ibid., 17r–17v.
5. For examples of the economic connections between a town and its hinterland, see the following cases about Asyūṭ and its surrounding villages: DWQ, Maḥkamat Asyūṭ 7, p. 20, case 42 (1 S 1208/7 Sept. 1793); DWQ, Maḥkamat Asyūṭ 7, p. 34, case 73 (28 R 1208/2 Dec. 1793); DWQ, Maḥkamat Asyūṭ 9, p. 85, case 196 (9 B 1219/13 Oct. 1804); DWQ, Maḥkamat Asyūṭ 9, p. 88, case 202 (20 Ş 1219/23 Nov. 1804); DWQ, Maḥkamat Asyūṭ 9, p. 178, case 402 (25 Ca 1219/31 Aug. 1804). For the case of Damanhūr and its hinterland, see DWQ, Maḥkamat al-Baḥayra 30, p. 27, case 50 (Evail L 1219/2–11 Jan. 1805); DWQ, Maḥkamat al-Baḥayra 30, p. 34, case 60 (8 Za 1219/8 Feb. 1805).
6. This form of land organization was known as the *çift-hane* system. Halil İnalcık, "The Ottoman State: Economy and Society, 1300–1600," in *An Economic and Social History of the Ottoman Empire*, ed. Halil İnalcık with Donald Quataert, 2 vols. (Cambridge: Cambridge University Press, 1994), 1: 143–54; idem., "The Emergence of Big Farms, *Çiftliks*: State, Landlords, and Tenants," in *Landholding and Commercial Agriculture in the Middle East*, ed. Çağlar Keyder and Faruk Tabak (Albany: State University of New York Press, 1991), 17–34; Barkey, *Bandits and Bureaucrats*, 108–13; Gilles Veinstein, "On the *Çiftlik* Debate," in *Landholding and Commercial Agriculture in the Middle East*, ed. Çağlar Keyder and Faruk Tabak (Albany: State University of New York Press, 1991), 35–53.
7. On the paper trade, see Hanna, *In Praise of Books*. On the Ottoman silk trade, see İnalcık, "Ottoman State," 218–55. On coffee see, for example, Raymond, "A Divided Sea"; idem., "Une famille de grands négociants en café au Caire." Generally on the wealth accumulation of artisans and merchants in Ottoman Egypt, see Raymond, *Artisans et commerçants au Caire*; Hanna, *Making Big Money*.
8. On the absence of animals in modern scholarship of the early modern period, see Fudge, *Brutal Reasoning*, 4–5.

9. On this point, see the following illuminating discussion: Jean and John L. Comaroff, "Goodly Beasts, Beastly Goods: Cattle and Commodities in a South African Context," *American Ethnologist* 17 (1990): 195–216.
10. This was of course true in other periods of Egyptian history as well. See, for example, Joseph G. Manning, "A Ptolemaic Agreement Concerning a Donkey with an Unusual Warranty Clause. The Strange Case of *P. dem. Princ.* 1 (inv. 7524)," *Enchoria* 28 (2002/2003): 46–61.
11. Most studies of agricultural labor in the Ottoman Empire have—perhaps not surprisingly—focused on the role of peasants in the rural economy. For general historiographical discussions of peasants in Ottoman studies, see Suraiya Faroqhi, "Agriculture and Rural Life in the Ottoman Empire (ca 1500–1878) (A Report on Scholarly Literature Published between 1970 and 1985)," *New Perspectives on Turkey* 1 (1987): 3–34; idem., "Ottoman Peasants and Rural Life: The Historiography of the Twentieth Century," *Archivum Ottomanicum* 18 (2000): 153–82. For studies of Ottoman peasants and labor during the early modern period, see idem., "The Peasants of Saideli in the Late Sixteenth Century," *Archivum Ottomanicum* 8 (1983): 215–50; idem., "Rural Society in Anatolia and the Balkans During the Sixteenth Century, I," *Turcica* 9 (1977): 161–95; idem., "Rural Society in Anatolia and the Balkans During the Sixteenth Century, II," *Turcica* 11 (1979): 103–53; İslamoğlu-İnan, *State and Peasant in the Ottoman Empire*. For the specific case of Egypt, see Cuno, *Pasha's Peasants*; ʿAbd al-Raḥīm, *al-Rīf al-Miṣrī*; Yāsir ʿAbd al-Minʿam Maḥārīq, *al-Minūfiyya fī al-Qarn al-Thāmin ʿAshar* (Cairo: al-Hayʾa al-Miṣriyya al-ʿĀmma lil-Kitāb, 2000).
12. Beyond the caloric sustenance of the empire's human populations, animals' labor and their transport capacities were obviously also crucial for provisioning the Ottoman military. On the use of animals in military provisioning, see Rhoads Murphey, *Ottoman Warfare, 1500–1700* (New Brunswick, NJ: Rutgers University Press, 1999), 65–83.
13. On the importance of animal capital in the Egyptian countryside, see Mutawallī, *Qānūn Nāmah Miṣr*, 31–32.
14. I use the term *peasant* advisedly to refer to the vast majority of people living in rural Egypt. This is the word that comes through most often in the original Arabic and Ottoman Turkish sources as *fallāḥ* or the collective plural *ahālī* (s. *ahl*). I will use the term interchangeably with *farmer* and *rural cultivator*. Because most residents of the countryside did not own large amounts of property, their choice to keep their small amounts of capital in animals is telling of just how crucial animals were to the entire economic base of rural Ottoman Egypt. Of course, not all animals used by Egyptian farmers were actually their own personal property. Even in the early modern period, peasants sometimes worked their own or others' fields using the animals of large property holders or other rural notables. For a discussion of the early modern usage of the term *ahālī* to refer collectively to the farmers of a particular village, see Mikhail, *Nature and Empire*, 262.
15. Domesticated animals were indeed such an important and common part of people's lives that interspecies sex was a well-known and extremely widespread phenomenon. According to the sixteenth-century Ottoman chronicler Deli Birader, "since this act is very famous and oft spoken of, there are many examples of it and stories about it, to such extent that nobody denies that such acts are being committed." Kuru, "Deli Birader," 252. Female donkeys were perhaps the animals with whom human males in Ottoman Egypt most often had sex. Deli Birader relates that Egypt had many donkey-houses in which men could pay for a few hours of sex with a female donkey. For a reproduction of a sixteenth-century Persian miniature painting depicting human sex with a camel, see Marianna Shreve Simpson, *Persian Poetry, Painting & Patronage: Illustrations in a Sixteenth-Century Masterpiece* (New Haven, CT: Yale University Press, 1998), 22–23. For comparative perspectives on early modern bestiality, see Susanne Hehenberger, "Dehumanised Sinners and their Instruments of Sin: Men and Animals in Early Modern Bestiality Cases, Austria 1500–1800," in *Early Modern Zoology: The Construction of Animals in Science, Literature and the Visual Arts*, ed. Karl A. E. Enenkel and Paul J. Smith, 2 vols. (Leiden: Brill, 2007), 2: 381–417; P. G. Maxwell-Stuart, " 'Wild, Filthie, Execrabill, Detestabill, and Unnatural Sin': Bestiality in Early Modern Scotland," in *Sodomy in Early Modern Europe*, ed. Tom Betteridge (Manchester: Manchester University Press, 2002), 82–93.

16. In considering the ubiquity of references to domesticated animals as property in the archival record of Islamic courts throughout Egypt in the early modern period, I focus on animal use in two regions of high agricultural output—the Nile Delta and Upper Egypt (al-Ṣaʿīd). The majority of the examples I give in this chapter come from the court records of cities in the former region (Damanhūr, al-Manṣūra, and Rosetta) rather than from cities in the latter (I mostly use the records of Isnā and Manfalūṭ), reflecting the Delta's higher levels of agricultural productivity and output during the Ottoman period. More peasants, more fields, and more canals in the Delta meant there were more animals there than in Upper Egypt. In this vein, consider that in the eighteenth century a full one-fourth of the Delta, as opposed to one-sixth of Upper Egypt (still a significant amount clearly), was planted in Egyptian clover (*birsīm*; *Trifolium alexandrinum* L.), used to raise and support draft animals for agricultural labor. Robert O. Collins, *The Nile* (New Haven, CT: Yale University Press, 2002), 132. For analyses and reproductions of a seventeenth-century Ottoman Turkish map of the Nile that was likely produced to accompany Evliya Çelebi's account of his travels in Egypt, see Robert Dankoff and Nuran Tezcan, eds., *Evliyâ Çelebi'nin Nil Haritası: "Dürr-i bî-misîl în Ahbâr-ı Nîl"* (Istanbul: Yapı Kredi Yayınları, 2011); Ettore Rossi, "A Turkish Map of the Nile River, about 1685," *Imago Mundi* 6 (1949): 73–75; John Alexander, "The Turks on the Middle Nile," *Archéologie du Nil Moyen* 7 (1996): 15–35; Intisar Elzein, "Ottoman Archaeology of the Middle Nile Valley in the Sudan," in *The Frontiers of the Ottoman World*, ed. A. C. S. Peacock (Oxford: Oxford University Press, 2009), 371–83.
17. Generally on animal wealth in Ottoman Egypt, see ʿAbd al-Raḥīm, *al-Rīf al-Miṣrī*, 208–10. For a comparative example of the economics of animal husbandry in the early modern Netherlands, see Jan de Vries, *The Dutch Rural Economy in the Golden Age, 1500–1700* (New Haven, CT: Yale University Press, 1974), 137–44.
18. On the role of livestock in constituting *urban* wealth in Ottoman Egypt, see Shaw, *Financial and Administrative Organization and Development*, 122, 140–41, 175–76.
19. For examples of these arrangements, see DWQ, Maḥkamat Isnā 6, p. 67, case 113 (11 L 1172/7 June 1759); DWQ, Maḥkamat Manfalūṭ 3, pp. 92–93, case 174 (n.d.); DWQ, Maḥkamat Isnā 8, pp. 111–12, case 177 (23 Ra 1173/13 Nov. 1759); DWQ, Maḥkamat al-Baḥayra 16, p. 89, case 155 (16 L 1200/11 Aug. 1786); DWQ, Maḥkamat Manfalūṭ 3, p. 62, case 117 (n.d.).
20. For a study of estate inventories from Bursa, Istanbul, and Edirne that makes this point as well, see Halil Inalcik, "Capital Formation in the Ottoman Empire," *Journal of Economic History* 29 (1969): 97–140. For an example of an eighteenth-century estate inventory from the Ottoman island of Andros that contained animals, see KLA, doc. 212 (25 Ca 1123/11 July 1711). In a wholly different context that nevertheless further evidences the economic importance of domesticated animals in the Ottoman Empire, the following case discusses the economic status of an Armenian church in Istanbul whose assets included a large number of livestock: BOA, Cedvet Adliye 3271 (4 M 1154/22 Mar. 1741). For the breakdown of revenues and expenses related to a garden, orchard, and *çiftlik* in Edirne that included many animals, see BOA, CS 8981 (14 Ş 1221/27 Oct. 1806).
21. DWQ, Maḥkamat al-Baḥayra 11, pp. 45–56, case 97 (1 Z 1196/7 Nov. 1782); DWQ, Maḥkamat Manfalūṭ 3, p. 111, case 216 (n.d.). In the cases that follow, I assume the animals in question to be healthy and still of an age fit for work. No statement in any of these cases suggests otherwise. Although the archival record is replete with references to animals as property in the countryside, it is difficult to determine precise prices for these animals and to trace how these prices fluctuated overtime and from place to place. As such, I focus in this section primarily on the *relative* prices of different species.

    A variety of buffalo cow similar to that found in Egypt was also well known in Mongol Iran. The fourteenth-century Persian scholar Qazvīnī describes this buffalo cow as "a strong-bodied animal, and very powerful. It has a worm in its brain which perpetually torments it, so that is sleeps little, and stays always in the water. It is an enemy of the lion and crocodile, and mostly gets the better of them, but it is helpless against gnats." Ḥamd Allāh Mustawfī Qazvīnī, *The Zoological Section of the Nuzhatu-l-Qulūb of Ḥamdullāh al-Mustaufī al-Qazwīnī*, ed. and trans. J. Stephenson (London: Royal Asiatic Society, 1928), 5.

22. See, for example, the following case that documents the relative numbers of animals in the possession of one Sulaymān Salām Dūjal at the time of his death: DWQ, Maḥkamat Isnā 3, p. 23, case 41 (8 L 1171/14 June 1758). Buffalo cows were clearly his most valuable possession. He owned six of them, in addition to four cows, a she-camel, and some sheep. For other cases that highlight the high value of the buffalo cow, see DWQ, Maḥkamat Rashīd 144, pp. 493–95, case 525 (25 Ra 1152/2 July 1739); DWQ, Maḥkamat Manfalūṭ 3, p. 74, case 135 (n.d.); DWQ, Maḥkamat Manfalūṭ 1, p. 223, case 549 (21 B 1098/2 June 1687).
23. DWQ, Maḥkamat al-Baḥayra 24, pp. 291–92, case 526 (20 B 1209/10 Feb. 1795). More than most animals, camels have received a good deal of attention from historians of the Middle East. See, for example, Richard W. Bulliet, *The Camel and the Wheel* (Cambridge, MA: Harvard University Press, 1975); idem., *Cotton, Climate, and Camels*; Halil İnalcık, "'Arab' Camel Drivers in Western Anatolia in the Fifteenth Century," *Revue d'Histoire Maghrebine* 10 (1983): 256–70; idem., "Ottoman State," 38–39 and 62–63; Suraiya Faroqhi, "Camels, Wagons, and the Ottoman State in the Sixteenth and Seventeenth Centuries," *International Journal of Middle East Studies* 14 (1982): 523–39; Roger S. Bagnall, "The Camel, the Wagon, and the Donkey in Later Roman Egypt," *Bulletin of the American Society of Papyrologists* 22 (1985): 1–6.
24. For an eleventh-century description of the physical characteristics of buffalo cows and camels, see Ibn Bakhtīshūʿ, *Ṭabāʾiʿ al-Ḥayawān wa al-Insān*, 13r–14v.
25. DWQ, Maḥkamat al-Baḥayra 11, pp. 45–56, case 97 (1 Z 1196/7 Nov. 1782).
26. DWQ, Maḥkamat al-Baḥayra 23, p. 198, case 357 (Awākhir L 1208/20–30 May 1794). On Bīwīṭ, see Ramzī, *al-Qāmūs al-Jughrāfī*, pt. 2, vol. 2: 268–69.
27. While the assets of this estate did indeed total 298 niṣf fiḍḍa, 134.61 niṣf fiḍḍa of this amount were used to repay debts and cover other expenses, therefore leaving only 163.39 niṣf fiḍḍa for the inheritors.
28. As this case illustrates, the buffalo cow was not always the most expensive animal in an estate's livestock collection. Nevertheless, the general point remains that it was the single most expensive domesticated animal in estates *more often* than any other animal.
29. See, for example, DWQ, Maḥkamat Rashīd 132, p. 278, case 419 (10 Ra 1138/15 Nov. 1725). The buffalo cow in this case was valued at 500 niṣf fiḍḍa and the horse at 720 niṣf fiḍḍa. In the following case, a buffalo cow was priced higher than a horse: DWQ, Maḥkamat Rashīd 139, pp. 107–08, case 172 (30 M 1146/12 July 1733). The buffalo cow was valued at 770 niṣf fiḍḍa, while the horse only at 210 niṣf fiḍḍa.
30. Muṣṭafā ibn al-Ḥājj Ibrāhīm tābiʿ al-Marḥūm Ḥasan Aghā ʿAzbān al-Damardāshī, *Tārīkh Waqāyiʿ Miṣr al-Qāhira al-Maḥrūsa: Kinānat Allāh fī Arḍihi*, ed. Ṣalāḥ Aḥmad Harīdī ʿAlī (Cairo: Dār al-Kutub wa al-Wathāʾiq al-Qawmiyya, 2002), 167, 201, 204, 308–09, 350–51. For a case involving the transport of horses to stables in Manfalūṭ, see DWQ, Maḥkamat Manfalūṭ 1, pp. 207–08, case 513 (14 B 1098/26 May 1687). For examples of the military seizure of horses, see Aḥmad Shalabī ibn ʿAbd al-Ghanī al-Ḥanafī al-Miṣrī, *Awḍaḥ al-Ishārāt fīman Tawallā Miṣr al-Qāhira min al-Wuzarāʾ wa al-Bāshāt: al-Mulaqqab bi-al-Tārīkh al-ʿAynī*, ed. ʿAbd al-Raḥīm ʿAbd al-Raḥmān ʿAbd al-Raḥīm (Cairo: Maktabat al-Khānjī, 1978), 468–69; idem., *Awḍaḥ al-Ishārāt fīman Tawallā Miṣr al-Qāhira min al-Wuzaraʾ wa al-Bāshāt*, Beinecke, Landberg MSS 3, 173v; al-Damardāshī, *Tārīkh Waqāyiʿ Miṣr al-Qāhira*, 293. Under the former Mamluk military system, heads of cavalry, known as emirs, were ranked according to the number of horsemen their plot of land could support. Hathaway, *Politics of Households*, 8–9. Halil Inalcik similarly suggests that part of the logic behind distributing land to cavalrymen in the Ottoman Empire was to give them the resources they needed to adequately support their horses. Inalcik, *Classical Age*, 107–08. On the social status attached to horses in medieval Europe, see Salisbury, *Beast Within*, 28–31 and 35. On various military uses of horses, see James Boyd, "Horse Power: The Japanese Army, Mongolia and the Horse, 1927-43," *Japan Forum* 22 (2010): 23–42; R. L. DiNardo and Austin Bay, "Horse-Drawn Transport in the German Army," *Journal of Contemporary History* 23 (1988): 129–42; Steven J. Corvi, "Men of Mercy: The Evolution of the Royal Army Veterinary Corps and the Soldier-Horse Bond During the Great War," *Journal of the Society for Army Historical Research* 76 (1998): 272–84; Jilly Cooper, *Animals in War* (London: Heinemann, 1983). More generally on the history of the human-horse relationship, see Pita Kelekna, *The Horse*

in Human History (Cambridge: Cambridge University Press, 2009); Elaine Walker, *Horse* (London: Reaktion Books, 2008).

31. Aḥmad al-Damurdāshī Katkhudā ʿAzabān, *Kitāb al-Durra al-Muṣāna fī Akhbār al-Kināna*, ed. ʿAbd al-Raḥīm ʿAbd al-Raḥman ʿAbd al-Raḥīm (Cairo: Institut français d'archéologie orientale, 1989), 16. For a comparative example of the gifting of birds in early modern Japan, see Martha Chaiklin, "Exotic-Bird Collecting in Early-Modern Japan," in *JAPANimals: History and Culture in Japan's Animal Life*, ed. Gregory M. Pflugfelder and Brett L. Walker (Ann Arbor: Center for Japanese Studies, University of Michigan, 2005), 132–39.
32. Shaw, *Niẓâmnâme-i Mıṣır*, 14–15.
33. Of course, other animals were also commonly given as gifts. On the gifting of animals other than horses, see ʿAbd al-Raḥman ibn Ḥasan al-Jabartī, *ʿAjāʾib al-Āthār fī al-Tarājim wa al-Akhbār*, ed. ʿAbd al-Raḥīm ʿAbd al-Raḥman ʿAbd al-Raḥīm, 4 vols. (Cairo: Maṭbaʿat Dār al-Kutub al-Miṣriyya, 1998), 1: 407. The very wealthy Bedouin shaykh al-Humām from the village of Akhmīm in Upper Egypt, for example, gave a gift of three hundred camels to the pilgrimage procession every year. Shaw, *Niẓâmnâme-i Mıṣır*, 41. On Shaykh al-Humām more generally, see ibid., 44. On the village of Akhmīm, see Ramzī, *al-Qāmūs al-Jughrāfī*, pt. 2, vol. 4: 89–90. For more on the use of camels in the annual pilgrimage caravan, see Kerr, "Letters," 13–14.
34. For a case about the collection and use of horses in the transport of emissaries from the Ottoman Empire to Iran, see BOA, Cevdet Hariciye 3089 (2 Ra 1154/18 May 1741). On the imperial acquisition of horses from Sofia, Edirne, Filibe (modern Plovdiv, Bulgaria), and Tatarpazarı, see BOA, CS 5088 (4 Ra 1197/7 Feb. 1783). Horses also played an essential role in the maintenance of systems of post in the Ottoman Empire. Colin Heywood, "The Ottoman *Menzilhane* and *Ulak* System in Rumeli in the Eighteenth Century," in *Türkiye'nin Sosyal ve Ekonomik Tarihi, 1071–1920: Birinci Uluslararası Türkiye'nin Sosyal ve Ekonomik Tarihi Kongresi Tebliğleri*, ed. Osman Okyar and Halil İnalcık (Ankara: Meteksan Şirketi, 1980), 179–86; idem., "Some Turkish Archival Sources for the History of the Menzilhane Network in Rumeli during the Eighteenth Century (Notes and Documents on the Ottoman Ulak, I)," *Boğaziçi Üniversitesi Dergisi* 4–5 (1976–1977): 39–54; idem., "The Via Egnatia in the Ottoman Period: The Menzilḫānes of the Ṣol Ḳol in the Late 17th/Early 18th Century," in *The Via Egnatia under Ottoman Rule (1380–1699): Halcyon Days in Crete II*, ed. Elizabeth Zachariadou (Rethymnon: Crete University Press, 1996), 129–44. On the use of horses and pigeons in the Mamluk post system, see Shehada, *Mamluks and Animals*, 59–67. On post in the Islamic world more generally, see Adam J. Silverstein, *Postal Systems in the Pre-Modern Islamic World* (Cambridge: Cambridge University Press, 2010); Gagan D. S. Sood, "'Correspondence is Equal to Half a Meeting': The Composition and Comprehension of Letters in Eighteenth-Century Islamic Eurasia," *Journal of the Economic and Social History of the Orient* 50 (2007): 172–214.
35. DWQ, Maḥkamat Rashīd 144, pp. 493–95, case 525 (25 Ra 1152/2 July 1739).
36. DWQ, Maḥkamat Manfalūṭ 3, p. 111, case 216 (n.d.). The adjective "red" here most likely refers to the reddish-brown hue of the animal's coat.
37. DWQ, Maḥkamat Manfalūṭ 1, p. 223, case 549 (21 B 1098/2 June 1687).
38. For a case involving the sharing of a she-camel, see DWQ, Maḥkamat Manfalūṭ 3, p. 46, case 93 (n.d.).
39. For further cases illustrating the relative values of these animals, see DWQ, Maḥkamat Isnā 6, pp. 73–79, case 131 (12 L 1172/8 June 1759); DWQ, Maḥkamat Manfalūṭ 3, pp. 92–93, case 174 (n.d.); DWQ, Maḥkamat Isnā 6, p. 67, case 113 (11 L 1172/7 June 1759); DWQ, Maḥkamat Isnā 8, pp. 111–12, case 177 (23 Ra 1173/13 Nov. 1759). One could similarly develop a hierarchy of the values of different animals in the Ottoman-Egyptian countryside based on the price of their skins. Indeed, a quite robust trade in different kinds of leather from Egypt developed during the Ottoman period. Buffalo cow leather was the most expensive, followed by ox and cow leather, with camel skins being the cheapest. For examples of the early nineteenth-century trade in animal skins, see DWQ, Maḥkamat al-Baḥayra 37, p. 203, case 423 (29 B 1232/15 June 1817); DWQ, Maḥkamat al-Baḥayra 37, p. 204, case 426 (17 B 1232/3 June 1817).

40. Human lactose toleration is a genetic anomaly observed in high rates among adults in northern Europe and parts of the Middle East and North Africa, including Egypt. High adult lactose digestion is likely the adaptive result of cultures of intense dairying and hence can serve as a kind of proxy for estimating the historic importance of human dairy consumption—and therefore human interactions with milk-producing animals. For some of the scientific evidence for this point, see Clare Holden and Ruth Mace, "Phylogenetic Analysis of the Evolution of Lactose Digestion in Adults," *Human Biology* 81 (2009): 597–619. See also the discussion in Bulliet, *Hunters, Herders, and Hamburgers*, 106–09.
41. According to Sufi tradition, it was wrong to use cows as beasts of burden. When riding a cow one day, a rider purportedly heard the animal exclaim, "I was not created for this!" Schimmel, *Islam and the Wonders of Creation*, 48. Schimmel does not provide a citation for this story or quotation. For an account that points to the importance of donkeys as means of conveyance in Ottoman Egypt, see Muḥammad ibn Abī Bakr Muḥibb al-Dīn al-Ḥamawī, *Ḥādī al-Aẓʿān al-Najdiyya ilā al-Diyār al-Miṣriyya*, ed. Muḥammad ʿAdnān al-Bakhīt (Muʾta: Jāmiʿat Muʾta ʿImādat al-Baḥth al-ʿIlmī wa al-Dirāsāt al-ʿUlyā, 1993), 28–29. See also the following version of the text: idem., *al-Durra al-Muḍīʾa fī al-Riḥla al-Miṣriyya*, Beinecke, Landberg MSS 427.
42. For evidence of a donkey market in Cairo in the second half of the seventeenth century, see Ibrāhīm ibn Abī Bakr al-Ṣawāliḥī al-ʿAwfī al-Ḥanbalī, *Tarājim al-Ṣawāʿiq fī Wāqiʿat al-Ṣanājiq*, ed. ʿAbd al-Raḥīm ʿAbd al-Raḥman ʿAbd al-Raḥīm (Cairo: Institut français d'archéologie orientale, 1986), 54. I also consulted the following manuscript version of this text: Ibrāhīm ibn Abī Bakr al-ʿAwfī, *Tarājim al-Ṣawāʿiq fī Wāqiʿat al-Ṣanājiq*, Beinecke, Landberg MSS 228.
43. DWQ, Maḥkamat Manfalūṭ 1, p. 69, case 144 (n.d.). For an example of wealth kept in sheep, see ʿAbd al-Ghanī, *Awḍaḥ al-Ishārāt*, 528.
44. Of these animals, donkeys were perhaps the most important for matters of prestige since, like horses, they were often used as riding mounts. On the prestige afforded to noble men *and* women who rode donkeys, see Andreas Tietze, *Muṣṭafā ʿĀlī's Description of Cairo of 1599: Text, Transliteration, Translation, Notes* (Vienna: Verlag Der Österreichischen Akademie Der Wissenschaften, 1975), 41. For further evidence of the importance of riding mounts as markers of status, see ibid., 81. For a useful global history of donkeys, see Jill Bough, *Donkey* (London: Reaktion Books, 2011).
45. These tables were compiled on the basis of DWQ, Maḥkamat Isnā 6, pp. 73–79, case 131 (12 L 1172/8 June 1759).
46. For additional probate cases involving shares in cows and other animals and their relative costs, see DWQ, Maḥkamat Manfalūṭ 3, pp. 92–93, case 174 (n.d.); DWQ, Maḥkamat Isnā 6, p. 67, case 113 (11 L 1172/7 June 1759); DWQ, Maḥkamat Isnā 8, pp. 111–12, case 177 (23 Ra 1173/13 Nov. 1759); DWQ, Maḥkamat Manfalūṭ 1, pp. 72–73, case 152 (8 Ş 1212/25 Jan. 1798). For a case involving the sale of a camel, see DWQ, Maḥkamat Manfalūṭ 3, p. 33, case 57 (n.d.).
47. DWQ, Maḥkamat Manfalūṭ 3, pp. 4–5, case 3 (4 R 1228/6 Apr. 1813). This case shows that animals did sometimes maintain their value into the early nineteenth century. This was true mostly in places like Manfalūṭ that were integrated relatively late into the domestic and global economies that changed so much in Ottoman Egypt.
48. DWQ, Maḥkamat Rashīd 154, p. 179, case 197 (15 Za 1162/27 Oct. 1749).
49. For another example of the comparative transactional values of animals and humans, see the following account involving horses and eunuchs: al-Damardāshī, *Tārīkh Waqāyiʿ Miṣr al-Qāhira*, 131.
50. DWQ, Maḥkamat Rashīd 148, pp. 67–68, case 86 (1 B 1154/11 Sept. 1741).
51. For another example of a financial transaction involving slaves and animals and their relative prices, see al-Damurdāshī, *al-Durra al-Muṣāna*, 44.
52. For a revealing comparative discussion of these points in the context of eighteenth-century Jamaica, see Philip D. Morgan, "Slaves and Livestock in Eighteenth-Century Jamaica: Vineyard Pen, 1750–1751," *William and Mary Quarterly* 52 (1995): 47–76.
53. There is no equivalent of *wergeld* in Islamic law. Historians of medieval Europe have usefully used the *wergeld* values of humans and animals to assess the relative valuations of humans to animals. Thus, for example, the Visigoths considered a nobleman to be five

hundred times the value of the most prized animal, and the Franks considered a stallion to be only half the value of a man's nose or ear. For the Burgundians, the tooth of an individual from the lowest classes was equal to the value of a noble hunting dog. Salisbury, *Beast Within*, 36–37.

54. Tietze, *Muṣṭafā ʿĀlī's Description of Cairo*, 43.
55. Ibid.
56. On the variety of dairy products made from the milk of different farm animals and their relative values, see al-Jabartī, *ʿAjāʾib al-Āthār* (1998), 1: 184, 338–39; 2: 205; 3: 332, 469; al-Damardāshī, *Tārīkh Waqāyiʿ Miṣr al-Qāhira*, 113.
57. This was also the case in medieval Europe where food animals were considered less valuable than laboring animals. Salisbury, *Beast Within*, 34.
58. For a case illustrating the high cost of she-camels, see DWQ, Maḥkamat Rashīd 130, pp. 172–76, case 245 (20 C 1136/16 Mar. 1724).
59. DWQ, Maḥkamat Isnā 6, pp. 1–3, case 1 (23 Ca 1172/23 Jan. 1759). In addition to this she-camel, ʿAlī ʿAbd al-Qādir al-Ḥamdanī owned three camels, two cows (one of which is described as being yellow), and a donkey (described as black). These animals were clearly a significant investment in capital and formed the bulk of his inheritable estate.
60. This despite the fact that the Quran seems to privilege cow over camel milk: "In cattle too you have a worthy lesson. We give you to drink of that which is in their bellies, between the bowels and the blood-streams: pure milk, pleasant for those who drink it." Quran 16: 66. See also in this regard Quran 23: 21 and 36: 73. Generally on the differences between cow and camel milk in Islamic tradition, see Schimmel, *Islam and the Wonders of Creation*, 47–48.
61. Nomadic populations in Anatolia regularly used camels to transport grain to Istanbul from villages at some distance from the imperial capital. Barkey, *Bandits and Bureaucrats*, 122. Egyptian Bedouin too no doubt often used their camels for similar sorts of (from the perspective of the imperial state at least) constructive social and economic purposes. For more on the importance of camels in Arab and Persian historical writings, see Schimmel, *Islam and the Wonders of Creation*, 46–47.
62. On the kinds and prices of meat products in eighteenth-century Ottoman Egypt, see al-Jabartī, *ʿAjāʾib al-Āthār* (1998), 1: 184, 338–39; 2: 198, 205, 274–75; 3: 332, 469, 551; al-Damardāshī, *Tārīkh Waqāyiʿ Miṣr al-Qāhira*, 113 and 117. For a very useful analysis of the provisioning of meat to Istanbul, see Antony Greenwood, "Istanbul's Meat Provisioning: A Study of the *Celepkeşan* System" (Ph.D. diss., University of Chicago, 1988). On meat consumption in the Ottoman palace and the palace kitchens more generally, see Arif Bilgin, *Osmanlı Saray Mutfağı (1453–1650)* (Istanbul: Kitabevi Yayınları, 2004); Ömer Lütfi Barkan, "İstanbul Saraylarına ait Muhasebe Defterleri," *Belgeler* IX/13 (1979): 1–380; Hedda Reindl-Kiel, "The Chickens of Paradise: Official Meals in the Seventeenth-Century Ottoman Palace," in *The Illuminated Table, the Prosperous House: Food and Shelter in Ottoman Material Culture*, ed. Suraiya Faroqhi and Christoph Neumann (Istanbul: Orient-Institut, 2003), 59–88; Joanita Vroom, "'Mr. Turkey Goes to Turkey,' Or: How an Eighteenth-Century Dutch Diplomat Lunched at Topkapı Palace," in *Starting with Food: Culinary Approaches to Ottoman History*, ed. Amy Singer (Princeton, NJ: Markus Wiener Publishers, 2011), 139–75; Tülay Artan, "Aspects of the Ottoman Elite's Food Consumption: Looking for 'Staples,' 'Luxuries,' and 'Delicacies' in a Changing Century," in *Consumption Studies and the History of the Ottoman Empire, 1550–1922: An Introduction*, ed. Donald Quataert (Albany: State University of New York Press, 2000), 107–200.
63. al-Jabartī, *ʿAjāʾib al-Āthār* (1998), 1: 545; 3: 300.
64. These festivities also included fireworks displays, a grand procession to the citadel, the firing of cannons, and a lavish banquet. al-Damurdāshī, *al-Durra al-Muṣāna*, 6, 9, 28, 40, 57, 69, 72, 75, 79–80, 84, 103, 113, 114, 122–23, 132–33, 142, 159, 162, 179, 199, 209, 212, 224, 237, 241, 248, 251, 252–54, 260.
65. Shaw, *Niẓâmnâme-i Mısır*, 47–48. The *okke* was an Ottoman dry measure equivalent to 400 *dirhem* or 1.2828 kg (2.8281 lbs). Walther Hinz, *Islamische Masse und Gewichte umgerechnet ins metrische System* (Leiden: Brill, 1955), 24.
66. DWQ, Maḥkamat Isnā 3, pp. 15–16, case 29 (8 L 1171/14 June 1758). On Aṣfūn, see Ramzī, *al-Qāmūs al-Jughrāfī*, pt. 2, vol. 4: 152. For another example of a *waqf* endowment that

included animals to be slaughtered for the poor and other guests of the *waqf*, see al-Jabartī, *ʿAjāʾib al-Āthār* (1998), 1: 612.

67. On the use of islands as natural enclosures to confine animals, see Anderson, *Creatures of Empire*, 160.

68. DWQ, Maḥkamat Rashīd 134, p. 332, case 443 (30 B 1140/11 Mar. 1728). Generally on Ottoman Rosetta, see Ṣalāḥ Aḥmad Harīdī ʿAlī, "al-Ḥayāh al-Iqtiṣādiyya wa al-Ijtimāʿiyya fī Madīnat Rashīd fī al-ʿAṣr al-ʿUthmānī, Dirāsa Wathāʾiqiyya," *Egyptian Historical Review* 30–31 (1983–1984): 327–78; Nāṣir ʿUthmān, "Maḥkamat Rashīd ka-Maṣdar li-Dirāsat Tijārat al-Nasīj fī Madīnat al-Iskandariyya fī al-ʿAṣr al-ʿUthmānī," *al-Rūznāma: al-Ḥauliyya al-Miṣriyya lil-Wathāʾiq* 3 (2005): 355–85.

69. al-Damurdāshī, *al-Durra al-Muṣāna*, 123; al-Damardāshī, *Tārīkh Waqāyiʿ Miṣr al-Qāhira*, 115.

70. al-Damurdāshī, *al-Durra al-Muṣāna*, 214; Shaw, *Financial and Administrative Organization and Development*, 127; idem., *Niẓâmnâme-i Mıṣır*, 47. For another example of a similar tax farm, see idem., *Financial and Administrative Organization and Development*, 135.

71. Idem., *Financial and Administrative Organization and Development*, 130 and 301.

72. Ibid.; idem., *Niẓâmnâme-i Mıṣır*, 47, n. 1.

73. Idem., *Niẓâmnâme-i Mıṣır*, 19.

74. Idem., *Financial and Administrative Organization and Development*, 139. For more on fish and fishing in Egypt, see *Laqis al-Ḥanak fīmā Warada fī al-Samak*, DKM, Makhṭūṭāt al-Zakiyya 373.

75. The tax revenue raised from this corporation between 1742–43 and 1760–61 was 1,620 paras. By way of comparison, the money collected during the same period from taxes on Cairo's sardine sellers was 1,800 paras. On the taxes levied on other corporations in the city, see Shaw, *Financial and Administrative Organization and Development*, 119.

76. For natural histories of the Nile and accounts of its role in Egyptian social life, see the following medieval and early modern accounts of the river: Muḥammad ibn ʿAbd al-Muʾmin ibn Muḥammad al-Jawharī, *Mūjaz fī Mabdāʾ al-Nīl wa Muntahāhu*, SK, Lâleli 3752; Muḥammad ibn Shuʿayyib, *Zahr al-Basāṭīn fī Faḍl al-Nīl wa Faḍl Miṣr wa al-Qāhira wa al-Qarāfa*, DKM, Buldān Taymūr 60; Ibn al-ʿImād, *Kitāb al-Nīl al-Saʿīd al-Mubārak*, DKM, Jughrāfiyā Ḥalīm 5; idem., *Risāla fī Manbaʿ al-Nīl*, al-Maktaba al-Baladiyya bi-al-Iskandariyya, Alexandria, ms. 1627 Bāʾ; idem., *Kitāb al-Qaul al-Mufīd fī al-Nīl al-Saʿīd*, al-Maktaba al-Baladiyya bi-al-Iskandariyya, Alexandria, ms. 4939 Dāl; Ibn Riḍwān, *Risāla fī Ziyādat al-Nīl wa Naqṣihi ʿalā al-Dawām*, DKM, Majāmīʿ Mīm 213; *Risāla fī al-Nīl wa Ziyādatihi*, al-Maktaba al-Baladiyya bi-al-Iskandariyya, Alexandria, ms. 1707 Bāʾ (4); Shihāb al-Dīn ibn al-ʿImād al-Aqfahsī, *Mabdāʾ Nīl Miṣr wa al-Ahrām wa Faḍīlat Miṣr wa al-Miqyās*, SK, Veliyüddin Efendi 3182, 38r–56v; Nūḥ Afandī al-Ṭahshawārī, *Tārīkh Miṣr wa al-Nīl wa Khabar man Malakahā min Ibtidāʾ al-Zamān*, DKM, Tārīkh Ṭalʿat 2037; Nūr al-Dīn ʿAlī ibn Dāwud ibn Ibrāhīm al-Jawharī al-Khaṭīb, *al-Durr al-Thamīn al-Manẓūm fīmā Wurida fī Miṣr wa Aʿmālihā bi-al-Khuṣūṣ wa al-ʿUlūm*, Bibliothèque nationale de France, Paris, ms. Fonds Arabe 1812, 34r–38r; *ʿAjāʾib al-Bilād wa al-Aqṭār wa al-Nīl wa al-Anhār wa al-Barārīy wa al-Biḥār*, DKM, Jughrāfiyā Mīm 7; *Dhikr Kalām al-Nās fī Manbaʿ al-Nīl wa Makhrajihi wa Ziyādatihi*, Beinecke, Landberg MSS 365; *Taʾrīkh Miṣr wa ʿAjāʾibuhā*, Beinecke, Landberg MSS 659, 175r–87r; Ibn Ẓahīra, *al-Faḍāʾil al-Bāhira*, 109r–16r; Rūmī, *Kitāb Taʾrīkh Miṣr wa al-Nīl*, 94r–101r; Celâl-zâde Sâlih Çelebi, *Tarih-i Mısr-ı Cedîd*, 80–83 and 249–71; Evliyâ Çelebi, *Evliyâ Çelebi Seyahatnâmesi*, 10: 182–86; Ibn Abī al-Surūr, *al-Nuzha al-Zahiyya* (1998), 241–54. See also the following four slightly different versions of the same text: Shihāb al-Dīn Aḥmad ibn Muḥammad ʿAlī al-Anṣārī al-Ḥijāzī, *al-Nīl al-Rāʾid fī al-Nīl al-Zāʾid*, SK, Ayasofya 3528; Badr al-Dīn ʿAlī ibn Aḥmad ibn Muḥammad al-Bulqīnī, *Kitāb al-Nīl al-Rāʾid fī al-Nīl al-Zāʾid*, SK, Fatih 4181; Shihāb al-Dīn Aḥmad al-Ḥijāzī al-Anṣārī, *al-Nīl al-Rāʾid min al-Nīl al-Zāʾid*, al-Khizāna al-Ḥasaniyya, Rabat, ms. 318; Shihāb al-Dīn Aḥmad ibn Muḥammad ʿAlī al-Ḥijāzī al-Anṣārī, *Nīl al-Rāʾid fī al-Nīl al-Zāʾid*, Khuda Bakhsh Oriental Public Library, Patna, ms. 1069. The following three versions of another text about the Nile: Jalāl al-Dīn al-Maḥallī, *Muqaddima fī Mabdāʾ al-Nīl wa Muntahāhu*, SK, Serez 3838/11; idem., *Risāla fī Mabdāʾ al-Nīl*, DKM, Majāmīʿ Ṭalʿat 972; idem., *Kitāb Mabdāʾ al-Nīl*, DKM, Jughrāfiyā 381. The two versions of the following text: idem., *Muqaddima fī Nīl Miṣr al-Mubārak*, SK, Ayasofya 3446; idem., *Muqaddima fī*

Dhikr Nīl Miṣr, SK, Kara Çelebi Zade 355/7, 52r–66r. And the following two versions of another text: Aḥmad ibn Muḥammad ibn Muḥammad ibn ʿAbd al-Salām, *al-Fayḍ al-Madīd fī Akhbār al-Nīl al-Saʿīd*, al-Maktaba al-Baladiyya bi-al-Iskandariyya, Alexandria, ms. 2591 Dāl; idem., *al-Fayḍ al-Madīd fī Akhbār al-Nīl al-Saʿīd*, Maʿhad al-Makhṭūṭāt al-ʿArabiyya, Cairo, al-Jughrāfiyā wa al-Buldān 39. For a useful comparative account of fishing in early modern Istanbul, see Suraiya Faroqhi, "Fish and Fishermen in Ottoman Istanbul," in *Water on Sand: Environmental Histories of the Middle East and North Africa*, ed. Alan Mikhail (New York: Oxford University Press, 2013), 91–109.

77. DWQ, Maḥkamat al-Baḥayra 8, p. 24, case 37 (19 S 1176/8 Sept. 1762); Kerr, "Letters," 8. On the use of messenger pigeons in Ottoman Egypt, see Tietze, *Muṣṭafā ʿĀlī's Description of Cairo*, 51.
78. Kerr, "Letters," 8.
79. Tietze, *Muṣṭafā ʿĀlī's Description of Cairo*, 32. According to Muṣṭafā ʿĀlī, other nondairy animal products that were both cheap and regularly consumed in Egypt included "indigestible dishes like ox heads, ox feet, lungs, and tripe." Ibid., 44.
80. For a comparative case of the history and current status of equine labor in the urban setting of Fez, Morocco, see Diana K. Davis and Denys Frappier, "The Social Context of Working Equines in the Urban Middle East: The Example of Fez Medina," *Journal of North African Studies* 5 (2000): 51–68.
81. al-Jabartī, *ʿAjāʾib al-Āthār* (1998), 1: 538–39. On the cultivation of sugarcane in Egypt, see Sato Tsugitaka, *State and Rural Society in Medieval Islam: Sultans, Muqtaʿs and Fallahun* (Leiden: Brill, 1997), 211–20.
82. DWQ, Maḍābiṭ al-Daqahliyya 20, pp. 110–11, no case no. (7 C 1188/15 Aug. 1774). For evidence of the transport of ten mules from Malta to the Bey of Cairo that same summer (a rather rare occurrence in this period), see ACM, LC 12, 48r–48v (6 July 1774). For an earlier instance of the use of animal labor to reinforce the embankments of the same canal of al-Baḥr al-Ṣaghīr in al-Daqahliyya, see DWQ, Maḥkamat al-Manṣūra 17, p. 383, no case no. (11 M 1119/14 Apr. 1707).
83. al-Damurdāshī, *al-Durra al-Muṣāna*, 131.
84. DWQ, Maḥkamat al-Manṣūra 3, p. 10, case 31 (19 S 1063/19 Jan. 1653). On the village of Sandūb, see Ramzī, *al-Qāmūs al-Jughrāfī*, pt. 2, vol. 1: 220.
85. DWQ, Maḥkamat al-Baḥayra 24, p. 56, case 102 (1 Ra 1209/26 Sept. 1794). One of these waterwheels was known as the waterwheel of Bāb al-Naṣr and the other was called al-Kitāniyya.
86. For another example of donkey labor, see Tietze, *Muṣṭafā ʿĀlī's Description of Cairo*, 51.
87. DWQ, al-Jusūr al-Sulṭāniyya 784, p. 129, no case no. (n.d.). On the village of Shūbar, see Ramzī, *al-Qāmūs al-Jughrāfī*, pt. 2, vol. 2: 101.
88. For more on these points, see Mikhail, *Nature and Empire*, 170–200.
89. See, for example, DWQ, Maḥkamat al-Baḥayra 10, p. 101, case 229 (1 Ra 1190/20 Apr. 1776).
90. DWQ, Maḥkamat Rashīd 132, pp. 200–01, case 311 (3 N 1137/16 May 1725). For a discussion of the use of camels in the transport of shipbuilding timbers to Ottoman Suez, see Alan Mikhail, "Anatolian Timber and Egyptian Grain: Things that Made the Ottoman Empire," in *Early Modern Things: Objects and Their Histories, 1500–1800*, ed. Paula Findlen (London: Routledge, 2013), 274–93.
91. See, for example, DWQ, Maḥkamat al-Baḥayra 5, p. 314, case 389 (10 Ş 1165/22 June 1752); DWQ, Maḥkamat Rashīd 145, p. 126, case 101 (30 Z 1151/9 Apr. 1739). In this second case, several donkeys were hired to remove dirt from the area around a public bath (*ḥammām*) under repair. These animal laborers cost more to hire for the project than the weigher (*qabbānī*), but were less expensive than the total cost of the human workers brought to the construction site. Because the relative numbers of animals and humans are not given in this case, it is impossible to determine the cost of a single donkey as compared to that of an individual human.
92. This was true not only in Egypt of course, but throughout the empire. For comparative examples of disputes over animal properties from the Ottoman Aegean island of Andros, see KLA, doc. 195 (N 1117/6 Dec. 1705); KLA, doc. 318 (7 N 1217/1 Jan. 1803).

93. Some of the many examples of such cases include the following: DWQ, Maḥkamat Isnā 3, p. 22, case 40 (3 L 1171/9 June 1758); DWQ, Maḥkamat Isnā 6, p. 31, case 42 (5 B 1172/5 Mar. 1759). The former case revolved around a dispute over the inheritance of a she-camel, while the latter centered around that of a buffalo cow. In the following case, the inheritors of an estate fought over the ownership of a cow, a donkey, and four ewes: DWQ, Maḥkamat al-Baḥayra 22, pp. 347–48, case 733 (15 Z 1207/24 July 1793). In a case from the same court in al-Baḥayra a few years later, three brothers fought over the ownership of a quarter of a buffalo cow calf that belonged to the deceased wife of one of the brothers. DWQ, Maḥkamat al-Baḥayra 25, p. 17, case 30 (26 Z 1211/22 June 1797). In another inheritance dispute from al-Baḥayra, Sitīta bint Sīdī Aḥmad Turkī from the village of Maḥallat Marḥūm and her maternal uncle (*khāluhā*) al-Ḥājj Yūsuf ibn al-Ḥājj Muḥammad al-ʿUrf went to court over the ownership of twelve buffalo cows, eight steers, one sheep, and a camel. DWQ, Maḥkamat al-Baḥayra 23, p. 45, case 87 (1 Ra 1208/7 Oct. 1793). On the village of Maḥallat Marḥūm, see Ramzī, *al-Qāmūs al-Jughrāfī*, pt. 2, vol. 2: 107.
94. DWQ, Maḥkamat Manfalūṭ 1, p. 296, case 738 (12 Z 1098/19 Oct. 1687).
95. DWQ, Maḥkamat al-Baḥayra 26, p. 13, case 24 (Awākhir M 1212/15–25 July 1797). The dispute in this case revolved around the sale by the amir Sulaymān Jurbajī Murād of a quarter of a calf and half a buffalo cow that belonged to his deceased wife. The quarter of the deceased's calf sold for three riyāls, while half of her buffalo cow sold for 38 riyāls—further evidence of the still relatively high value of the buffalo cow in comparison to other animals in rural Ottoman Egypt at the end of the eighteenth century.
96. For example, in a case from the court of al-Baḥayra from the year 1793, a woman named Sālima bint Ḥasan ʿĀmir brought her former husband Muḥammad Khalīf to court to assert her right to a fourth of a buffalo cow and an *ardabb* of wheat. In response to Sālima's claims, Muḥammad asserted that when he divorced her she gave up her right to this property in exchange for his payment of three riyāls to her. Continuing to press her claims, Sālima asked that Muḥammad produce evidence (*thubūt*) of this transaction. He in turn showed the court a *ḥukm* witnessed and signed by one Sulaymān al-Laqqānī and one al-Sayyid Muṣṭafā ʿUlwān and also another unspecified written legal instrument (*wathīqa*). Satisfied with this evidence, the judge in the case ruled in favor of the defendant Muḥammad and ordered Sālima to drop any claims against him for the property. DWQ, Maḥkamat al-Baḥayra 22, p. 343, case 724 (13 Z 1207/22 July 1793). For an example of the transfer of animal wealth from parents to a son, see the following case from Andros: KLA, doc. 11 (Evasıt Ş 993/8–17 Aug. 1585).
97. DWQ, Maḥkamat al-Baḥayra 5, p. 183, case 319 (26 Ş 1165/8 July 1752).
98. For examples of the forcible military seizure of camels, see al-Damardāshī, *Tārīkh Waqāyiʿ Miṣr al-Qāhira*, 75 and 301.
99. The kind of proof brought to bear in this case is not specified.
100. Furthermore, the language often used to describe theft (whether of livestock or other goods) again points to the centrality of animals in Egyptians' daily lives. The Ottoman writer Muṣṭafā ʿĀlī, for instance, likens imperial bureaucrats who steal from the Egyptian treasury to "hungry wolves and jackals that have broken into a flock of sheep." Tietze, *Muṣṭafā ʿĀlī's Description of Cairo*, 62. As this and numerous other examples show, the repertoire of available linguistic descriptors in Ottoman Egypt was heavily infused with animal imagery.
101. Bedouin raiders not only stole animals but also rode them effectively in their attacks. Indeed, this was one of the reasons for their skill and swiftness in battle. They are described as being those who ride horses and use spears (*ʿurbān sevārī ve sahib-i mezrāk olup*). Shaw, *Niẓâmnâme-i Mıṣır*, 26–27. On the general history of various Bedouin groups in Egypt, see Aḥmad ibn ʿAlī al-Maqrīzī, *al-Bayān wa al-Iʿrāb ʿammā bi-Arḍ Miṣr min al-Aʿrāb lil-Maqrīzī, maʿa Dirāsāt fī Tārīkh al-ʿUrūba fī Wādī al-Nīl*, ed. ʿAbd al-Majīd ʿĀbidīn (Cairo: ʿĀlam al-Kutub, 1961). I also consulted the following manuscript version of this text: idem., *al-Bayān wa al-Iʿrāb ʿammā bi-Arḍ Miṣr min al-Aʿrāb*, Beinecke, Landberg MSS 492.
102. For an example of bandits (*ashqiyāʾ* in Arabic, *eşkıya* in Ottoman Turkish) stealing animals, see DWQ, Maḥkamat al-Manṣūra 18, p. 267, no case no. (3 M 1120/26 Mar. 1708). This case is an Ottoman buyuruldu sent from Istanbul to the court of al-Manṣūra. As evidenced by this order and numerous other imperial firmans and decrees sent to the courts of Ottoman Egypt, the imperial administration was quite concerned about stopping the theft of animals

from Egyptian villages. For some of the many examples of the plunder of animals during battle, see al-Jabartī, *ʿAjāʾib al-Āthār* (1998), 1: 190, 543; 2: 150.

103. For other examples of Bedouin raids involving the theft of animals, see al-Damurdāshī, *al-Durra al-Muṣāna,* 7–8, 41, 43–44, 129. Bedouin were not always (or only) enemies of local animal economies. Indeed, they very often played integral roles in them. Nomadic groups facilitated the trade in cattle between villages, for example, and also regularly used their camels to transport grain and other foodstuffs from rural communities to large towns and cities. Barkey, *Bandits and Bureaucrats,* 122.
104. On the role of medieval European animals in increasing their owners' social status, see Salisbury, *Beast Within,* 27–32.
105. DWQ, Maḥkamat al-Baḥayra 16, p. 232, case 403 (27 R 1201/16 Feb. 1787). On al-Raḥmāniyya, see Ramzī, *al-Qāmūs al-Jughrāfī,* pt. 2, vol. 2: 305.
106. Again, the description of a steer as "red" probably indicates that the animal had a reddish-brown coat.
107. For another example of the social status attached to the possession of animals, see al-Jabartī, *ʿAjāʾib al-Āthār* (1998), 3: 272.
108. They identified the steers led by the defendant at the wedding as "the two aforementioned described steers" (*al-thaurayn al-madhkūrayn al-mawṣūfayn*).
109. Many questions are raised by this case. For instance, why did it only come to court seventeen days after the wedding in which the defendant was seen with the animals? Was this simply a function of the timing of the general convening of the court? Did the plaintiff perhaps wait to form a case? This seems unlikely given the matter's simplicity and the fact that the plaintiff's key piece of evidence, the testimony of his two witnesses, was probably immediately available after the night of the wedding. Perhaps the plaintiff approached the defendant before coming to court to attempt to solve the case between them without the need to make recourse to a formal legal institution. On these and related issues of legal procedure, see Ergene, *Local Court, Provincial Society and Justice*; Peirce, *Morality Tales.*
110. DWQ, Maḥkamat Manfalūṭ 3, p. 62, case 117 (n.d.). On the village of Jalda, see Ramzī, *al-Qāmūs al-Jughrāfī,* pt. 2, vol. 4: 46–47.
111. "Green" here most likely refers to a greenish-gray or grayish-brown coat color. The term *khaḍrāʾ* was often used to denote browns or grays with a tinge of green. For example, half-tanned leather, certain varieties of sun-dried mud brick, and camel leather were often described as being "green." El-Said Badawi and Martin Hinds, *A Dictionary of Egyptian Arabic* (Beirut: Librairie du Liban, 1986), s.v. *khāʾ, ḍād, rāʾ.*
112. This case does not offer any details as to how these two witnesses evidenced their knowledge of the hereditary link between the two donkeys. Perhaps they witnessed the young donkey's birth, or perhaps they had known Rushdān's donkey to be pregnant before she wandered away. Jenny pregnancies last approximately eleven months—which fits well with the timeline of this case.
113. For a similar court case about establishing the legal ownership of a disputed wandering donkey, see DWQ, Maḥkamat Manfalūṭ 3, pp. 86–87, case 161 (n.d.).
114. DWQ, Maḥkamat al-Baḥayra 16, p. 89, case 155 (16 L 1200/11 Aug. 1786).
115. For a comparative example of a dispute centered around ownership rights to a newborn calf, see the following case from the island of Andros involving a priest and a group of monks: KLA, doc. 191 (1115/6 May 1703).
116. DWQ, Maḥkamat al-Baḥayra 23, p. 129, case 241 (21 C 1208/23 Jan. 1794). On Sunṭīs, see Ramzī, *al-Qāmūs al-Jughrāfī,* pt. 2, vol. 2: 287.
117. On the village of Zāwiyat Naʿīm, see Ramzī, *al-Qāmūs al-Jughrāfī,* pt. 2, vol. 2: 240–41.
118. On Ruzzāfa, see ibid., pt. 2, vol. 2: 261.
119. For those wanting to press their claims to certain plots of land, however, other people's wandering animals became a source of much concern, and there were indeed many court cases in which domestic animals became embroiled in such property disputes. See, for example, the following case from the island of Andros concerning attempts by the monastery of Hagios Nikolaos to prevent individual farmers' sheep from using the monastery's pastureland: KLA, doc. 354 (n.d.). For similar cases from Andros involving wandering sheep and pastureland, see KLA, doc. 82 (Evail C 1030/23 Apr.–2 May 1621); KLA, doc. 277 (28 Ra 1177/6 Oct.

1763). On sheep and goat raising on other islands in the Aegean during the Ottoman period, see Yasemin Demircan, "XV–XVII. Yüzyıllarda Ege Adaları'nda Çeşitli Yönleriyle Küçükbaş Hayvancılık," *Ankara Üniversitesi Osmanlı Tarihi Araştırma ve Uygulama Merkezi Dergisi* 24 (2008): 71–95.

120. On the prohibitive costs of feeding domestic work animals in medieval Europe, see Salisbury, *Beast Within*, 20.
121. For a comparative example of this phenomenon in colonial America, see Anderson, *Creatures of Empire*, 165–66.
122. On this kind of "communal oversight" of livestock, see ibid., 163.
123. In medieval Europe as well, a wandering animal nevertheless always remained its owner's legal property. Salisbury, *Beast Within*, 32.

2. Unleashing the Beast

1. Kasaba, *Ottoman Empire and World Economy*, 23–27.
2. Richards, "Global System," 71. On these points, see also idem., *Unending Frontier*, 617–22.
3. The Brenner debate is usefully summarized, with contributions by many of the scholars involved in these discussions, in Aston and Philpin, *Brenner Debate*. The literature on divergence of course began with Pomeranz, *Great Divergence*. For a recent consideration of the subject, see Rosenthal and Wong, *Before and Beyond Divergence*.
4. In 1735–36, for example, cattle and sheep populations were decimated by disease. On the epizootics of these years, see Abū al-Fayḍ Aḥmad ibn Qara Kamāl, *Jawāhir al-Bayān fī Dawlat Āl ʿUthmān*, Biblioteca Apostolica Vaticana, Vatican Arabo 870, 20v–21r; ʿAbd al-Raḥman al-Jabartī, *ʿAbd al-Raḥman al-Jabartī's History of Egypt: ʿAjāʾib al-Āthār fī al-Tarājim wa al-Akhbār*, ed. Thomas Philipp and Moshe Perlmann, 4 vols. (Stuttgart: Franz Steiner Verlag, 1994), 1: 242–43 and 273–74; al-Damurdāshī, *al-Durra al-Muṣāna*, 202. For other earlier instances of human and animal diseases and population reductions, see Nāṣir Aḥmad Ibrāhīm, *al-Azamāt al-Ijtimāʿiyya fī Miṣr fī al-Qarn al-Sābiʿ ʿAshar* (Cairo: Dār al-Āfāq al-ʿArabiyya, 1998); Stuart J. Borsch, *The Black Death in Egypt and England: A Comparative Study* (Austin: University of Texas Press, 2005); idem., "Environment and Population: The Collapse of Large Irrigation Systems Reconsidered," *Comparative Studies in Society and History* 46 (2004): 451–68; Karl W. Butzer, *Early Hydraulic Civilization in Egypt: A Study in Cultural Ecology* (Chicago: University of Chicago Press, 1976); Michael W. Dols, *The Black Death in the Middle East* (Princeton, NJ: Princeton University Press, 1977); idem., "The General Mortality of the Black Death in the Mamluk Empire," in *The Islamic Middle East, 700–1900: Studies in Social and Economic History*, ed. Abraham L. Udovitch (Princeton, NJ: Darwin Press, 1981), 397–428; William F. Tucker, "Natural Disasters and the Peasantry in Mamlūk Egypt," *Journal of the Economic and Social History of the Orient* 24 (1981): 215–24.
5. Crecelius, *Roots of Modern Egypt*.
6. Ibid., 79–91 and 159–68; Kerr, "Letters," 12–13.
7. Stanford J. Shaw, "Landholding and Land-Tax Revenues in Ottoman Egypt," in *Political and Social Change in Modern Egypt: Historical Studies from the Ottoman Conquest to the United Arab Republic*, ed. P. M. Holt (London: Oxford University Press, 1968), 91–103; Holt, *Egypt and the Fertile Crescent*, 85–101. For a recent rethinking of the role of Ottoman provincial elites at the end of the eighteenth century, see Yaycıoğlu, "Provincial Power-Holders."
8. Kenneth M. Cuno, "Commercial Relations between Town and Village in Eighteenth and Early Nineteenth-Century Egypt," *Annales Islamologiques* 24 (1988): 111–35; Alan R. Richards, "Primitive Accumulation in Egypt, 1798–1882," in *The Ottoman Empire and the World-Economy*, ed. Huri İslamoğlu-İnan (Cambridge: Cambridge University Press, 1987), 203–43.
9. Richard W. Bulliet identifies four phases in the history of human-animal relations: the separation of humans from other animals, predomesticity, domesticity, and postdomesticity. For his full characterization of these four periods and his history of the transitions between them, see Bulliet, *Hunters, Herders, and Hamburgers*. Although at first glance the economic transitions I am discussing here would seem to map onto Bulliet's study of changes in human-animal relations from the phase of domesticity to the first moments of postdomesticity, it should

be noted that he specifically distances his analysis from any connection to modernization. Ibid., 36–37.

10. This is one of the primary points of contrast between the seemingly similar situations of Egypt at the end of the eighteenth century and northern Europe in the first half of the fourteenth century. As William Chester Jordan has shown, apart from potentially improving nutritional standards, famine and its attendant consequences in Europe between 1315 and 1322 did *not* alter social and economic relations in any lasting way as they did in late Ottoman Egypt. William Chester Jordan, *The Great Famine: Northern Europe in the Early Fourteenth Century* (Princeton, NJ: Princeton University Press, 1996). Philip Slavin has recently attempted to link the Great Famine and the Black Death through what he terms the Great Bovine Pestilence. Philip Slavin, "The Great Bovine Pestilence and its Economic and Environmental Consequences in England and Wales, 1318–50," *Economic History Review* 65 (2012): 1239–66. In this regard, see also Timothy P. Newfield, "A Cattle Panzootic in Early Fourteenth-Century Europe," *Agricultural History Review* 57 (2009): 155–90.
11. Cuno, *Pasha's Peasants*; Richards, "Primitive Accumulation in Egypt."
12. Kasaba, *Ottoman Empire and World Economy*, 23–27.
13. Mikhail, *Nature and Empire*, 170–200.
14. On the beginnings of rail in Egypt, see Omar Abdel-Aziz Omar, "Anglo-Egyptian Relations and the Construction of the Alexandria-Cairo-Suez Railway (1833–1858)" (D.Phil. thesis, University of London, 1966). If the American example is any indication, it seems safe to assume that rail *initially* actually increased the use of horses in Egypt—both in the original construction of rail lines and in the movement of people and goods to, from, and between trains. Ann Norton Greene, *Horses at Work: Harnessing Power in Industrial America* (Cambridge, MA: Harvard University Press, 2008), 43–45 and 75–82. On horse labor in America more generally, see Clay McShane and Joel A. Tarr, *The Horse in the City: Living Machines in the Nineteenth Century* (Baltimore: Johns Hopkins University Press, 2007).
15. J. R. McNeill offers the following definition of an "energy regime": "the collection of arrangements whereby energy is harvested from the sun (or uranium atoms), directed, stored, bought, sold, used for work or wasted, and ultimately dissipated." J. R. McNeill, *Something New under the Sun: An Environmental History of the Twentieth-Century World* (New York: W. W. Norton, 2000), 297. More generally on the relationship between energy and economy, see ibid., 296–324. For a biological explanation of the relationship between animals and energy, see Brian K. McNab, *Extreme Measures: The Ecological Energetics of Birds and Mammals* (Chicago: University of Chicago Press, 2012).
16. On some of the consequences of a lack of wood in Ottoman Egypt, see Mikhail, *Nature and Empire*, 124–69.
17. For studies that make this point in some detail, see Vaclav Smil, *Energy in World History* (Boulder: Westview Press, 1994); idem., *Energy in Nature and Society: General Energetics of Complex Systems* (Cambridge: Massachusetts Institute of Technology Press, 2008); Stephen J. Pyne, *World Fire: The Culture of Fire on Earth* (Seattle: University of Washington Press, 1997); idem., *Vestal Fire: An Environmental History, Told through Fire, of Europe and Europe's Encounter with the World* (Seattle: University of Washington Press, 1997); Warde, *Ecology, Economy and State Formation*; Edmund Burke III, "The Big Story: Human History, Energy Regimes, and the Environment," in *The Environment and World History*, ed. Edmund Burke III and Kenneth Pomeranz (Berkeley: University of California Press, 2009), 33–53; Mary C. Stiner and Gillian Feeley-Harnik, "Energy and Ecosystems," in *Deep History: The Architecture of Past and Present*, ed. Andrew Shryock and Daniel Lord Smail (Berkeley: University of California Press, 2011), 78–102.
18. This point is made convincingly in the following: Richards, *Unending Frontier*; Fernand Braudel, *The Wheels of Commerce*, trans. Siân Reynolds, vol. 2 of *Civilization and Capitalism, 15th–18th Century* (London: Collins, 1982); Richard W. Bulliet, "History and Animal Energy in the Arid Zone," in *Water on Sand: Environmental Histories of the Middle East and North Africa*, ed. Alan Mikhail (New York: Oxford University Press, 2013), 51–69.
19. On 1800 as the turning point between energy regimes see, for example, Bruce Podobnik, "Toward a Sustainable Energy Regime: A Long-Wave Interpretation of Global Energy Shifts," *Technological Forecasting and Social Change* 62 (1999), 155–61; Burke, "The Big Story," 35.

20. On plague in Egypt in 1781, see ACM, LC 12, 84r–84v (15 Apr. 1781); ACM, LC 12, 88r–88v (25 Sep. 1781). Throughout the eighteenth century, certain areas of Europe also experienced successive waves of epizootics, suggesting that this was likely a Mediterranean-wide phenomenon in the early modern period. On the European epizootics, see Karl Appuhn, "Ecologies of Beef: Eighteenth-Century Epizootics and the Environmental History of Early Modern Europe," *Environmental History* 15 (2010): 268–87.
21. BOA, HAT 29/1361 (13 Ş 1198/1 Jul. 1784). I use the term "plague" throughout as the translation for both *taun* (the Ottoman Turkish version of the Arabic *ṭāʿūn*) and *veba* (or *wabāʾ* in Arabic). The original sources use both words to refer to diseases of humans and animals alike. The human diseases termed "plague" no doubt included a wide variety of ailments beyond just *Yersinia pestis*—very likely anthrax, typhus, or any number of other parasitic diseases. As for the epizootics that are also termed *taun* and *veba*, they could have been anthrax, rinderpest, bovine pleuropneumonia, foot and mouth disease, or something else entirely. The larger and more important point to make—and probably part of the reason the original sources confusingly use disease terms so interchangeably—is that the same conditions made both humans vulnerable to epidemic diseases and animals vulnerable to epizootics. Following the sources, and with really no other logical choice, I have necessarily replicated this general and wide-ranging, if nonspecific, use of the word "plague" to refer to human diseases. I also sometimes term epizootics "plague" when the sources direct this usage, but more generally simply use the catchall descriptor "epizootic." It would be several decades after this period before veterinary medicine would make possible clearer delineations of specific animal diseases. For a very useful treatment of some of the difficulties involved in identifying human diseases in Ottoman Turkish and Arabic source materials, see Sam White, "Rethinking Disease in Ottoman History," *International Journal of Middle East Studies* 42 (2010), 555–58. For their part, Venetians as well were not always sure how to categorize their animals' diseases. On this point, see Appuhn, "Ecologies of Beef," 270 and 279–81. In the case of the Great Bovine Pestilence of the fourteenth century, the scholarly consensus is that the disease was rinderpest. Newfield, "Cattle Panzootic," 188; Slavin, "Great Bovine Pestilence," 1240. For a useful analysis of the history of anthrax that touches on both the disease's Ottoman past and its relationship to other animal maladies, see Susan D. Jones, *Death in a Small Package: A Short History of Anthrax* (Baltimore: Johns Hopkins University Press, 2010). For some of the current literature on the history of veterinary medicine, see idem., *Valuing Animals*; Brown and Gilfoyle, *Healing the Herds*; Louise Hill Curth, *The Care of Brute Beasts: A Social and Cultural Study of Veterinary Medicine in Early Modern England* (Leiden: Brill, 2010); Swabe, *Animals, Disease, and Human Society*; Shehada, *Mamluks and Animals*; Diana K. Davis, "Brutes, Beasts, and Empire: Veterinary Medicine and Environmental Policy in French North Africa and British India," *Journal of Historical Geography* 34 (2008): 242–67. For a useful study of camel diseases in modern Jordan, see Odeh F. al-Rawashdeh, Falah K. al-Ani, Labib A. Sharrif, Khaled M. al-Qudah, Yasin al-Hami, and Nicolas Frank, "A Survey of Camel (*Camelus dromedarius*) Diseases in Jordan," *Journal of Zoo and Wildlife Medicine* 31 (2000): 335–38.
22. BOA, HAT 28/1354 (7 Za 1198/22 Sep. 1784).
23. Ismāʿīl ibn Saʿd al-Khashshāb, *Khulāṣat mā Yurād min Akhbār al-Amīr Murād*, ed. and trans. Ḥamza ʿAbd al-ʿAzīz Badr and Daniel Crecelius (Cairo: al-ʿArabī lil-Nashr wa al-Tawzīʿ, 1992), 24–25.
24. For the official registering of flood levels by the Ottoman high divan in Cairo in the period between 1741 and 1804, see DWQ, al-Dīwān al-ʿĀlī 1; DWQ, al-Dīwān al-ʿĀlī 2. For a study of these sources, see Jīhān Aḥmad ʿUmrān, "Dirāsa Diblūmātiyya li-Wathāʾiq Wafāʾ al-Nīl bi-Sijilāt al-Dīwān al-ʿĀlī maʿa Nashr Namādhij minhā," *Waqāʾiʿ Tārīkhiyya: Dauriyya ʿIlmiyya Muḥakkama* (2004): 347–81.
25. BOA, HAT 28/1354 (7 Za 1198/22 Sep. 1784); al-Jabartī, *ʿAjāʾib al-Āthār* (1994), 2: 139, 156–57.
26. This situation of prolonged food shortages led to riots in Cairo in the spring of 1785. al-Jabartī, *ʿAjāʾib al-Āthār* (1994), 2: 210; 3: 100.
27. Ibid., 2: 139. These years' desperate circumstances of drought, famine, and disease resulted in part from the eruption of the Laki Fissure in Iceland in 1783 and 1784, a climate event of truly

global proportions that led to reductions in monsoon levels over the Indian Ocean, decreasing temperatures around the Mediterranean basin, and dramatically low flood levels for the Nile and other riverine systems in Europe and the Middle East. It contributed to a decade (1785 to 1795) of cold winters, inadequate floods, depressed cultivation levels, and disease outbreaks throughout Egypt. On the eruption, see Luke Oman, Alan Robock, Georgiy L. Stenchikov, and Thorvaldur Thordarson, "High-Latitude Eruptions Cast Shadow over the African Monsoon and the Flow of the Nile," *Geophysical Research Letters* 33 (2006): L18711. For observations that make clear the relationships between weather, plague, and food supplies and their combined effects on populations in Egypt (and elsewhere) in 1791, see TNA, PC 1/19/24 (19. Mar.–23 Dec. 1791); TNA, FO 24/1, 191r–196v (12 July 1791) and 197r–200v (7 Oct. 1791).

28. al-Jabartī, *ʿAjāʾib al-Āthār* (1994), 2: 228–29.
29. Ibid., 2: 229.
30. Ibid., 2: 228. And, al-Jabartī continues, "other losses were on a similar scale." Ibid.
31. Ibid., 2: 241.
32. Ibid., 2: 229.
33. Ibid., 2: 232.
34. Ibid.
35. Ibid., 2: 241.
36. For a recent treatment of the role of plague in Ottoman history, see White, "Rethinking Disease."
37. al-Jabartī, *ʿAjāʾib al-Āthār* (1994), 2: 260 and 263. For references to some of the human victims of this plague, see ibid., 2: 275, 280, 282. For more on the plague of 1788, see ACM, LC 12, 115r–16r (26 June 1788).
38. BOA, Cevdet Dahiliye, 1722 (Evasıt N 1205/15–24 May 1791); al-Jabartī, *ʿAjāʾib al-Āthār* (1994), 2: 315; al-Khashshāb, *Akhbār al-Amīr Murād*, 33–34.
39. BOA, HAT 1399/56283 (29 Z 1205/29 Aug. 1791). There is no internal evidence for the date of this case. The date given is the one assigned by the BOA. Although this high number of deaths is surely exaggerated, it indicates that this plague was considered of unmatched severity. For a discussion of how to interpret reported numbers of human plague causalities, see Alan Mikhail, "The Nature of Plague in Late Eighteenth-Century Egypt," *Bulletin of the History of Medicine* 82 (2008), 254–57.
40. BOA, HAT 1412/57500 (29 Z 1205/29 Aug. 1791). There is no internal evidence for the date of this case. The date given is the one assigned by the BOA.
41. Ismāʿīl ibn Saʿd al-Khashshāb, *Akhbār Ahl al-Qarn al-Thānī ʿAshar: Tārīkh al-Mamālīk fī al-Qāhira*, ed. ʿAbd al-ʿAzīz Jamāl al-Dīn and ʿImād Abū Ghāzī (Cairo: al-ʿArabī lil-Nashr wa al-Tawzīʿ, 1990), 58. The extreme brutality and lethality of this plague epidemic is attested by the fact that it is mentioned in almost all of the narrative sources of the period, whereas other instances of disease appear in only one or some accounts.
42. A set of thirty-seven letters sent by British consuls in several of these locales describe the desperate situation of plague throughout 1791. TNA, PC 1/19/24 (19. Mar.–23 Dec. 1791). For more British correspondence about plague in Egypt in 1791 and 1792, see for example the following reports sent by the British consul in Egypt George Baldwin to London: TNA, FO 24/1, 183r–185v (4 July 1791), 191r–196v (12 July 1791), 211r–212v (21 June 1792).
43. On the 1792 plague, see BOA, HAT 209/11213 (29 Z 1206/18 Aug. 1792). As in previous cases, the date given here is that assigned by the BOA. There is no internal evidence for the date of this case.
44. For British concerns about food and water shortages in Egypt in this period, see the following letter from Consul Baldwin to London: TNA, FO 24/1, 197r–200v (7 Oct. 1791).
45. al-Jabartī, *ʿAjāʾib al-Āthār* (1994), 2: 374. For more on the problem of worms in the countryside in this period, see ibid., 2: 397.
46. Ibid., 2: 374.
47. BOA, HAT 209/11213 (29 Z 1206/18 Aug. 1792). As in previous cases, the date given here is that assigned by the BOA. There is no internal evidence for the date of this case.
48. al-Jabartī, *ʿAjāʾib al-Āthār* (1994), 2: 374 and 397.

49. Ibid., 2: 374–75.
50. Ibid., 2: 397.
51. Ibid.
52. Ibid.
53. BOA, HAT 245/13801A (3 Za 1213/9 Apr. 1799); BOA, HAT 240/13451 (29 N 1214/24 Feb. 1800).
54. al-Jabartī, *ʿAjāʾib al-Āthār* (1994), 3: 149–50. On the role of fodder shortages in bovine malnutrition and death in early fourteenth-century England and Wales, see Slavin, "Great Bovine Pestilence," 1244–47.
55. For another instance of attempts to sell off starving animals before they died, see al-Jabartī, *ʿAjāʾib al-Āthār* (1994), 3: 402.
56. Ibid., 3: 459. In the fall of 1791, for example, food and fodder supplies were so low that the price of grain tripled. TNA, FO 24/1, 197r–200v (7 Oct. 1791).
57. al-Jabartī, *ʿAjāʾib al-Āthār* (1994), 3: 158.
58. ʿAbd Allāh Sharqāwī, *Tuḥfat al-Nāẓirīn fī man Waliya Miṣr min al-Mulūk wa al-Salāṭīn*, ed. Riḥāb ʿAbd al-Ḥamīd al-Qārī (Cairo: Maktabat Madbūlī, 1996), 124–25. For other examples of how the monopolization of food supplies led to animal starvation and death, see al-Jabartī, *ʿAjāʾib al-Āthār* (1994), 3: 402.
59. BOA, HAT 86/3520 (29 N 1216/3 Feb. 1802).
60. For nineteenth-century French discussions of the etiology and prevention of human plague in Egypt, see Antoine Barthélemy Clot-Bey, *De la peste observée en Égypte: recherches et considérations sur cette maladie* (Paris: Fortin, Masson, 1840); M. Hamont, *Destruction de la peste: lazarets et quarantaines* (Paris: Imprimerie de Bourgogne et Martinet, 1844).
61. al-Jabartī, *ʿAjāʾib al-Āthār* (1994), 3: 480.
62. This major storm hit when Egypt was already suffering from "various diseases" (*enva-ı emraz*) throughout the countryside and Cairo. BOA, HAT 88/3601 (1 Za 1218/12 Feb. 1804).
63. al-Jabartī, *ʿAjāʾib al-Āthār* (1994), 4: 117.
64. The description of hailstones as the size of hen's eggs is used here as a standard trope to convey that they were exceptionally large. Even bigger hailstones were sometimes described as being the size of "grinding stones." Ibid., 4: 126.
65. Ibid.
66. Ibid., 4: 240.
67. See, for example, ibid., 4: 412. For a wider discussion of the role of natural disasters in Ottoman history, see Elizabeth Zachariadou, ed., *Natural Disasters in the Ottoman Empire* (Rethymnon: Crete University Press, 1999). On earthquakes in particular, see N. N. Ambraseys and C. F. Finkel, *The Seismicity of Turkey and Adjacent Areas: A Historical Review, 1500–1800* (Istanbul: Eren, 1995); N. N. Ambraseys, C. P. Melville, and R. D. Adams, *The Seismicity of Egypt, Arabia and the Red Sea: A Historical Review* (Cambridge: Cambridge University Press, 1994); Nicholas Ambraseys, *Earthquakes in the Eastern Mediterranean and Middle East: A Multidisciplinary Study of Seismicity up to 1900* (Cambridge: Cambridge University Press, 2009); N. N. Ambraseys and C. P. Melville, *A History of Persian Earthquakes* (Cambridge: Cambridge University Press, 1982); Mohamed Reda Sbeinati, Ryad Darawcheh, and Mikhail Mouty, "The Historical Earthquakes of Syria: An Analysis of Large and Moderate Earthquakes from 1365 B.C. to 1900 A.D.," *Annals of Geophysics* 48 (2005): 347–435. For a late fifteenth-century Arabic treatise on earthquakes, see Jalāl al-Dīn al-Suyūṭī, *Kashf al-Ṣalṣala ʿan Waṣf al-Zalzala*, al-Maktaba al-Waṭaniyya lil-Mamlaka al-Maghribiyya, Rabat, ms. 1027 Kāf. This text was published as idem., *Kashf al-Ṣalṣala ʿan Waṣf al-Zalzala*, ed. ʿAbd al-Laṭīf al-Saʿdānī (Rabat: Wizārat al-Dawla al-Mukallafa bi-al-Shuʾūn al-Thaqāfiyya wa al-Taʿlīm al-Aṣlī, 1971).
68. Data from the Great Bovine Pestilence in England and Wales in the early fourteenth century suggest that it took about twenty years to recover about 60 percent of the bovid population lost in that crisis and that about 20 percent of the population never recovered. Slavin, "Great Bovine Pestilence," 1249–54.
69. For a comparative analysis of the relationships between epizootics and meat supply in early modern Venice, see Appuhn, "Ecologies of Beef."

70. For a very useful study of the history of French animal slaughter and meat consumption, see Noëlie Vialles, *Animal to Edible*, trans. J. A. Underwood (Cambridge: Cambridge University Press, 1994). For the American case, see Roger Horowitz, *Putting Meat on the American Table: Taste, Technology, Transformation* (Baltimore: Johns Hopkins University Press, 2006).
71. al-Jabartī, *ʿAjāʾib al-Āthār* (1994), 4: 380–81.
72. For accounts of the reign of Mehmet ʿAli, see Fahmy, *All the Pasha's Men*; idem., "The Era of Muhammad ʿAli Pasha, 1805–1848," in *Modern Egypt, from 1517 to the End of the Twentieth Century*, vol. 2 of *The Cambridge History of Egypt*, ed. M. W. Daly (Cambridge: Cambridge University Press, 1998), 139–79; idem., *Mehmed Ali*; Marsot, *Egypt in the Reign of Muhammad Ali*; al-Rāfʿī, *ʿAṣr Muḥammad ʿAlī*.
73. al-Jabartī, *ʿAjāʾib al-Āthār* (1994), 4: 380.
74. Ibid., 4: 381.
75. On egg shortages in November 1817, see ibid., 4: 405.
76. Ibid., 4: 382.
77. Ibid., 4: 382–84.
78. Ibid., 4: 395.
79. Ibid.
80. Owners of some of these livestock were paid for their animals' meat, but in these cases they were forced to also give over the heart, liver, neck, head, skin, genitals, intestines, and offal for free. Ibid., 4: 383.
81. Ibid., 4: 390.
82. Ibid., 4: 384.
83. Ibid., 4: 397.
84. Ibid., 2: 232.
85. For a comparative example of the animal labor lost during bovine epizootics in early modern Venice, see Appuhn, "Ecologies of Beef," 279.
86. al-Khashshāb, *Akhbār al-Amīr Murād*, 24–26.
87. al-Jabartī, *ʿAjāʾib al-Āthār* (1994), 3: 345–46.
88. This led to an increase in the price of waterskins, since they now had to be carried on a human's rather than a donkey's back. Ibid., 3: 346. On the history of *al-Khalīj*, see André Raymond, *Cairo: City of History*, trans. Willard Wood (Cairo: American University in Cairo Press, 2001), 123–27 and 221–24; Suʿād Māhir, "Majrī Miyāh Famm al-Khalīj," *Egyptian Historical Review* 7 (1958): 134–57; *ʿAwāʾid al-Miṣriyyīn ʿand Izdiyād al-Nīl*, DKM, Makhṭūṭāt al-Zakiyya 584. For an example of the regular work needed to clean and dredge the canal, see DWQ, Maḥkamat Miṣr al-Qadīma 90, pp. 386–87, case 1932 (13 Z 969/14 Aug. 1562).
89. Except for these sorts of accentuating circumstances, donkeys were rarely kept inside homes, especially in Cairo. It was, however, quite common for rural peoples to keep water buffalo inside their homes because these animals were rather sedentary and docile and because their large bodies were useful sources of heat during the winter. Donkeys, by contrast, did not give off as much heat as water buffalo and were generally more active and fidgety, making it much more difficult to keep them indoors.
90. For other examples of attempts by soldiers to steal animals hidden inside people's homes, see al-Jabartī, *ʿAjāʾib al-Āthār* (1994), 4: 163–64.
91. Bruce McGowan, "The Age of the Ayans, 1699–1812," in *An Economic and Social History of the Ottoman Empire*, ed. Halil İnalcık with Donald Quataert, 2 vols. (Cambridge: Cambridge University Press, 1994), 2: 692.
92. About this situation at the end of the eighteenth century, Reşat Kasaba writes that rural cultivators in the Ottoman Empire "were transformed from being peasant producers whose freedom and protection was institutionally guaranteed to the status of, at best, sharecroppers—but more commonly, indebted tenants or dispossessed wage laborers." Kasaba, *Ottoman Empire and World Economy*, 26.
93. al-Jabartī, *ʿAjāʾib al-Āthār* (1994), 2: 241.
94. In certain parts of England in the late sixteenth century, land consolidation based on enclosure was similarly part of the commercial development of the countryside. In the English

case, however, there is much evidence to suggest that animal populations, especially those of sheep, actually increased markedly in lockstep with the concentration and commercialization of land. John Martin, "Sheep and Enclosure in Sixteenth-Century Northamptonshire," *Agricultural History Review* 36 (1988): 39–54.

95. On the history of land use and land ownership in Egypt see, for example, Richards, "Primitive Accumulation in Egypt"; Shaw, "Landholding and Land-Tax Revenues in Ottoman Egypt"; idem., *Financial and Administrative Organization and Development,* 12–97; Cuno, *Pasha's Peasants;* ʿAbd al-Raḥīm, *al-Rīf al-Miṣrī,* 83–143; Gabriel Baer, *A History of Landownership in Modern Egypt, 1800–1950* (London: Oxford University Press, 1962).
96. On the rule of these two leaders, see Crecelius, *Roots of Modern Egypt;* ACM, LC 12, 192r–93r (9 Mar. 1771).
97. Marsot, *Egypt in the Reign of Muhammad Ali,* 14–19; Fahmy, *Mehmed Ali,* 23–24.
98. The early nineteenth-century estates in which animals do not play a role are many. While absence of evidence in thousands of cases is not evidence of complete absence, animals appear far less frequently in the estate inventories of members of all classes of rural society between 1780 and 1820 than they did earlier. For a discussion of the increasing role of land as property in the first few decades of the nineteenth century, see Cuno, *Pasha's Peasants,* 103–17.
99. On usufruct as effective, though of course not legal, ownership, see ibid., 74–84; Gabriel Baer, "The Dissolution of the Egyptian Village Community," *Die Welt des Islams* 6 (1959), 59.
100. This kind of confiscation was known as *müsadere* in the Ottoman Empire. For a discussion of this phenomenon in the early modern period, see Rifaʿat ʿAli Abou-El-Haj, *Formation of the Modern State: The Ottoman Empire, Sixteenth to Eighteenth Centuries,* 2nd ed. (Syracuse, NY: Syracuse University Press, 2005), 48–49.
101. Cuno, *Pasha's Peasants,* 33–47.
102. The creation of large rural estates in Egypt further exacerbated animal shortages in the countryside by reducing the amount of available pastureland. As competing amirs, beys, and other Egyptian elites continued to seize land and other properties, less and less pastureland was left for animals to graze, making it much more difficult to maintain sizeable herds in and around major towns and cities. al-Jabartī, *ʿAjāʾib al-Āthār* (1994), 3: 477–78; 4: 380.
103. On the emergence of regularity and reproducibility as paradigms of labor and natural resource management in Ottoman Egypt, see Mikhail, *Nature and Empire.* For a study of similar developments in the United States and Europe, see Jennifer Karns Alexander, *The Mantra of Efficiency: From Waterwheel to Social Control* (Baltimore: Johns Hopkins University Press, 2008). See also Scott, *Seeing Like a State.*
104. Rudolph Peters, "'For His Correction and as a Deterrent Example for Others': Meḥmed ʿAlī's First Criminal Legislation (1829–1830)," *Islamic Law and Society* 6 (1999): 164–92.
105. For a sustained discussion of the processes involved in the simplification and de-skilling of human labor, see Harry Braverman, *Labor and Monopoly Capital: The Degradation of Work in the Twentieth Century* (New York: Monthly Review Press, 1974). See also the helpful analyses in Daron Acemoglu, "Technical Change, Inequality, and the Labor Market," *Journal of Economic Literature* 40 (2002): 7–72.
106. For a general discussion of corvée in Egypt in the seventeenth and eighteenth centuries, see ʿAbd al-Ḥamīd Sulaymān, "al-Sukhra fī Miṣr fī al-Qarnayn al-Sābiʿ ʿAshar wa al-Thāmin ʿAshar, Dirāsa fī al-Asbāb wa al-Natāʾij," in *al-Rafḍ wa al-Iḥtijāj fī al-Mujtamaʿ al-Miṣrī fī al-ʿAṣr al-ʿUthmānī,* ed. Nāṣir Ibrāhīm and Raʾūf ʿAbbās (Cairo: Markaz al-Buḥūth wa al-Dirāsāt al-Ijtimāʿiyya, 2004), 89–126.
107. al-Shirbīnī, *Hazz al-Quḥūf,* 2: 328–29. For other descriptions of corvée in al-Shirbīnī's text, see ibid., 2: 13, 49, 79, 333.
108. This land was again known as *ūsya* land and was generally found only in the Delta and Middle Egypt. The figure of 10 percent comes from the end of the eighteenth century. Cuno, *Pasha's Peasants,* 36–37.
109. On this point, al-Shirbīnī offers the following verse: "And on the day when the corvée descends on the people in the village / Umm Waṭīf hides me in the oven." al-Shirbīnī, *Hazz al-Quḥūf,* 2: 327. For extended commentary on this verse, see ibid., 2: 327–31.

110. For these definitions, see *Lisān al-ʿArab*, 4 vols. (Beirut: Dār Lisān al-ʿArab, 1970), s.v. *ʿawana*; Edward William Lane, *An Arabic-English Lexicon*, 8 vols. (Beirut: Librairie du Liban, 1968), s.v. *ʿawana*.
111. *Lisān al-ʿArab*, s.v. *sakhara*; Lane, *An Arabic-English Lexicon*, s.v. *sakhara*.
112. Mikhail, *Nature and Empire*, 170–200.
113. al-Jabartī, *ʿAjāʾib al-Āthār* (1994), 4: 289. For a discussion of this important passage, see Cuno, *Pasha's Peasants*, 5–6 and 37.
114. Helen Anne B. Rivlin, *The Agricultural Policy of Muḥammad ʿAlī in Egypt* (Cambridge, MA: Harvard University Press, 1961), 248.
115. Cuno, *Pasha's Peasants*, 115. Between 1805 and 1812, Mehmet ʿAli's administration undertook a large-scale survey of the countryside to determine the percentage of agricultural lands watered and unwatered. The results are summarized by subprovince in the following registers. For the subprovince of al-Sharqiyya, see DWQ, Taḥrīr Aṭyān al-Sharāqī wa al-Riyy 3019 (1220/1805 and 1806). For al-Minūfiyya, see DWQ, Taḥrīr Aṭyān al-Sharāqī wa al-Riyy 3020 (1220/1805 and 1806) and 3027 (1226/1811 and 1812). For al-Qalyūbiyya, see DWQ, Taḥrīr Aṭyān al-Sharāqī wa al-Riyy 3021 (1222/1807 and 1808), 3024 (1225/1810 and 1811), and 3025 (1226/1811 and 1812). For al-Baḥayra, see DWQ, Taḥrīr Aṭyān al-Sharāqī wa al-Riyy 3022 (1222/1807 and 1808). For al-Manṣūra, see DWQ, Taḥrīr Aṭyān al-Sharāqī wa al-Riyy 3023 (1222/1807 and 1808) and 3028 (1226/1811 and 1812). For al-Gharbiyya, see DWQ, Taḥrīr Aṭyān al-Sharāqī wa al-Riyy 3026 (1226/1811 and 1812).
116. Rivlin, *Agricultural Policy of Muḥammad ʿAlī*, 247.
117. Details of these and other canal repairs are given in the following: ibid., 213–41; al-Rāfʿī, *ʿAṣr Muḥammad ʿAlī*, 487–95.
118. Peters, "Meḥmed ʿAlī's First Criminal Legislation."
119. Ibid., 170 and 180.
120. For example, one important outcome of the changing economy of animals in this period was what corvée meant for the laboring agricultural family as a whole. As men were taken away from their home villages to work on infrastructural projects, women and other familial relations often replaced men in the laboring economy of their home communities. For a discussion of this point, see Tucker, *Women in Nineteenth-Century Egypt*, 29. For a useful recent analysis of the important role of women in rural labor in early modern France, see Chandra Mukerji, *Impossible Engineering: Technology and Territoriality on the Canal du Midi* (Princeton, NJ: Princeton University Press, 2009).
121. Rivlin, *Agricultural Policy of Muḥammad ʿAlī*, 243–45.
122. For later examples of the principle that the organization of corvée labor was the responsibility of village heads, see Baer, "Egyptian Village Community," 63–64 and 66–68.
123. According to M. A. Linant de Bellefonds's estimates, in each year between 1820 and 1831 about sixty-seven thousand men were used to build canals and another four hundred thousand were employed to clean them. Cuno, *Pasha's Peasants*, 122.
124. On the population of Cairo, see Panzac, "Alexandrie: évolution d'une ville cosmopolite," 147. On the population of Egypt as a whole, see idem., *La Peste dans l'Empire Ottoman, 1700–1850* (Louvain: Association pour le développement des études turques, 1985), 271; Raymond, "La population du Caire et de l'Égypte."
125. This was true, for example, during certain periods of repair work on the Maḥmūdiyya Canal. Marsot, *Egypt in the Reign of Muhammad Ali*, 151. For earlier examples, see DWQ, Maḥkamat al-Manṣūra 1, p. 84, case 197 (20 Z 1055/6 Feb. 1646); DWQ, Maḥkamat al-Manṣūra 7, p. 134, case 340 (7 Za 1091/29 Nov. 1680); DWQ, Maḥkamat al-Manṣūra 11, p. 8, case 16 (3 Ra 1113/8 Aug. 1701).
126. Rivlin, *Agricultural Policy of Muḥammad ʿAlī*, 227–28.
127. Ibid., 231.
128. Marsot, *Egypt in the Reign of Muhammad Ali*, 152.
129. Rivlin, *Agricultural Policy of Muḥammad ʿAlī*, 232. On the intimate connection between corvée and Mehmet ʿAli's military conscription efforts, see Fahmy, "The Era of Muhammad ʿAli," 163 and 166.
130. Marsot, *Egypt in the Reign of Muhammad Ali*, 150.
131. 2 al-Maʿiyya al-Saniyya (1 N 1251/20 Dec. 1835), cited in ibid., 150–51.

132. 1 al-Maʿiyya al-Saniyya (1 Ş 1253/30 Oct. 1837), cited in Marsot, *Egypt in the Reign of Muhammad Ali*, 151.
133. 2 al-Maʿiyya al-Saniyya (23 Ra 1251/18 July 1835), cited in Marsot, *Egypt in the Reign of Muhammad Ali*, 151.
134. For a detailed discussion and analysis of this construction project, see Mikhail, *Nature and Empire*, 242–96.
135. Rivlin, *Agricultural Policy of Muḥammad ʿAlī*, 219–20 and 353, n. 15. The estimate of 360,000 comes from M. A. Linant de Bellefonds, *Mémoires sur les principaux travaux d'utilité publiqué éxécutés en Egypte depuis la plus haute antiquité jusqu'à nos jours: accompagné d'un atlas renfermant neuf planches grand in-folio imprimées en couleur* (Paris: Arthus Bertrand, 1872–73), 351. Without citation, Marsot writes that one quarter million peasants were brought to work on the canal. Marsot, *Egypt in the Reign of Muhammad Ali*, 151.
136. al-Jabartī, *ʿAjā'ib al-Āthār* (1994), 4: 408.
137. Most of what they cleared from the city's streets was dumped into Cairo's main urban canal (*al-Khalīj*). Since this waterway also served as one of the city's main trash receptacles, the massive amounts of dirt, sand, and garbage deposited in it only clogged it up even more. Ibid.
138. For a discussion of this number of dead, see Mikhail, *Nature and Empire*, 281–82 and 289–90.
139. al-Jabartī, *ʿAjā'ib al-Āthār* (1994), 4: 427. For examples of subsequent problems with shipping on the canal deriving from its narrowness and shallow depth, see DWQ, Maḥkamat al-Baḥayra 38, p. 336, case 789 (8 M 1236/16 Oct. 1820); DWQ, Maḥkamat al-Baḥayra 38, p. 329, case 772 (8 Za 1237/27 Jul. 1822); DWQ, Maḥkamat al-Baḥayra 38, p. 336, case 791 (28 M 1236/5 Nov. 1820); DWQ, Maḥkamat al-Baḥayra 38, p. 335, case 788 (16 Za 1237/4 Aug. 1822).
140. The utility of animal transport labor in moving goods between the Red and Mediterranean Seas was so specialized, unique, and seemingly indispensible that even during late nineteenth-century discussions about the use of rail, animals were still cited as the best means of transport to cover this distance. In a letter dated September 20, 1850, the British consul in Cairo, Alfred S. Walne, wrote the following to the British consul general in Egypt, Sir Charles Murray: "His Highness [ʿAbbas I] proposes when the single line has been laid down between Alexandria and Cairo, to carry a single tram road from the latter place to Suez, so that in due time there may be one almost continued line of communication by rail from one port to the other; the only difference being that on the flatter and paying part of the road locomotives would be employed, whilst on the less frequented and hilly part the carriages would be drawn by horses." TNA, FO 78/841 (20 Sept. 1850), reproduced in Helen Anne B. Rivlin, "The Railway Question in the Ottoman-Egyptian Crisis of 1850–1852," *Middle East Journal* 15 (1961), 387.
141. For accounts of various Ottoman plans for a Suez Canal that ultimately never materialized, see Casale, *Ottoman Age of Exploration*, 135–37, 159–70, 201–02; Colin Imber, *The Ottoman Empire, 1300–1650: The Structure of Power* (New York: Palgrave Macmillan, 2002), 62; Mustafa Bilge, "Suez Canal in the Ottoman Sources," in *Proceedings of the International Conference on Egypt during the Ottoman Era: 26–30 November 2007, Cairo, Egypt*, ed. Research Centre for Islamic History, Art and Culture (Istanbul: IRCICA, 2010), 89–113.
142. For Ottoman orders to Egypt about the organization of the pilgrimage and pilgrims' security, see BOA, MM, 5: 164 (Evail C 1147/29 Oct.–7 Nov. 1734); BOA, MM, 6: 448 (Evasıt Ra 1161/11–20 Mar. 1748). More generally on Ottoman Egypt's central role in the financing and provisioning of the annual *ḥajj*, see Mikhail, *Nature and Empire*, 113–18.
143. For example, in October 1779 soldiers were sent out across Egypt with instructions to collect—by any means necessary—the camels, mules, and donkeys needed for the pilgrimage caravan set to leave Cairo on the twenty-ninth of that month. Whoever was found riding a donkey or mule was stopped on the road, and his animal was forcibly taken from him. If the person was a notable of some importance, he was financially compensated for the animal. If not, no money was paid. al-Jabartī makes special note of the fact that this massive seizure of animals for the pilgrimage caravan—the likes of which, he says, had never been seen before—greatly decreased the number of domestic animals available for labor and sale in Egypt and led to massive inflation. al-Jabartī, *ʿAjā'ib al-Āthār* (1994), 2: 82.

During the pilgrimage season of 1811, a somewhat more forceful seizure of animals was undertaken by Egyptian authorities. Troops were dispatched with the specific aim of collecting animals from Bedouin communities since it was known that they possessed large numbers of donkeys, horses, and mules. Moreover, as in 1779, troops were stationed on all the roads leading in and out of Cairo (and other major cities) to seize any donkeys, mules, camels, horses, and other beasts of burden that were ridden or otherwise moved into or out of cities. In contrast to 1779, however, no consideration was made in 1811 for the status of the person on the animal. This collection of animals for the pilgrimage in the early nineteenth century was so extensive and efficient that people stopped riding their animals altogether for fear that they would be taken from them. Ibid., 4: 188. As these cases show, the annual pilgrimage was one of the important factors exerting pressure on Ottoman Egypt's animal supply.

144. Janet L. Abu-Lughod, *Cairo: 1001 Years of* The City Victorious (Princeton, NJ: Princeton University Press, 1971), 83–97; Raymond, *Cairo*, 293–304.
145. On Barak, "Egyptian Times: Temporality, Personhood, and the Technopolitical Making of Modern Egypt, 1830–1930" (Ph.D. diss., New York University, 2009), 36. John Galloway writes that Mehmet ʿAli began seeking foreign consultation, including from Galloway's brother, on the feasibility of a Cairo-Suez rail line ten years before the publication of his account in 1844. John Alexander Galloway, *Communication with India, China, &c.: Observations on the Proposed Improvements in the Overland Route via Egypt, with Remarks on the Ship Canal, the Boulac Canal, and the Suez Railroad* (London: John Weale, 1844), 22. Thus, even though construction of the first rail lines in Egypt began only in 1851, it was in the middle of the 1830s (and probably even somewhat earlier) that ideas to link Cairo and Suez by rail began in earnest. Khedive ʿAbbas, Mehmet ʿAli's grandson and eventual successor, hired the well-known British civil engineer and member of parliament Robert Stephenson to oversee the construction of Egypt's first rail link between Cairo and Alexandria. For a copy of the contract for this work, see TNA, CO 1069/855 (12 July 1851). On the life and career of Robert Stephenson, see Michael R. Bailey, ed., *Robert Stephenson—The Eminent Engineer* (Aldershot, UK: Ashgate, 2003); Derrick Beckett, *Stephensons' Britain* (Newton Abbot, UK: David and Charles, 1984); L. T. C. Rolt, *George and Robert Stephenson: The Railway Revolution* (London: Longmans, 1960).
146. On some of the diplomatic wrangling involved in efforts to build railways in Egypt, see Rivlin, "Railway Question."
147. For a discussion of some of the economics of rail in nineteenth-century America see, for instance, Richard White, *Railroaded: The Transcontinentals and the Making of Modern America* (New York: W. W. Norton, 2011); William Cronon, *Nature's Metropolis: Chicago and the Great West* (New York: W. W. Norton, 1991).
148. Galloway, *Proposed Improvements in the Overland Route*, 18.
149. Ibid., 16. To avoid any possible confusion, it should be made clear that the "fellow-creatures" Galloway had in mind were human, not animal. For many different reasons, rail in Egypt, not surprisingly, had numerous supporters among British observers. For example, Thomas Waghorn, a strong advocate of funneling through Egypt post and other transport services connecting Britain and India, wrote in the 1830s "that the powers of the steam engine, will, under Providence, be one of the means of bringing the unlettered and darkened millions of the East to Christianity." Thomas Waghorn, *Egypt as it is in 1837* (London: Smith, Elder, and Co., 1837), 27–28. On Waghorn's advocacy of rail in Egypt, see also idem., *Truths Concerning Mahomet Ali, Egypt, Arabia, and Syria: Addressed to the Five Powers or to their Representatives in the Contemplated Congress* (London: Smith, Elder, and Co., 1840). Nearly fifty years later, Lord Cromer, the head British colonial official in Egypt after 1882, wrote in July 1885 that he thought rail potentially useful for the reconquest of the Sudan. TNA, FO 633/5 (3–4 July 1885). For other British statements about the utility and importance of rail in Egypt, see TNA, FO 141/131 (14 May 1869); TNA, FO 926/15 (24 Nov. 1875); TNA, FO 633/3, p. 79 (27 Jul. 1882).
150. Galloway, *Proposed Improvements in the Overland Route*, 6.
151. For examples of British discussions about the Egyptian postal service in the 1870s, see TNA, FO 143/10 (1874–78).

152. Needless to say, British interests in rail in Egypt stemmed largely from the desire to increase the speed and efficiency of moving Indian Army troops and supplies. For discussions in this regard, see the following from the 1860s and 1870s: TNA, MT 23/4 (12 Nov. 1866); TNA, MT 23/13 (1867); TNA, MT 23/18 (1868); TNA, MT 23/27 (1871).
153. Galloway, *Proposed Improvements in the Overland Route*, 3.
154. For added commentary on the need to improve "the celerity and comfort of those who travel across the Desert" between Cairo and Suez, see ibid., 22.
155. For a critical discussion of rail as a quintessential symbol of modernity, see Marian Aguiar, *Tracking Modernity: India's Railway and the Culture of Mobility* (Minneapolis: University of Minnesota Press, 2011).
156. British officials in Egypt were very conscious of the need to maintain coal and other energy supplies in the country to fuel both rail and ship. Egypt's energy regime in the 1860s and 1870s indeed became a crucial engine of the British Empire. See, for example, TNA, FO 141/70 (13 Apr. 1869); TNA, FO 141/131 (1879); TNA, MT 23/13 (1867). For more on the coal energy regime, see Timothy Mitchell, *Carbon Democracy: Political Power in the Age of Oil* (London: Verso, 2011), 12–42.
157. Galloway, *Proposed Improvements in the Overland Route*, 5. Emphasis in original.
158. Another example of the intersecting histories of animals and machines in the Ottoman Empire, and particularly of the perceived disutility of animals in the nineteenth century, was the opening of an underground rail tunnel to move passengers up and down the steep hill between Galata and Pera in Istanbul. Built by a British company, the tunnel was opened in late 1874. To prove the tunnel train's safety and efficiency, its first runs were carried out with only animals aboard. Before this rail link, animals provided the indispensible labor needed to move loads uphill between financial centers in the Ottoman capital. After its construction, however, the only productive labor they were deemed capable of carrying out was to serve as expendable first passengers—crash dummies—on a new, unknown, and possibly dangerous means of conveyance. I take this information about animals as the train's first passengers from the station plaque about the tunnel's history.
159. Galloway, *Proposed Improvements in the Overland Route*, 18.
160. Tuan, *Dominance and Affection*, 15.
161. Ibid.
162. For the British, this preference for machine over both human and nonhuman animal labor was expressed in the late 1870s and early 1880s through the language of both economic efficiency and the purported alleviation of human suffering. Earlier in the nineteenth century, Mehmet ʿAli sought to utilize human labor for its efficacy and perceived superiority over animal labor. By contrast, in the last quarter of the nineteenth century the British came to see corvée as an obstacle to both financial gain and moral uprightness. They thus advocated for the abolition of the practice in favor of machines of all kinds—steam pumps, rail, dredging devices, and telegraphs. For expressions of this desire for machine labor over that of humans and other animals, see the "Annual Report of the Controllers General to His Highness the Khedive, 1880" in TNA, T 1/12928 (21 Feb. 1881). For further statements of British concern about the moral and financial hazards of corvée in this period, see TNA, FO 926/14 (1879); TNA, FO 633/5, 77r–78r (16 May 1887), 79r–79v (23 May 1887), 81v–83v (29 May 1887); TNA, FO 633/45, p. 23 (15 Feb. 1879).
163. Podobnik, "Toward a Sustainable Energy Regime," 155–61; Burke, "The Big Story," 35.
164. Numerous examples of this process from around the globe are given in Richards, "Global System."
165. For an illuminating analysis in this regard, see Bulliet, "History and Animal Energy in the Arid Zone"; idem., "The Camel and the Watermill," *International Journal of Middle East Studies* 42 (2010): 666–68.

3. In-Between

1. The literature on the history of dogs and human-dog relations is therefore vast and varied. Most useful for my thinking on these subjects have been the following: Kete, *Beast in the Boudoir*; Susan McHugh, *Dog* (London: Reaktion Books, 2004); Harriet Ritvo, "Pride and

Pedigree: The Evolution of the Victorian Dog Fancy," *Victorian Studies* 29 (1986): 227–53; Aaron Herald Skabelund, *Empire of Dogs: Canines, Japan, and the Making of the Modern Imperial World* (Ithaca, NY: Cornell University Press, 2011); idem., "Breeding Racism: The Imperial Battlefields of the 'German' Shepherd," *Society and Animals* 16 (2008): 354–71; Kenneth Stow, *Jewish Dogs: An Image and Its Interpreters: Continuity in the Catholic-Jewish Encounter* (Stanford, CA: Stanford University Press, 2006); Donna Haraway, *The Companion Species Manifesto: Dogs, People, and Significant Otherness* (Chicago: Prickly Paradigm Press, 2003); idem., *When Species Meet*; Robert J. Lilly and Michael B. Puckett, "Social Control and Dogs: A Sociohistorical Analysis," *Crime and Delinquency* 43 (1997): 123–47; Sarah Cheang, "Women, Pets, and Imperialism: The British Pekingese Dog and Nostalgia for Old China," *Journal of British Studies* 45 (2006): 359–87; Jacob Tropp, "Dogs, Poison, and the Meaning of Colonial Intervention in the Transkei, South Africa," *Journal of African History* 43 (2002): 451–72; Jan Bondeson, *Amazing Dogs: A Cabinet of Canine Curiosities* (Ithaca, NY: Cornell University Press, 2011); Helena Pycior, "The Public and Private Lives of 'First Dogs': Warren G. Harding's Laddie Boy and Franklin D. Roosevelt's Fala," in *Beastly Natures: Animals, Humans, and the Study of History*, ed. Dorothee Brantz (Charlottesville: University of Virginia Press, 2010), 176–203; Mark Derr, *A Dog's History of America: How Our Best Friend Explored, Conquered, and Settled a Continent* (New York: North Point Press, 2004); Amy Nelson, "The Legacy of Laika: Celebrity, Sacrifice, and the Soviet Space Dog," in *Beastly Natures: Animals, Humans, and the Study of History*, ed. Dorothee Brantz (Charlottesville: University of Virginia Press, 2010), 204–24; Marion Schwartz, *A History of Dogs in the Early Americas* (New Haven, CT: Yale University Press, 1977); James Boyce, "Canine Revolution: The Social and Environmental Impact of the Introduction of the Dog to Tasmania," *Environmental History* 11 (2006): 102–29; John Grier Varner and Jeannette Johnson Varner, *Dogs of the Conquest* (Norman: University of Oklahoma Press, 1983); John K. Walton, "Mad Dogs and Englishmen: The Conflict over Rabies in Late Victorian England," *Journal of Social History* 13 (1979): 219–39; Lance van Sittert and Sandra Swart, eds., *Canis Africanis: A Dog History of Southern Africa* (Leiden: Brill, 2008).

2. On dog domestication see, for example, Raymond Coppinger and Lorna Coppinger, *Dogs: A Startling New Understanding of Canine Origin, Behavior, and Evolution* (New York: Scribner, 2001); Carles Vilà et al., "Multiple and Ancient Origins of the Domestic Dog," *Science* 276 (1997): 1687–89; Ádám Miklósi, *Dog Behaviour, Evolution, and Cognition* (New York: Oxford University Press, 2008); James Serpell, ed., *The Domestic Dog: Its Evolution, Behaviour, and Interactions with People* (Cambridge: Cambridge University Press, 1995); Horowitz, *Inside of a Dog*; Mark Derr, *How the Dog Became the Dog: From Wolves to Our Best Friends* (New York: Overlook Press, 2011).
3. His answer to this question is Michael Pollan, *The Botany of Desire: A Plant's-Eye View of the World* (New York: Random House, 2001).
4. Katy M. Evans and Vicki J. Adams, "Proportion of Litters of Purebred Dogs Born by Caesarean Section," *Journal of Small Animal Practice* 51 (2010): 113–18.
5. Sultan inspired the title of the following, which is sadly not really about him at all: David Edmonds and John Eidinow, *Rousseau's Dog: Two Great Thinkers at War in the Age of Enlightenment* (New York: HarperCollins, 2006). As regards the early modern French connection to the Ottoman Empire in the realm of things animal, it is also noteworthy that the animal combat arena constructed in the royal gardens of Vincennes in the 1650s was dubbed the *sérial des bêtes sauvages* (the seraglio of wild beasts), a name—according to Peter Sahlins—purposefully chosen in part to evoke "the exotic world of the Ottoman palace." Peter Sahlins, "The Royal Menageries of Louis XIV and the Civilizing Process Revisited," *French Historical Studies* 35 (2012), 240–41.
6. On Blondie, Hitler, and his other dogs, see Boria Sax, *Animals in the Third Reich: Pets, Scapegoats, and the Holocaust* (New York: Continuum, 2000), 87–89; McHugh, *Dog*, 27 and 117.
7. The movie star Rin Tin Tin was apparently so popular that he garnered the most votes for the first-ever Best Actor Award from the Academy of Motion Pictures in 1929. It was thought, however, that offering the inaugural prize to a nonhuman was not a good precedent for the new award, so the prize was taken away from Rin Tin Tin and instead given to Emil

Jannings. Susan Orlean, *Rin Tin Tin: The Life and the Legend* (New York: Simon and Schuster, 2011), 88–89.

8. On these conflicting views of dogs see, for example, James Serpell, "From Paragon to Pariah: Some Reflections on Human Attitudes to Dogs," in *The Domestic Dog: Its Evolution, Behaviour, and Interactions with People*, ed. James Serpell (Cambridge: Cambridge University Press, 1995), 245–56; Belk, "Metaphoric Relationships with Pets."
9. The following blanket statement is typical: "To the Egyptian the dog is an unclean, abhorrent animal." Afaf Lutfi Al-Sayyid Marsot, "The Cartoon in Egypt," *Comparative Studies in Society and History* 13 (1971), 5. See also Shehada, *Mamluks and Animals*, 75. For a comparative perspective on Islam's supposed canine hostility, see Sophia Menache, "Dogs: God's Worst Enemies?" *Society and Animals* 5 (1997): 23–44.
10. As in many societies, the black dog in particular was regarded by Muslim writers as more troublesome than any other kind of dog. *Encyclopedia of Religion and Nature* (London: Thoemmes Continuum, 2005), s.v. "Dogs in the Islamic Tradition and Nature" (Khaled Abou El Fadl).
11. For comparative purposes, consider that from antiquity until at least the nineteenth century there was a very active dog trade between Tibet and what is today Uttar Pradesh in north central India. Maheshwar P. Joshi and C. W. Brown, "Some Dynamics of Indo-Tibetan Trade through Uttarākhaṇḍa (Kumaon-Garhwal), India," *Journal of the Economic and Social History of the Orient* 30 (1987), 308–09. On some of the utility of dogs in T'ang China, see Edward H. Schafer, "The Conservation of Nature under the T'ang Dynasty," *Journal of the Economic and Social History of the Orient* 5 (1962): 279–308.
12. None of the evidence for the human-dog relationship in Ottoman Egypt suggests that dogs had individual names. In ancient Egypt, by contrast, there is a record of dog names, with about eighty having been identified. Most of the names evoke the animal's physical attributes: color, speed, size, and so on. Brewer, "Hunting, Animal Husbandry and Diet in Ancient Egypt," 451–52; Kasia Szpakowska, *Daily Life in Ancient Egypt: Recreating Lahun* (Malden: Blackwell, 2008), 59–60. See also Angela M. J. Tooley, "Coffin of a Dog from Beni Hasan," *Journal of Egyptian Archaeology* 74 (1988): 207–11. The presence of dog names in ancient Egypt was likely a function of the fact that canines were sometimes kept as pets, a phenomenon that, as I will discuss later, was quite rare in the Ottoman period.
13. Abou El Fadl, "Dogs in the Islamic Tradition and Nature"; Ze'ev Maghen, "Dead Tradition: Joseph Schacht and the Origins of 'Popular Practice,'" *Islamic Law and Society* 10 (2003), 297–313.
14. Maghen, "Dead Tradition," 298.
15. The scholar of Islamic law Joseph Schacht thought that the notion of dog impurity in Islam was borrowed from Judaism. Joseph Schacht, *The Origins of Muḥammadan Jurisprudence* (Oxford: Clarendon Press, 1950), 216. Ze'ev Maghen disputes this idea, stating that nowhere in the Talmud are dogs cited as impure and that no evidence exists for the presence of this idea in Jewish practice between the seventh and the tenth centuries CE. Maghen believes that Schacht found notions of canine impurity in the writings of Sephardic scholars, who were themselves influenced by Islamic ideas of dog impurity. Thus, for Maghen, Schacht mistakenly took these borrowed ideas as representative of normative Jewish beliefs. According to Maghen, the evidence indeed overwhelmingly points to the exact opposite phenomenon: the influence of Islamic ideas about dogs on Sephardic Jews. Maghen, "Dead Tradition," 297–313. For a philosophical take on artistic representations of Jews and animals, see Andrew Benjamin, *Of Jews and Animals* (Edinburgh: Edinburgh University Press, 2010).
16. Quran 7: 176.
17. For a study comparing ideas about the impurity of pigs and dogs, see Maghen, "Dead Tradition," 297–313.
18. The most likely disease referent here is rabies. For the identification of certain parasites found in street dogs in Egypt during an anti-rabies campaign in the late twentieth century, see E. M. Mikhail, N. S. Mansour, and H. N. Awadalla, "Identification of *Trichinella* Isolates from Naturally Infected Stray Dogs in Egypt," *Journal of Parasitology* 80 (1994): 151–54.
19. Abou El Fadl, "Dogs in the Islamic Tradition and Nature"; Maghen, "Dead Tradition," 298–99.
20. Abou El Fadl, "Dogs in the Islamic Tradition and Nature."

21. Maghen, "Dead Tradition," 300; Ignaz Goldziher, "Islamisme et Parsisme," *Revue de L'Histoire des Religions* 43 (1901), 18. Goldziher's discussion of dogs during the Prophet's lifetime comes as part of a comparative study between Islam and Zoroastrianism. For a study of dogs in Zoroastrianism, see Mahnaz Moazami, "The Dog in Zoroastrian Religion: *Vidēvdād* Chapter XIII," *Indo-Iranian Journal* 49 (2006): 127–49. More generally on animals in Zoroastrianism, see idem., "Evil Animals in the Zoroastrian Religion," *History of Religions* 44 (2005): 300–17; Richard Foltz, "Zoroastrian Attitudes toward Animals," *Society and Animals* 18 (2010): 367–78; S. K. Mendoza Forrest, *Witches, Whores, and Sorcerers: The Concept of Evil in Early Iran* (Austin: University of Texas Press, 2011), 92–93, 104–06, 110–11, 118–20.
22. Abou El Fadl, "Dogs in the Islamic Tradition and Nature."
23. Ibid.
24. Ibid.
25. Quran 18: 18; *Encyclopaedia of Islam*, 2nd ed. (Leiden: Brill Online, 2013), s.v. "Kalb" (François Viré).
26. Perlo, *Kinship and Killing*, 195–96.
27. *Encyclopaedia of the Qur'ān* (Leiden: Brill Online, 2013), s.v. "Dog" (Bruce Fudge); Viré, "Kalb."
28. Robert Dankoff, "Animal Traits in the Army Commander," *Journal of Turkish Studies* 1 (1977), 99–104.
29. For a comparative perspective, see G. H. Bousquet, "Des animaux et de leur traitement selon le judaïsme, le christianisme et l'islam," *Studia Islamica* 9 (1958): 31–48.
30. For a capacious discussion of the Arabic literary works on animals by al-Jāḥiẓ, Qazvīnī, and al-Damīrī, see M. V. McDonald, "Animal-Books as a Genre in Arabic Literature," *Bulletin (British Society for Middle Eastern Studies)* 15 (1988): 3–10.
31. Abī ʿUthmān ʿAmr ibn Baḥr al-Jāḥiẓ, *Kitāb al-Ḥayawān*, ed. Muḥammad Bāsil ʿUyūn al-Sūd, 4 vols. (Beirut: Dār al-Kutub al-ʿIlmiyya, 1998); idem., *Kitāb al-Ḥayawān*, SK, Reisülküttab 584; idem., *Kitāb al-Ḥayawān*, pt. I, Beinecke, Landberg MSS 236.
32. For general discussions of the book, see Saʿīd Ḥ. Manṣūr, *The World-View of al-Jāḥiẓ in* Kitāb al-Ḥayawān (Alexandria: Dar al-Maareff, 1977); Charles Pellat, *The Life and Works of Jāḥiẓ, Translations of Selected Texts*, trans. D. M. Hawke (Berkeley: University of California Press, 1969), 21–22.
33. Qazvīnī's text is the following: Ḥamd Allāh Mustawfī Qazvīnī, *Nuzhat al-Qulūb* (Tehran: Dunyā-yi Kitāb, 1983/1984). The relevant animal portions of this text have been translated as idem., *Zoological Section of the Nuzhatu-l-Qulūb*. For al-Damīrī's text, see the following versions: Muḥammad ibn Mūsā al-Damīrī, *Ḥayāt al-Ḥayawān al-Kubrā*, ed. Ibrāhīm Ṣāliḥ, 4 vols. (Damascus: Dār al-Bashā'ir lil-Ṭibāʿa wa al-Nashr wa al-Tawzīʿ, 2005); idem., *Ḥayāt al-Ḥayawān al-Kubrā*, 2 vols., Beinecke, Salisbury MSS 78–79; idem., *Ḥayāt al-Ḥayawān*, SK, Harput 372; idem., *Ḥayāt al-Ḥayawān*, SK, Damad İbrahim 859; idem., *Ḥayāt al-Ḥayawān*, SK, Pertevniyal 763; idem., *Ad-Damîrî's Ḥayât al-Ḥayawân (A Zoological Lexicon)*, trans. A. S. G. Jayakar, 2 vols. (London: Luzac, 1906–08).
34. Pellat, *Life and Works of Jāḥiẓ*, 173.
35. Viré, "Kalb."
36. Pellat, *Life and Works of Jāḥiẓ*, 143.
37. For a discussion of the reverse phenomenon—human feeding dog—see Tuan, *Dominance and Affection*, 92–94. According to Tuan, Pacific Islanders in the nineteenth century considered the meat of dogs nursed by human women to be the best possible kind of dog meat.
38. This theme of animal-suckling human is not an uncommon one in Islamic literature. One of the most prominent examples is the story of Ḥayy Ibn-Yaqẓān, who was nursed in his infancy by a doe. Muḥammad ibn ʿAbd al-Malik Ibn Ṭufayl, *Ibn Tufayl's Hayy Ibn Yaqzān: A Philosophical Tale*, trans. Lenn Evan Goodman, 5th ed. (Los Angeles: Gee Tee Bee, 2003), 106, 109. For a recent study of this story as it relates to autodidacticism, see Avner Ben-Zaken, *Reading* Ḥayy Ibn-Yaqẓān: *A Cross-Cultural History of Autodidacticism* (Baltimore: Johns Hopkins University Press, 2011). For a broader view of interspecies suckling, see Londa Schiebinger, *Nature's Body: Gender in the Making of Modern Science* (New Brunswick, NJ: Rutgers University Press, 2004), 53–65.

39. Pellat, *Life and Works of Jāḥiẓ*, 143.
40. Ibid.
41. Muḥammad ibn Khalaf ibn al-Marzubān, *Faḍl al-Kilāb ʿalā Kathīr mimman Labisa al-Thiyāb*, Beinecke, Landberg MSS 350. I also consulted the following published Arabic version of the text and its English translation: idem., *The Book of the Superiority of Dogs over Many of Those who Wear Clothes*, trans. and ed. G. R. Smith and M. A. S. Abdel Haleem (Warminster: Aris and Phillips, 1978). All page number citations refer to the published Arabic text. Ibn al-Marzubān died in 921.
42. Ibid., 24.
43. Ibid.
44. Ibid., 33–35.
45. Ibid., 33. Unfortunately, we are not told the dog's name. Interestingly, the major account of the reign of the Mughal emperor Akbar in the second half of the sixteenth century states explicitly that Akbar loved dogs, imported them to India from many different locations, and gave names to all of them. Abul Fazl ʾAllami, *The Ain i Akbari*, trans. H. Blochmann, 3 vols. (Frankfurt am Main: Institute for the History of Arabic-Islamic Science at the Johann Wolfgang Goethe University, 1993), 1: 290.
46. Ibn al-Marzubān, *Superiority of Dogs*, 34.
47. For a useful comparative study that sheds light on the role of poisonous creatures in certain Islamic traditions, see Jürgen Wasim Frembgen, "The Scorpion in Muslim Folklore," *Asian Folklore Studies* 63 (2004): 95–123. For mention of a book about poisonous animals in Greco-Roman Egypt authored by Hermes the Second, see Ṣāʿid al-Andalusī, *Science in the Medieval World: "Book of the Categories of Nations,"* trans. and ed. Semaʿan I. Salem and Alok Kumar (Austin: University of Texas Press, 1991), 36.
48. This part of the story evokes earlier associations made between dogs and women—associations that were strongly repudiated. Abou El Fadl, "Dogs in the Islamic Tradition and Nature."
49. The dog-as-hero is of course a trope that exists in multiple traditions. For the case of twentieth-century Japan, for example, see Skabelund, *Empire of Dogs*, 130–70.
50. In al-Jāḥiẓ's borrowed Aristotelian schema, for example, animals who walk are the most esteemed and those who crawl or slither the most lowly. Aristotle writes that "to understand the world, we must understand mean and lowly things." For a very creative and illuminating account of human-animal relations that uses this aphorism as its organizing thread, see Kate Jackson, *Mean and Lowly Things: Snakes, Science, and Survival in the Congo* (Cambridge, MA: Harvard University Press, 2008).
51. In the story of the king and his pudding, the dog dies so that the king may live. If the dog is indeed between human and animal, then perhaps he also exists, as Christ did, between God and man. If the Christian idiom holds—an idiom I do not want to push too far—then the snake is clearly a symbol of the devil and his trickery in tempting and attempting to kill man. On this symbolism, see Frembgen, "The Scorpion in Muslim Folklore."
52. Ibn al-Marzubān, *Superiority of Dogs*, 55.
53. Jean-Claude Schmitt, *The Holy Greyhound: Guinefort, Healer of Children since the Thirteenth Century*, trans. Martin Thom (Cambridge: Cambridge University Press, 1983), 40–43.
54. For an Arabic version of this compendium, see Bīdpāʾī, *Kitāb Kalīla wa Dimna: Taʾlīf Bīdpā al-Faylasūf al-Hindī, Tarjamahu ilā al-ʿArabiyya fī Ṣadr al-Dawla al-ʿAbbāsiyya ʿAbd Allah ibn al-Muqaffaʿ* (Bombay: Sharaf al-Dīn al-Kutubī, 1968).
55. The legend and its role in medieval French peasant culture is the direct subject of Jean-Claude Schmitt's extremely provocative and useful work *The Holy Greyhound*.
56. Ibn al-Marzubān, *Superiority of Dogs*, 57.
57. For a useful study of the human ethics of the burial of dogs and other companionate species, see Philip Howell, "A Place for the Animal Dead: Pets, Pet Cemeteries and Animal Ethics in Late Victorian Britain," *Ethics, Place and Environment* 5 (2002): 5–22.
58. By the fourteenth century, the Persian polymath Qazvīnī had the following positive take on dogs as integral and constructive members of human societies: "It is a faithful beast, patient in undergoing hardships and in enduring hunger, in rendering service and in keeping off enemies. It follows up game with quick intelligence, and although kept hungry it is faithful and

will not leave its master, acknowledging the duty of obedience." Qazvīnī, *Zoological Section of the Nuzhatu-l-Qulūb*, 34. For an eleventh-century description of canines' physical characteristics, see Ibn Bakhtīshūʿ, *Ṭabāʾiʿ al-Ḥayawān wa al-Insān*, 32v–34r.

59. On the village of al-Ajhūr, see Ramzī, *al-Qāmūs al-Jughrāfī*, pt. 2., vol. 1: 53.
60. R. Y. Ebied and M. J. L. Young, "An Unpublished Legal Work on a Difference between the Shāfiʿites and Mālikites," *Orientalia Lovaniensia Periodica* 8 (1977), 252. This article includes the Arabic text of an (as of 1977) unpublished and uncatalogued manuscript from the University of Leeds, along with its English translation. All references are to the Arabic text.
61. On some of the many students al-Ajhūrī trained while at al-Azhar, see al-Jabartī, *ʿAjāʾib al-Āthār* (1994), 1: 107–10; 2: 168 and 242.
62. Ebied and Young, "A Difference between the Shāfiʿites and Mālikites," 254.
63. Ibid.
64. Ibid.
65. Ibid. In his eighth point, al-Ajhūrī asserts that the Prophet allowed the drinking and use for purification purposes of water from ponds between Mecca and Medina, despite the well-known fact that wild beasts (*al-sibāʿ*) regularly drank from them. Ibid., 255.
66. Ibid.
67. The line is: *al-kilāb luhātha lā yakhlū famuhā ʿan al-ruṭūbāt wa tasāquṭ al-luʿāb*. Ibid., 256.
68. Ibid., 256–57.
69. Ibid., 254.
70. Maghen, "Dead Tradition," 307.
71. Ebied and Young, "A Difference between the Shāfiʿites and Mālikites," 254.
72. Ibid., 255.
73. Ibid., 256.
74. Elsewhere in the Ottoman Empire, dogs were so integral to social life that they even played a role in place names. For example, the name of a village in the vicinity of Erzurum in eastern Anatolia was simply Köpek (Dog). For a case about the collection of taxes from Köpek in the early eighteenth century, see BOA, İbnülemin Askeriye 7691 (14 R 1118/26 July 1706).
75. Antonius Gonzales, *Voyage en Égypte du Père Antonius Gonzales, 1665–1666*, trans. Charles Libois, 2 vols. (Cairo: Institut français d'archéologie orientale, 1977), 2: 556. For more on this Franciscan in Egypt, see Alastair Hamilton, *The Copts and the West, 1439–1822: The European Discovery of the Egyptian Church* (Oxford: Oxford University Press, 2006), 80. For roughly contemporaneous descriptions of dogs elsewhere in the Ottoman Empire, see İrvin Cemil Schick, "Evliya Çelebi'den Köpeklere Dair," *Toplumsal Tarih* 202 (2010): 34–44.
76. It was not just Cairo that had a large number of dogs, but other Egyptian cities as well. About Damietta and its dogs, see Gonzales, *Voyage en Égypte*, 2: 556.
77. Ellis Veryard, *An Account of Divers Choice Remarks, as well Geographical, as Historical, Political, Mathematical, Physical, and Moral; Taken in a Journey through the Low-Countries, France, Italy, and Part of Spain; with the Isles of Sicily and Malta. As also, a Voyage to the Levant: A Description of Candia, Egypt, the Red Sea, the Desarts of Arabia, Mount-Horeb, and Mount-Sinai; the Coasts of Palestine, Syria, and Asia-Minor; the Hellespont, Propontis, and Constantinople; the Isles of the Carpathian, Egean, and Ionian Seas* (Exon: Sam. Farley, 1701), 319.
78. For a comparative example of the use of hogs for sanitation purposes in nineteenth-century New York City, see Catherine McNeur, "The 'Swinish Multitude': Controversies over Hogs in Antebellum New York City," *Journal of Urban History* 37 (2011): 639–60.
79. Gonzales, *Voyage en Égypte*, 1: 182.
80. Ibid.
81. Ibid., 1: 210; 2: 557.
82. This was also the case in Istanbul. Catherine Pinguet, "Istanbul's Street Dogs at the End of the Ottoman Empire: Protection or Extermination," in *Animals and People in the Ottoman Empire*, ed. Suraiya Faroqhi (Istanbul: Eren, 2010), 354–55.
83. Antonius Gonzales carefully notes that while dogs were allowed in mosques, they (and small children) were not allowed in churches for fear that these smallish humans and canines would disturb worshipers. Gonzales, *Voyage en Égypte*, 1: 222. For a study that shows how discourses about anti-cruelty in late nineteenth-century America drew commonalities

between animals and children, see Susan J. Pearson, *The Rights of the Defenseless: Protecting Animals and Children in Gilded Age America* (Chicago: University of Chicago Press, 2011).

84. al-Damardāshī, *Tārīkh Waqāyiʿ Miṣr al-Qāhira*, 115; Gonzales, *Voyage en Égypte*, 1: 22.
85. Gonzales, *Voyage en Égypte*, 2: 570.
86. In the Ottoman capital as well, dogs and humans were in such frequent contact in the early modern period that canines regularly came to serve as symbols and similes for individual humans and human communities. When in the middle of the seventeenth century, for example, a male page (*lala*) in the palace was nowhere to be found, the sultan issued a firman describing the servant's wanderings as being like those of a dog (*köpek gibi*). BOA, HAT 1446/34 (18 B 1058/8 Aug. 1648).
87. al-Shirbīnī, *Hazz al-Quḥūf*, 2: 238–40.
88. Ibid., 2: 238.
89. Ibid., 2: 239.
90. Ibid., 2: 240.
91. Viré, "Kalb." For a useful discussion of the social and economic importance of dogs as herders in the Indian Himalayas, see Joshi and Brown, "Dynamics of Indo-Tibetan Trade," 308–09.
92. al-Shirbīnī, *Hazz al-Quḥūf*, 2: 84.
93. Gonzales, *Voyage en Égypte*, 2: 557 and 614.
94. For examples of dog vomit imagery in Ottoman Egypt, see al-Shirbīnī, *Hazz al-Quḥūf*, 2: 367; al-Jabartī, *ʿAjāʾib al-Āthār* (1994), 4: 143.
95. Also in the realm of the medicinal, al-Jāḥiẓ reports that "dog dirt" is highly effective in the treatment of stab wounds. Pellat, *Life and Works of Jāḥiẓ*, 146. For examples of the medicinal uses of dog excrement in twentieth-century Yemen, see Hanne Schönig, "Reflections on the Use of Animal Drugs in Yemen," *Quaderni di Studi Arabi* 20/21 (2002–03), 169.
96. Chapter Five addresses hunting more thoroughly.
97. Edward William Lane, *An Account of the Manners and Customs of the Modern Egyptians: The Definitive 1860 Edition*, intro. Jason Thompson (Cairo: American University in Cairo Press, 2003), 95.
98. Gonzales, *Voyage en Égypte*, 2: 553.
99. Allsen, *Royal Hunt in Eurasian History*, 54–56.
100. For an analysis of the cultural representations of foxes in various traditions, see Hans-Jörg Uther, "The Fox in World Literature: Reflections on a 'Fictional Animal,'" *Asian Folklore Studies* 65 (2006): 133–60; Martin Wallen, *Fox* (London: Reaktion Books, 2006).
101. Allsen, *Royal Hunt in Eurasian History*, 239.
102. These breeds later became known as *leporarius* and *veltres* in the Latin West. Today's European whippets and wolfhounds are the direct descendants of these early Middle Eastern imports.
103. Ibn al-Marzubān, *Superiority of Dogs*, xxxii.
104. On Mamluk treatments of dog diseases, see: Shehada, *Mamluks and Animals*, 305–07.
105. Allsen, *Royal Hunt in Eurasian History*, 55–56.
106. Ibid., 56.
107. Ibid., 240–41.
108. Ibid., 240.
109. Elisabetta Borromeo, "The Ottomans and Hunting, according to Julien Bordier's Travelogue (1604–1612)," in *Animals and People in the Ottoman Empire*, ed. Suraiya Faroqhi (Istanbul: Eren, 2010), 227.
110. Allsen, *Royal Hunt in Eurasian History*, 240. This quote points again to the centrality of Iran to the global early modern animal trade.
111. Tülay Artan, "A *Book of Kings* Produced and Presented as a Treatise on Hunting," *Muqarnas* 25 (2008), 301.
112. al-Shirbīnī, *Hazz al-Quḥūf*, 2: 283.
113. Borromeo, "The Ottomans and Hunting," 225–26.
114. For more on these two hunting breeds, see François Viré, "À propos des chiens de chasse *salûqî* et *zagârî*," *Revue des études islamiques* 41 (1973): 331–40.
115. Each handler was responsible for the daily walking and grooming of two dogs.
116. *Polis à merveille*. Borromeo, "The Ottomans and Hunting," 226.

117. On the many uses of dogs in multiple arenas of warfare, see Michael G. Lemish, *War Dogs: Canines in Combat* (Washington, DC: Brassey's, 1996); Skabelund, *Empire of Dogs*, 130–70; Charles L. Dean, *Soldiers and Sled Dogs: A History of Military Dog Mushing* (Lincoln: University of Nebraska Press, 2005); Thomas R. Buecker, "The Fort Robinson War Dog Reception and Training Center, 1942–46," *Military History of the West* 38 (2008): 115–40; Cooper, *Animals in War*; Marilyn W. Seguin, *Dogs of War: And Stories of Other Beasts of Battle in the Civil War* (Boston: Branden, 1998); Abel A. Alves, *The Animals of Spain: An Introduction to Imperial Perceptions and Human Interaction with Other Animals, 1492–1826* (Leiden: Brill, 2011), 150–57; William W. Putney, *Always Faithful: A Memoir of the Marine Dogs of WWII* (New York: Free Press, 2001); Fairfax Davis Downey, *Dogs for Defense: American Dogs in the Second World War, 1941–45* (New York: Daniel P. McDonald, 1955).
118. *Saymāniyya* is the plural Arabized form of the Persian singular *segban* (it comes to Turkish as *sekban*), with *seg* meaning "dog" and *ban* "keeper." For a discussion of the etymology of various words for "dog" in Turkish, see Ahmet Caferoğlu, "Türk Onomastiğinde 'Köpek' Kültü," *Belleten* 25 (1961): 1–11.
119. Midhat Sertoğlu, *Osmanlı Tarih Lûgatı* (Istanbul: Enderun Kitabevi, 1986), 309; Daniel Crecelius and ʿAbd al-Wahhab Bakr, trans., *al-Damurdashi's Chronicle of Egypt, 1688–1755: al-Durra al-Musana fi Akhbar al-Kinana* (Leiden: Brill, 1991), 85, n. 251.
120. This use of dogs continues. In the 2011 American attack that killed Osama bin Laden in Pakistan, the commandos that stormed bin Laden's compound were aided by a group of canines who advanced them to sniff out any hidden explosives and to lead the way to the targets of the raid. Gardiner Harris, "A Bin Laden Hunter on Four Legs," *New York Times*, May 4, 2011. For an account of the U.S. military's use of dogs in Afghanistan and elsewhere, see Maria Goodavage, *Soldier Dogs: The Untold Story of America's Canine Heroes* (New York: Dutton, 2012).
121. For an early nineteenth-century example of *sekban* regiments using various animals for different military purposes, see TSMA, E. 1173/72 (13 C 1215/1 Nov. 1800).
122. Artan, "*A Book of Kings* as a Treatise on Hunting," 300. Into the early nineteenth century, the sixty-fourth and seventy-first regiments of the Ottoman janissary corps still regularly kept large numbers of mastiffs (s. *sansun*), terriers (s. *zağar*), and other dogs as part of their fighting forces. BOA, CS 6637 (13 M 1216/26 May 1801).
123. al-Damurdāshī, *al-Durra al-Muṣāna*, 109–12.
124. Ibid., 88–89.
125. For a slightly different version of the story of this dog attack, see al-Damardāshī, *Tārīkh Waqāyiʿ Miṣr al-Qāhira*, 160–61.
126. Pinguet, "Istanbul's Street Dogs," 370.
127. Gonzales, *Voyage en Égypte*, 1: 182.

4. Evolution in the Streets

1. For comparative examples of how urban dogs challenged state authority, see Walton, "Mad Dogs and Englishmen"; Jesse S. Palsetia, "Mad Dogs and Parsis: The Bombay Dog Riots of 1832," *Journal of the Royal Asiatic Society* 11 (2001): 13–30; İrvin Cemil Schick, "İstanbul'da 1910'da Gerçekleşen Büyük Köpek İtlâfı: Bir Mekan Üzerinde Çekişme Vakası," *Toplumsal Tarih* 200 (2010): 22–33; Jeffrey C. Sanders, "Animal Trouble and Urban Anxiety: Human-Animal Interaction in Post-Earth Day Seattle," *Environmental History* 16 (2011): 226–61.
2. For an account of nearly contemporaneous attempts to remove hogs from New York City, see McNeur, "'Swinish Multitude.'"
3. The historical literature on dogs as companionate species—pets—is enormous. Studies that have been most useful to me include Kete, *Beast in the Boudoir*; Fudge, *Pets*; Grier, *Pets in America*; Ritvo, "Pride and Pedigree"; Tuan, *Dominance and Affection*; Haraway, *When Species Meet*.
4. On dog evolution, see the studies cited in note 2 of Chapter Three.

5. Edmund Russell, *Evolutionary History: Uniting History and Biology to Understand Life on Earth* (New York: Cambridge University Press, 2011), 18.
6. For a discussion of some of the political and civilizational anxieties associated with street dogs, see McHugh, *Dog*, 127–45.
7. For some of the history of and conceptual problems involved in using dogs as guardians of other animals, see the following practical guide: Janet Vorwald Dohner, *Livestock Guardians: Using Dogs, Donkeys, and Llamas to Protect Your Herd* (North Adams, MA: Storey Publishing, 2007).
8. For a discussion and translation of this law, see Hiroshi Kato, "Egyptian Village Community under Muḥammad ʿAlī's Rule: An Annotation of 'Qānūn al-Filāḥa,'" *Orient* 16 (1980): 183–222.
9. As Rudolph Peters rightly directs, the Law of Agriculture should be considered in the context of the criminal legislation promulgated by Mehmet ʿAli's government that same year. Peters, "Meḥmed ʿAlī's First Criminal Legislation."
10. Kato, "Qānūn al-Filāḥa," 205.
11. A similar section of the 1830 law punished Bedouin who intentionally let their animals into peasants' fields. The punishment in such cases was a fine of 100 riyāls per faddan damaged. Ibid., 199.
12. Ibid., 195.
13. Ibid., 204–05.
14. Ibid., 206. This section appears nearly verbatim in the criminal law of 1829. Peters, "Meḥmed ʿAlī's First Criminal Legislation," 188.
15. Kato, "Qānūn al-Filāḥa," 198.
16. Peters, "Meḥmed ʿAlī's First Criminal Legislation," 165.
17. For a useful analysis of the logic and history of these kinds of state practices, see Scott, *Seeing Like a State.*
18. For accounts of Napoleon's invasion and occupation of Egypt, see Cole, *Napoleon's Egypt*; Raymond, *Égyptiens et Français.*
19. al-Jabartī, *ʿAjāʾib al-Āthār* (1994), 3: 51–52.
20. Visitors to Egypt often opined that it seemed dogs were especially aggressive toward them as foreigners and Christians. Visiting Rosetta in the early 1870s, James Morris Morgan, a former American Confederate naval officer, wrote: "I went out on to the street to look at the moon and take a little stroll. The only living thing I met was a pariah dog that snarled and disappeared through the entrance of a handsome house. While perfectly harmless to natives I knew that these wild dogs, especially when in packs, had a great aversion for Christians, and where one was met it was certain that there were many more near by, so I determined to return to my palace." James Morris Morgan, *Recollections of a Rebel Reefer* (Boston and New York: Houghton Mifflin, 1917), 294. For other examples of these ideas about Egyptian dogs' views of foreigners, see Frederick Henniker, *Notes during a Visit to Egypt, Nubia, the Oasis, Mount Sinai, and Jerusalem* (London: John Murray, 1823), 135; James Augustus St. John, *Egypt and Mohammed Ali; or, Travels in the Valley of the Nile*, 2 vols. (London: Longman, Rees, Orme, Brown, Green, and Longman, 1834), 2: 112. Dogs clearly know when they are encountering a person for the first time. Their abilities to perceive that person's religious affiliation, however, seem less acute. For a comparative example from T'ang China of the use of dogs to warn humans of strangers in their midst, see Schafer, "Conservation of Nature," 303.
21. For a comparative example of the use of poisoned meat to kill dogs in South Africa, see Tropp, "Dogs, Poison, and the Meaning of Colonial Intervention."
22. Dogs were also frequently victims of the various skirmishes between French and Ottoman soldiers that periodically erupted during the three years of the French occupation. In the spring of 1800, for example, there was a period of a few weeks that saw heavy fighting between French and Ottoman forces in Cairo. In addition to widespread plundering and looting in markets and elsewhere, many humans, dogs, and other animals perished as a result of these clashes. It was poor humans that suffered most from this violence. The chronicler al-Jabartī is, however, very clear in drawing an equivalence between these weaker members of

human society and the dogs and other animals that died alongside them. Thus we see in this case the persistence of the idea that certain kinds of humans and certain kinds of animals—in this example, the poor and dogs—were conceptualized as being very close to one another on a spectrum of living things. al-Jabartī, *ʿAjāʾib al-Āthār* (1994), 3: 158.

23. Aspects of this story bear resemblance to "the great cat massacre" in 1730s Paris analyzed by Robert Darnton. Especially important in both these instances of violence against animals was that the dogs and cats made noise at night, thereby robbing humans of precious sleep and preventing nighttime security marches. Robert Darnton, *The Great Cat Massacre and Other Episodes in French Cultural History* (New York: Basic Books, 1999), 75–104. For more on the problem of dogs as night, see Craig Koslofsky, *Evening's Empire: A History of the Night in Early Modern Europe* (Cambridge: Cambridge University Press, 2011), 202 and 218. On the history of the night, see also Wolfgang Schivelbusch, *Disenchanted Night: The Industrialization of Light in the Nineteenth Century*, trans. Angela Davies (Berkeley: University of California Press, 1988).
24. For a history of the phenomenon of urban dog killings in Europe, see McHugh, *Dog*, 130–34.
25. For an earlier example of the overcrowding of Cairo's streets with horses and other animals, see Rūmī, *Kitāb Taʾrīkh Miṣr wa al-Nīl*, 91r–91v. For a later example, see Letter from Arthur Rodgers to His Father, March 30, 1881, Arthur Rodgers Correspondence (1865–1902), BANC MSS 2004/224 cz, The Bancroft Library, University of California, Berkeley.
26. al-Jabartī, *ʿAjāʾib al-Āthār* (1998), 4: 437; idem., *ʿAjāʾib al-Āthār* (1994), 4: 396.
27. See, for example, Marsot, *Egypt in the Reign of Muhammad Ali*, 162–95.
28. On urban dogs and noise, see Alan M. Beck, *The Ecology of Stray Dogs: A Study of Free-Ranging Urban Animals* (West Lafayette, IN: NotaBell Books, 2002), 68–69.
29. Viré, "Kalb."
30. Pinguet, "Istanbul's Street Dogs," 355–56.
31. Some of Istanbul's residents interpreted the ship's sinking as divine retribution for attempts to rid the city of its dogs. A similar operation to relocate Istanbul's strays to an island near the city occurred during the reign of Sultan Abdülaziz (r. 1861–76). On this occasion, the dogs were successfully moved to the island of Sivriada. Still, in this instance too, a fire that destroyed parts of Istanbul shortly after this dog removal was interpreted by some as divine punishment for human cruelty to dogs. Ibid.
32. Beck, *Ecology of Stray Dogs*, 45–69.
33. al-Jabartī, *ʿAjāʾib al-Āthār* (1994), 4: 396.
34. On the emergence of this separation between species, see the following very important work: Keith Thomas, *Man and the Natural World: Changing Attitudes in England, 1500–1800* (London: Allen Lane, 1983).
35. On Seattle, see Sanders, "Animal Trouble and Urban Anxiety." On Istanbul, see Pinguet, "Istanbul's Street Dogs."
36. For a partly Freudian analysis of the "redefinition of dog shit as litter," see McHugh, *Dog*, 179–82.
37. For a history of sanitation in modern American cities, see Martin V. Melosi, *Garbage in the Cities: Refuse, Reform, and the Environment, 1880–1980* (College Station: Texas A&M University Press, 1981).
38. Catherine Pinguet, *Les chiens d'Istanbul: des rapports entre l'homme et l'animal de l'Antiquité à nos jours* (Saint-Pourçain-sur-Soule: Bleu autour, 2008); idem., "Istanbul's Street Dogs"; Schick, "Büyük Köpek İtlâfı"; Ekrem Işın, "Street Dogs of Istanbul: The Four-Legged Municipality," in *Everyday Life in Istanbul: Social Historical Essays on People, Culture and Spatial Relations*, trans. Virginia Taylor Saçlıoğlu (Istanbul: Yapı Kredi Yayınları, 2001), 219–26; Taner Timur, "XIX. Yüzyılda İstanbul'un Köpekleri," *Tarih ve Toplum* 20: 117 (September 1993): 10–14.
39. Abu-Lughod, *Cairo*, 83–97; Raymond, *Cairo*, 291–308; Khaled Fahmy, "An Olfactory Tale of Two Cities: Cairo in the Nineteenth Century," in *Historians in Cairo: Essays in Honor of George Scanlon*, ed. Jill Edwards (Cairo: American University in Cairo Press, 2002), 155–59. In the second half of the century, the segregation of Cairo (and Alexandria) into two quarters—one European and elite, the other native and poor—emerged as the most fundamental force driving the expansion of urban Egypt. Abu-Lughod, *Cairo*, 98–117; Raymond, *Cairo*,

309–39; Fahmy, "An Olfactory Tale," 155–87. On Alexandria, see idem., "Towards a Social History of Modern Alexandria," in *Alexandria, Real and Imagined*, ed. Anthony Hirst and Michael Silk (Aldershot, UK: Ashgate, 2004), 281–306; idem., "For Cavafy, with Love and Squalor: Some Critical Notes on the History and Historiography of Modern Alexandria," in *Alexandria, Real and Imagined*, ed. Anthony Hirst and Michael Silk (Aldershot, UK: Ashgate, 2004), 263–80.

40. Abu-Lughod, *Cairo*, 91–92 and 95–97; Raymond, *Cairo*, 303.
41. Mikhail, *Nature and Empire*, 230–41.
42. Abu-Lughod, *Cairo*, 91–93; Raymond, *Cairo*, 302–03.
43. For an illuminating study of urban dogs, see Beck, *Ecology of Stray Dogs*.
44. Sometimes in Cairo, but more commonly in the countryside, donkeys were also used to haul garbage. Tietze, *Muṣṭafā ʿĀlī's Description of Cairo*, 51.
45. Similar things happened to hog populations in New York City in the first half of the nineteenth century. McNeur, "'Swinish Multitude.'"
46. George Fleming, *Travels on Horseback in Mantchu Tartary: Being a Summer's Ride Beyond the Great Wall of China* (London: Hurst and Blackett, 1863), 233–34. Also cited in Skabelund, *Empire of Dogs*, 19.
47. About Istanbul's dogs, see in this regard Pinguet, "Istanbul's Street Dogs," 353–54.
48. Edward William Lane, *Cairo Fifty Years Ago*, ed. Stanley Lane-Poole (London: J. Murray, 1896), 35.
49. Ibid., 62; Abu-Lughod, *Cairo*, 86.
50. Abu-Lughod, *Cairo*, 86.
51. Lane, *Cairo Fifty Years Ago*, 37.
52. In Alexandria as well, trash mounds around the city's walls were understood to be the sovereign territory of the city's dogs. Moreover, turf wars over these mounds seem to have been a regular occurrence in Alexandria, with certain packs of dogs defending their mounds from other dogs and humans too. Henniker, *Notes during a Visit to Egypt*, 9–10. For a conceptual discussion of the use of the term "pack" to refer to dogs, see Horowitz, *Inside of a Dog*, 57–61.
53. On the competition among intramural dogs for food from urban trash in various Egyptian cities, see St. John, *Egypt and Mohammed Ali*, 2: 280; Henniker, *Notes during a Visit to Egypt*, 24. For the comparative example of dogs eating garbage in Ottoman Istanbul, see Pinguet, "Istanbul's Street Dogs," 354–55.
54. Abu-Lughod, *Cairo*, 92; Raymond, *Cairo*, 302–03; Fahmy, "An Olfactory Tale," 174.
55. For more on the removal of these mounds, see Lane, *Cairo Fifty Years Ago*, 37.
56. Thus as in many cities in the nineteenth century, Cairo's expansion was literally built on top of garbage. For the case of New York City, see Edward Humes, *Garbology: Our Dirty Love Affair with Trash* (New York: Avery, 2012), 45.
57. St. John, *Egypt and Mohammed Ali*, 1: 140–41.
58. Abu-Lughod, *Cairo*, 92.
59. For another example of clearing land for expanding urban construction projects, see al-Jabartī, *ʿAjāʾib al-Āthār* (1994), 4: 442–44.
60. St. John, *Egypt and Mohammed Ali*, 1: 141–42.
61. Consider, for example, what James Augustus St. John had to say about Cairo in December 1832 after being away from the city for a period. "I have not been many days in Cairo, and yet I discover that many changes have taken place in its appearance even since the descriptions of the very latest travellers were written. The streets, formerly disgustingly filthy, are now remarkable in general for their cleanliness, being all swept three times a day." Ibid., 1: 140.
62. For a useful analysis of changing understandings of disease and environment in the nineteenth century, see Linda Nash, *Inescapable Ecologies: A History of Environment, Disease, and Knowledge* (Berkeley: University of California Press, 2006).
63. For a useful comparative study of these issues, see the following examination of dogs in twentieth-century Baltimore: Beck, *Ecology of Stray Dogs*.
64. As garbage was increasingly moved outside of Cairo's walls, new extramural mounds developed and many dogs came to live near them. Decades later Cairo would grow beyond its walls and take over these spaces as well, pushing garbage and the city's dwindling dog population even farther away from the old city center. Following from Virginia DeJohn Anderson's

work on cattle in colonial America, Aaron Skabelund suggests that canines may usefully be thought of "as an 'advance guard' to expansion and as 'agents of empire.'" Skabelund, *Empire of Dogs*, 13. Indeed, in the case of Cairo's rapid nineteenth-century growth, the city certainly expanded toward areas conquered first by canines. Dogs were pushed outside the city's walls into new territory by a bulging urban population, only to be pushed even farther as the city followed them and eventually took over that land as well.

65. Abu-Lughod, *Cairo*, 86.
66. On competing Egyptian and European views of disease in nineteenth-century Egypt, see LaVerne Kuhnke, *Lives at Risk: Public Health in Nineteenth-Century Egypt* (Berkeley: University of California Press, 1990).
67. Other potentially problematic sources of disease, filth, and putridity were cesspools, slaughterhouses, fishmongers, cemeteries, and sites of regular public urination and defecation. Later in the late 1860s and 1870s, ʿAlī Mubārak, chief minister of public works, set about to remake Cairo to remove all of these danger zones. Fahmy, "An Olfactory Tale"; Mitchell, *Colonising Egypt*, 63–94.
68. Fahmy, "An Olfactory Tale," 166.
69. The medicalization of urban canine populations in the Ottoman Empire would continue to develop over the course of the nineteenth century and would eventually result in the establishment of, among other things, the Institute for Dog Medical Treatment (Darü'l Kelb Tedavihanesi) in Istanbul, a facility that fell under the administrative purview of the empire's general medical services. On the history of this canine institution, see BOA, Dahiliye İdare 55.5 (5 L 1328/ 10 Oct. 1910). Similar dog veterinary facilities would be established elsewhere in the empire as well. For example, see the following on the founding of a Canine Health Treatment Center (Köpek Hastalıkları Tedavihanesi) in the eastern Anatolian city of Erzincan in 1916: BOA, Dahiliye Nezareti Kalem-i Mahsûs Müdüriyeti 43.36 (14 Ca 1334/19 Mar. 1916). Moreover, like humans and other animal populations, dogs would increasingly become subject to quarantine efforts in Egypt and other parts of the empire in the late nineteenth and early twentieth centuries. For example, in 1923, dogs coming to Alexandria from parts of Europe were kept in quarantine for ten days due to fears of the spread of an unknown disease that was killing European dogs at the time. BOA, Hariciye Nezareti İstanbul Murahhaslığı 82.42 (1 Sept. 1923).
70. Already by the fall of 1820, Cairo's housing shortage was described as "critical." al-Jabartī, *ʿAjāʾib al-Āthār* (1994), 4: 444.
71. Lane, *Manners and Customs*, 284–85.
72. Indeed, at the linguistic and rhetorical levels as well, dogs seem to have been recoded in the early nineteenth century as particularly worthy of derision, disgust, hatred, and ultimately violence. Dog imagery was rarely used before the nineteenth century to insult or to illustrate the depravity, lowness, or disgusting character of humans. By the first few decades of the nineteenth century, however, we find quite a different story, with al-Jabartī, to take but one example, frequently using dog imagery and symbolism to critique human society. For instance, he describes those seeking customs concessions in the winter of 1812 as a bunch of "greedy dogs." al-Jabartī, *ʿAjāʾib al-Āthār* (1994), 4: 218. A few years later at the end of 1816, taxes on market goods were suspended for a short period during which people descended on the marketplace "like mad dogs" to secure meat, fruits, and vegetables. Idem., 4: 379. Elsewhere we read of treacherous soldiers as "fighting dogs" or of distrustful peasants as a pack of dogs. Idem., 3: 361; 4: 289. Later in the first half of the twentieth century, Egyptian social reformers sometimes referred to male youths milling about on the street as "stray dogs." Omnia El Shakry, "Youth as Peril and Promise: The Emergence of Adolescent Psychology in Postwar Egypt," *International Journal of Middle East Studies* 43 (2011), 599. The identification of people with dogs is, of course, now a common means of insult in Egypt and in many other (perhaps almost all?) societies.
73. Lane, *Manners and Customs*, 285.
74. Earlier instances of keeping dogs as pets in Egypt show that it was almost exclusively a practice of the foreign-acculturated elite. For instance, a Turkish woman was observed in her harem in the 1840s with a pug that was obviously kept as her pet. According to the British observer of this companionate relationship, who very clearly believed that all Muslims reviled dogs,

this woman kept "a little pug-dog, running about unconstrained all over the apartment, even upon the leewa'n, which is considered sacred, as being the usual place of prayer; but notwithstanding this little piece of uncleanness, was allowed full liberty, and every moment polluted the clothes of his mistress by rubbing up against them. She even patted him several times with her fair hand, and laughed at his barks and antics." James Augustus Saint John, *Egypt and Nubia* (London: Chapman and Hall, 1845), 239. For a statistical analysis of petkeeping attitudes in contemporary Kuwait, see Ghenaim Al-Fayez, Abdelwahid Awadalla, Donald I. Templer, and Hiroko Arikawa, "Companion Animal Attitude and its Family Pattern in Kuwait," *Society and Animals* 11 (2003): 17–28.

75. The British orientalist painter John Frederick Lewis (1804–1876) depicts a human-animal affective relationship in Egypt in his painting *A Lady Receiving Visitors (The Reception)*, 1873. This is an otherwise unremarkable rendition of the European male imaginary of the inside of an Egyptian harem. Important for our purposes though is that the scene shows one of the women in the harem keeping a pet gazelle. The woman and animal look at one another with a clearly knowing gaze. I have been unable to find further evidence of the claim in the caption of this painting's home institution, the Yale Center for British Art, that the gazelle was "a favorite Egyptian house pet." John Frederick Lewis, *A Lady Receiving Visitors (The Reception)*, 1873, Yale Center for British Art, Paul Mellon Collection, B1981.25.417.

    John Frederick Lewis lived in Egypt from 1841 to 1851. Much of his early career was devoted to depicting animal life, which he studied beginning at an early age during visits to menageries in Exeter Change and Windsor Great Park. The title of his first exhibited work was *A Donkey's Head*. After his time in Egypt, however, his career came to be defined mostly by his orientalist paintings of desert scenes, the harem, life in Cairo, and Egyptian antiquities. *Oxford Dictionary of National Biography* (Oxford: Oxford University Press, 2004), s.v. "Lewis, John Frederick (1804–1876)" (Kenneth Bendiner).

76. Lane, *Manners and Customs*, 286–87.

77. With no significant familial ties, this single woman was marked as socially marginal. Drawing connections between dogs and the socially liminal or oppressed (single people, the homeless, criminals, and so forth) is a common trope in literature, art, and film. McHugh, *Dog*, 157–70.

78. There was precedent in ancient Egypt for honoring dogs with proper burials. George A. Reisner, "The Dog which was Honored by the King of Upper and Lower Egypt," *Bulletin of the Museum of Fine Arts* 34 (1936): 96–99; Tooley, "Coffin of a Dog from Beni Hasan."

79. The human mourning and lamentation of dead dogs also had ancient antecedents in Egypt. In the fifth century BCE, Herodotus observed that "in whoever's house a cat dies naturally, those who dwell in the house all shave their eyebrows, but only these; if the dead animal is a dog, they shave all their body and head." Herodotus, *The History*, trans. David Grene (Chicago: University of Chicago Press, 1987), 2.66.

80. More generally on connections between human understandings of death and dogs, see McHugh, *Dog*, 18 and 39–48.

81. Lane, *Manners and Customs*, 286.

82. In true orientalist fashion, Lane offers the following sexist side comment about this: "For an Egyptian woman to keep a secret, and such a secret, was impossible." Ibid., 287.

83. On the commemorative practices of humans for their companion animals and what these practices reveal about the affective relationships between humans and animals, see Stanley Brandes, "The Meaning of American Pet Cemetery Gravestones," *Ethnology* 48 (2009): 99–118; Howard Williams, "Ashes to Asses: An Archaeological Perspective on Death and Donkeys," *Journal of Material Culture* 16 (2011): 219–39; Richard Chalfen, "Celebrating Life after Death: The Appearance of Snapshots in Japanese Pet Gravesites," *Visual Studies* 18 (2003): 144–56; Ingrid H. Tague, "Dead Pets: Satire and Sentiment in British Elegies and Epitaphs for Animals," *Eighteenth-Century Studies* 41 (2008): 289–306; Anna Chur-Hansen, "Cremation Services upon the Death of a Companion Animal: Views of Service Providers and Service Users," *Society and Animals* 19 (2011): 248–60; Howell, "A Place for the Animal Dead."

84. For a short history of dog cemeteries in Europe and the United States, see Bondeson, *Amazing Dogs*, 250–58. On the first American pet cemetery, see Edward C. Martin, Jr., *Dr. Johnson's*

Apple Orchard: The Story of America's First Pet Cemetery (Hartsdale, NY: Hartsdale Canine Cemetery, 1997).

85. Never mind that the ancient Egyptian god of the dead, Anubis, had the *canis* head of a jackal. McHugh, *Dog*, 40–41.
86. Generally on the problems of urban dog feces and urine, see Beck, *Ecology of Stray Dogs*, 53–59.
87. On this point, Thorstein Veblen writes in his classic 1899 study of bourgeois sensibilities that "he [the dog] is the filthiest of the domestic animals in his person and the nastiest in his habits. For this he makes up in a servile, fawning attitude towards his master, and a readiness to inflict damage and discomfort on all else. The dog, then, commends himself to our favor by affording play to our propensity for mastery, and as he is also an item of expense, and commonly serves no industrial purpose [in the capitalism of the nineteenth century], he holds a well-assured place in men's regard as a thing of good repute. The dog is at the same time associated in our imagination with the chase—a meritorious employment and an expression of the honorable predatory impulse." Thorstein Veblen, *The Theory of the Leisure Class* (Boston: Houghton Mifflin, 1973), 103.
88. On human interests in dog breeding and efforts to create canine breeds to fulfill certain human needs and desires, see McHugh, *Dog*, 58–126; Margaret E. Derry, *Bred for Perfection: Shorthorn Cattle, Collies, and Arabian Horses since 1800* (Baltimore: Johns Hopkins University Press, 2003), 48–102.
89. On dog domestication, see Horowitz, *Inside of a Dog*, 38–41; Russell, *Evolutionary History*, 63–66; Coppinger and Coppinger, *Dogs*; Vilà et al., "Multiple and Ancient Origins of the Domestic Dog"; Miklósi, *Dog Behaviour, Evolution, and Cognition*; Serpell, *Domestic Dog*; Derr, *How the Dog Became the Dog*.
90. On these points, in addition to the work cited in the previous note, see also McHugh, *Dog*, 19–25.
91. This point is made convincingly in Vilà et al., "Multiple and Ancient Origins of the Domestic Dog."
92. For an analysis of one particular Central Asian tale about wolves, see Namu Jila, "Myths and Traditional Beliefs about the Wolf and the Crow in Central Asia: Examples from the Turkic Wu-Sun and the Mongols," *Asian Folklore Studies* 65 (2006): 161–77. For a general cultural history of wolves, see Garry Marvin, *Wolf* (London: Reaktion Books, 2012).
93. Estimates for when dogs emerged as a distinct species range widely. Vilà et al., "Multiple and Ancient Origins of the Domestic Dog"; McHugh, *Dog*, 13.
94. D. K. Belyaev, "Destabilizing Selection as a Factor in Domestication," *Journal of Heredity* 70 (1979): 301–08; D. K. Belyaev, A. O. Ruvinsky, and L. N. Trut, "Inherited Activation-Inactivation of the Star Gene in Foxes: Its Bearing on the Problem of Domestication," *Journal of Heredity* 72 (1981): 267–74. For useful discussions of some of the important ramifications of Belyaev's experiments, see Lyudmila N. Trut, "Early Canid Domestication: The Farm-Fox Experiment," *American Scientist* 87 (1999): 160–69; Horowitz, *Inside of a Dog*, 35–37.
95. For a general discussion of canine genetics and its implications for both dogs and humans, see McHugh, *Dog*, 176–91.
96. Russell, *Evolutionary History*, 18.
97. Will Steffen, Paul J. Crutzen, and John R. McNeill, "The Anthropocene: Are Humans Now Overwhelming the Great Forces of Nature?" *Ambio* 36 (2007): 614–21.
98. Russell, *Evolutionary History*, 49; Paul J. Crutzen, "Geology of Mankind," *Nature* 415 (2002): 23.
99. For a historian's reflection on what the emergence of humans as a geological force means for the writing of history of, see Dipseh Chakrabarty, "The Climate of History: Four Theses," *Critical Inquiry* 35 (2009): 197–222.
100. Crutzen, "Geology of Mankind," 23.
101. Ibid.
102. Oman, Robock, Stenchikov, and Thordarson, "High-Latitude Eruptions."
103. One could, of course, make a similar point about other parts of the globe that witnessed massive social and political upheavals at the end of the eighteenth century. Events like the

French, Haitian, and American Revolutions; the Russo-Ottoman War; the first European settlement of Australia; the emergence of the Qajar Dynasty in Iran; the start of a fifty-year civil war in Tonga; and much else were all likely influenced—at least in part—by geophysical, climatic, and energetic changes associated with the beginnings of the Anthropocene.

104. In Edmund Russell's words, "We have accidentally shaped the evolution of populations by altering environments." Russell, *Evolutionary History*, 43.

105. Khālid Fahmī, "al-Ra'īs Mursī wa Akhlāqiyāt al-Naẓāfa," *Akhbār al-Adab*, August 14, 2012. Available at: http://www.jadaliyya.com/pages/index/6864/ (accessed January 17, 2013); Sarah El Deeb, "Egypt's Garbage Problem Continues to Grow," *Huffington Post*, September 1, 2012. http://www.huffingtonpost.com/2012/09/01/egypt-garbage-problem_n_1849254.html (accessed January 17, 2013).

5. Enchantment

1. For an illuminating discussion of this and other accounts of conversion, see Tijana Krstić, *Contested Conversions to Islam: Narratives of Religious Change in the Early Modern Ottoman Empire* (Stanford, CA: Stanford University Press, 2011), 75–84.
2. For another story of imperial rivalry involving a horse, see the following account of equestrian competition among the ruler of the island sultanate of Aceh, the Portuguese, and—through his gift of a horse—the Ottoman sultan: Casale, *Ottoman Age of Exploration*, 180–81.
3. Osman II (r. 1618–22) seems to have been the Ottoman sultan who loved horses most. He regularly hunted on horseback, wrote poetry about horses, and was often depicted in Ottoman paintings atop a horse. Advice literature written during his reign invoked equestrian themes and connected his political career to his hunting life. Most tellingly, his horse was buried on the grounds of the royal palace in Üsküdar complete with a headstone commemorating the animal's life. Tezcan, *Second Ottoman Empire*, 118–19. For a comparative perspective on the use of animal tombstones, see Brandes, "Meaning of American Pet Cemetery Gravestones."
4. On the emergence and history of the Ottoman-Safavid rivalry, see Adel Allouche, *The Origins and Development of the Ottoman-Ṣafavid Conflict (906–962/1500–1555)* (Berlin: K. Schwarz Verlag, 1983); Jean-Louis Bacqué-Grammont, *Les Ottomans, les Safavides et leurs voisins: contribution à l'histoire des relations internationales dans l'Orient islamique de 1514 à 1524* (Istanbul: Nederlands Historisch-Archaeologisch Instituut te Istanbul, 1987); Maḥmūd Ḥasan ʿAbd al-ʿAzīz al-Ṣarrāf, *Maʿrakat Chāldārān, 920 H/1514 M: Ūlā Ṣafaḥāt al-Ṣirāʿ al-ʿUthmānī al-Fārisī: al-Asbāb wa al-Natāʾij* (Cairo: Maktabat al-Nahḍa al-Miṣriyya, 1991).
5. Generally on Ottoman-Mughal relations, see Naimur Rahman Farooqi, *Mughal-Ottoman Relations: A Study of Political and Diplomatic Relations between Mughal India and the Ottoman Empire, 1556–1748* (Delhi: Idarah-i Adabiyat-i Delli, 1989); Bernard Lewis, "The Mughals and the Ottomans," in *From Babel to Dragomans: Interpreting the Middle East* (New York: Oxford University Press, 2004), 108–14; Muzaffar Alam and Sanjay Subrahmanyam, *Indo-Persian Travels in the Age of Discoveries, 1400–1800* (Cambridge: Cambridge University Press, 2007), 298–303. On the related topics of the Ottomans in the Indian Ocean and Ottoman-Portuguese rivalry, see Özbaran, *Ottoman Expansion towards the Indian Ocean*; idem., *Ottoman Response to European Expansion*; idem., "A Turkish Report on the Red Sea and the Portuguese in the Indian Ocean"; idem., "Ottoman Naval Power in the Indian Ocean in the 16th Century"; Casale, *Ottoman Age of Exploration*; idem., "Ottoman Administration of the Spice Trade"; Reid, "Sixteenth-Century Turkish Influence in Western Indonesia"; Tuchscherer, "La flotte impériale de Suez."
6. For a useful holistic treatment of the three major empires of the early modern Muslim world, see Stephen F. Dale, *The Muslim Empires of the Ottomans, Safavids, and Mughals* (Cambridge: Cambridge University Press, 2010).
7. The literature on Indian Ocean trade is obviously enormous. Most indispensible for me have been K. N. Chaudhuri, *Trade and Civilisation in the Indian Ocean: An Economic History from the Rise of Islam to 1750* (Cambridge: Cambridge University Press, 1985); idem., *The Trading*

World of Asia and the English East India Company, 1660–1760 (Cambridge: Cambridge University Press, 1978); Sanjay Subrahmanyam, *The Political Economy of Commerce: Southern India, 1500–1650* (Cambridge: Cambridge University Press, 1990); idem., ed., *Merchants, Markets and the State in Early Modern India* (Delhi: Oxford University Press, 1990); Ashin Das Gupta, *The World of the Indian Ocean Merchant, 1500–1800: Collected Essays of Ashin Das Gupta*, comp. Uma Das Gupta (New Delhi: Oxford University Press, 2001); idem., *Malabar in Asian Trade: 1740–1800* (Cambridge: Cambridge University Press, 1967); Pius Malekandathil, "Maritime Malabar and a Mercantile State: Polity and State Formation in Malabar under the Portuguese, 1498–1663," in *Maritime Malabar and the Europeans, 1500–1962*, ed. K. S. Mathew (Gurgaon: Hope India Publications, 2003), 197–227; James C. Boyajian, *Portuguese Trade in Asia under the Habsburgs, 1580–1640* (Baltimore: Johns Hopkins University Press, 1993); C. A. Bayly, *Imperial Meridian: The British Empire and the World, 1780–1830* (London: Longman, 1989); Willem Floor, *The Persian Gulf: A Political and Economic History of Five Port Cities, 1500–1730* (Washington, DC: Mage Publishers, 2006).

8. My thinking on gift-giving has been most influenced by Marcel Mauss, *The Gift: The Form and Reason for Exchange in Archaic Socieities*, trans. W. D. Halls (New York: W. W. Norton, 1990); Natalie Zemon Davis, *The Gift in Sixteenth-Century France* (Madison: University of Wisconsin Press, 2000); Gadi Algazi, Valentin Groebner, and Bernhard Jussen, eds., *Negotiating the Gift: Pre-Modern Figurations of Exchange* (Göttingen: Vandenhoeck and Ruprecht, 2003). See also the following manual on gifting: Aḥmad ibn al-Rashīd ibn al-Zubayr, *Book of Gifts and Rarities = Kitāb al-Hadāyā wa al-Tuḥaf*, trans. Ghāda al-Ḥijjāwī al-Qaddūmī (Cambridge, MA: Center for Middle Eastern Studies of Harvard University, 1996).
9. For a general discussion of ancient menageries, see R. J. Hoage, Anne Roskell, and Jane Mansour, "Menageries and Zoos to 1900," in *New Worlds, New Animals: From Menagerie to Zoological Park in the Nineteenth Century*, ed. R. J. Hoage and William A. Deiss (Baltimore: Johns Hopkins University Press, 1996), 8–15. For a broad perspective on the history of various diplomatic roles played by charismatic megafauna, see Liliane Bodson, ed., *Les animaux exotiques dans les relations internationales: espèces, fonctions, significations* (Liège: Université de Liège, 1998). For a popular account of the political history of charismatics across time and space, see Marina Belozerskaya, *The Medici Giraffe: And Other Tales of Exotic Animals and Power* (New York: Little, Brown and Company, 2006).
10. For a history of the cheetah in India, see Divyabhanusinh, *The End of a Trail: The Cheetah in India* (New Delhi: Banyan Books, 1995).
11. A. R. Williams, "Animal Mummies," *National Geographic* (November 2009): http://ngm.nationalgeographic.com/2009/11/animal-mummies/williams-text/1 (accessed March 5, 2012).
12. Hoage, Roskell, and Mansour, "Menageries and Zoos to 1900," 9.
13. Ibid.
14. On the history of animals in ancient Rome, see George Jennison, *Animals for Show and Pleasure in Ancient Rome* (Manchester: Manchester University Press, 1937).
15. On the Ptolemaic zoo, see Harry M. Hubbell, "Ptolemy's Zoo," *The Classical Journal* 31 (1935): 68–76.
16. Marco Polo, *The Travels of Marco Polo the Venetian*, trans. W. Marsden, rev. T. Wright, ed. Peter Harris (New York: Alfred A. Knopf, 2008), 98–99 and 137–44.
17. This collection included an elephant, a white bear, leopards, a giraffe, camels, lions, monkeys, and many other animals. Hoage, Roskell, and Mansour, "Menageries and Zoos to 1900," 12–13.
18. On the history of the Tower of London's zoological collection, see Daniel Hahn, *The Tower Menagerie: The Amazing 600-Year History of the Royal Collection of Wild and Ferocious Beasts Kept at the Tower of London* (New York: Jeremy P. Tarcher/Penguin, 2004).
19. Valerie Hansen, *The Open Empire: A History of China to 1600* (New York: W. W. Norton, 2000), 379–81.
20. Like many things in the Americas, this menagerie apparently came to an end with the Spanish conquests. Hoage, Roskell, and Mansour, "Menageries and Zoos to 1900," 12.
21. Ibid., 14–15. On Versailles, see also Sahlins, "Royal Menageries of Louis XIV." For a collection of animal images produced at the Versailles menagerie in the seventeenth century, see

Paola Gallerani, *The Menagerie of Pieter Boel: Animal Painter in the Age of Louis XIV*, trans. Catherine Bolton (Milan: Officina Libraria, 2011).

22. For an account of this elephant and the papal world he inhabited, see Silvio A. Bedini, *The Pope's Elephant* (Manchester: Carcanet Press, 1997). For the story of an albino elephant born to a human virgin in the Anatolian town of Turhal in the middle of the seventeenth century, see Evliyâ Çelebi, *Evliyâ Çelebi Seyahatnâmesi*, 3: 129–30.
23. For a short and very general sketch of the collection, see Metin And, *16. Yüzyılda İstanbul: Kent—Saray—Günlük Yaşam* (Istanbul: Yapı Kredi Yayınları, 2011), 129–32.
24. Allsen, *Royal Hunt in Eurasian History*, 204.
25. And, *16. Yüzyılda İstanbul*, 129.
26. For various travelers' comments about Istanbul's animal collection, see ibid., 129–32.
27. John Sanderson, *The Travels of John Sanderson in the Levant, 1584–1602: With his Autobiography and Selections from his Correspondence*, ed. Sir William Foster (London: Hakluyt Society, 1931), 57. Also cited in Gerald MacLean, "The Sultan's Beasts: Encountering Ottoman Fauna," in *Looking East: English Writing and the Ottoman Empire before 1800* (New York: Palgrave Macmillan, 2007), 152. For more on giraffes in Istanbul in the sixteenth century, see And, *16. Yüzyılda İstanbul*, 132. For a seventeenth-century Arabic travel account to Ethiopia commissioned by the Amir of Yemen al-Mutawakkil ʿAla Allah Ismāʿīl ibn al-Qāsim, see Sharaf al-Dīn al-Ḥasan ibn Aḥmad ibn Ṣāliḥ al-Yūsufī al-Jamālī al-Ḥaymī, *Riḥlat al-Ḥaymī ilā Arḍ al-Ḥabasha*, Veneranda Biblioteca Ambrosiana, Milan, ms. 35 B.
28. Sanderson, *Travels in the Levant*, 59–60. Also cited in MacLean, "The Sultan's Beasts," 153.
29. MacLean, "The Sultan's Beasts," 153.
30. For an analysis of India's central role in the early modern Eurasian economy, see Stephen Frederic Dale, *Indian Merchants and Eurasian Trade, 1600–1750* (Cambridge: Cambridge University Press, 1994).
31. See, for instance, S. D. Goitein and Mordechai Akiva Friedman, *India Traders of the Middle Ages: Documents from the Cairo Geniza, "India Book"* (Leiden: Brill, 2008); Ian Shaw, "Egypt and the Outside World," in *The Oxford History of Ancient Egypt*, ed. Ian Shaw (Oxford: Oxford University Press, 2000), 308–23; R. Stephen Humphreys, "Egypt in the World System of the later Middle Ages," in *Islamic Egypt, 640–1517*, vol. 1 of *The Cambridge History of Egypt*, ed. Carl F. Petry (Cambridge: Cambridge University Press, 1998), 445–61; Bernard Lewis, "The Fatimids and the Route to India," *Revue de la Faculté des Sciences Économiques de l'Université d'Istanbul* 11 (1949–1950): 50–54.
32. See, for example, the following cases about Venetian merchants in Egypt working to secure transport of spices, gold, and textiles from South Asia: ASV, Turchi, 554 (Evahir N 952/26 Nov.–5 Dec. 1545); ASV, Turchi, 714 (Evahir L 961/19–28 Sept. 1554); ASV, Algeri, Egitto, Marocco, 1, Documenti Egitto 1 (10 Ş 877/10 Jan. 1473). The following case concerns Armenian merchants in Egypt and their trading links across the eastern Mediterranean: ASV, Turchi, 1050 (Evahir R 1001/25 Jan.–3 Feb. 1593). For an analysis of Armenian trading networks in the early modern period, see Sebouh David Aslanian, *From the Indian Ocean to the Mediterranean: The Global Trade Networks of Armenian Merchants from New Julfa* (Berkeley: University of California Press, 2011). For an example of the commerce in salt between Venice and the Ottoman Balkans, see the Ottoman letter included in ASV, Senato, Deliberazioni, Costantinopoli, 1 (Evahir L 965/6–15 Aug. 1558).
33. al-Jabartī, *ʿAjāʾib al-Āthār* (1994), 4: 283; Hanna, *Making Big Money*, 34–35, 58, 77.
34. On the horse trade from the Arabian Peninsula and Persian Gulf to South Asia, see Ralph Kauz, "Horse Exports from the Persian Gulf until the Arrival of the Portuguese," in *Pferde in Asien: Geschichte, Handel Und Kultur*, ed. Bert G. Fragner, Ralph Kauz, Roderich Ptak, and Angela Schottenhammer (Vienna: Österreichischen Akademie der Wissenschaften, 2009), 137–52; Rudi Manuel Loureiro, "Portuguese Involvement in Sixteenth Century Horse Trade through the Arabian Sea," in *Pferde in Asien: Geschichte, Handel Und Kultur*, ed. Bert G. Fragner, Ralph Kauz, Roderich Ptak, and Angela Schottenhammer (Vienna: Österreichischen Akademie der Wissenschaften, 2009), 203–30; Ranabir Chakravarti, "Horse Trade and Piracy at Tana (Thana, Maharashtra, India): Gleanings from Marco Polo," *Journal of the Economic and Social History of the*

Orient 34 (1991): 159–82; Hedda Reindl-Kiel, "No Horses for the Enemy: Ottoman Trade Regulations and Horse Gifting," in *Pferde in Asien: Geschichte, Handel Und Kultur*, ed. Bert G. Fragner, Ralph Kauz, Roderich Ptak, and Angela Schottenhammer (Vienna: Österreichischen Akademie der Wissenschaften, 2009), 43–49; William G. Clarence-Smith, "Elephants, Horses, and the Coming of Islam to Northern Sumatra," *Indonesia and the Malay World* 32 (2004): 271–84.

On some of the military and symbolic uses of Arabian and Persian horses in the Vijayanagara Empire, whose capital was the largest and most important city in pre-colonial peninsular India, see Carla M. Sinopoli, "From the Lion Throne: Political and Social Dynamics of the Vijayanagara Empire," *Journal of the Economic and Social History of the Orient* 43 (2000), 377–80; John M. Fritz, "Vijayanagara: Authority and Meaning of a South Indian Imperial Capital," *American Anthropologist* 88 (1986): 44–55; Velcheru Narayana Rao and Sanjay Subrahmanyam, "Notes on Political Thought in Medieval and Early Modern South India," *Modern Asian Studies* 43 (2009): 175–210; A. Rangasvami Sarasvati, "Political Maxims of the Emperor Poet, Krishnadeva Raya," *Journal of Indian History* 6 (1925), 69 and 72, cited in Catherine B. Asher and Cynthia Talbot, *India before Europe* (Cambridge: Cambridge University Press, 2006), 77. On the history and importance of the city of Vijayanagara, see Anna Libera Dallapiccola, ed., with Stephanie Zingel-Avé Lallemant, *Vijayanagara, City and Empire: New Currents of Research*, 2 vols. (Stuttgart: Steiner Verlag Wiesbaden, 1985); Anila Verghese and Anna L. Dallapiccola, eds., *South India under Vijayanagara: Art and Archaeology* (New York: Oxford University Press, 2011); Fritz, "Authority and Meaning"; Asher and Talbot, *India before Europe*, 59–60 and 63; Alam and Subrahmanyam, *Indo-Persian Travels*, 67–79.

The major account of the reign of the sixteenth-century Mughal emperor Akbar, *Ain-i-Akbari*, makes clear that horses from the Ottoman and Persian worlds continued to play a central role in Mughal warfare throughout the sixteenth century. Abul Fazl 'Allami, *Ain i Akbari*, 1: 132–42, 215, 233–34. Horses from Iraq and Turan were well-known in Mughal India into the seventeenth century as well. See the following discussion with emperor Jahangir just before his death in 1627: Mutribi al-Asamm Samarqandi, *Conversations with Emperor Jahangir*, trans. Richard C. Foltz (Costa Mesa: Mazda Publishers, 1998), 37. Horses were also effectively employed for didactic and symbolic purposes in several South Indian histories written in this period. Sanjay Subrahmanyam, "Reflections on State-Making and History-Making in South India, 1500–1800," *Journal of the Economic and Social History of the Orient* 41 (1998): 382–416. For an analysis of the horse trade in the Indian Ocean and Persian Gulf in the eighteenth and nineteenth centuries, see Hala Fattah, *The Politics of Regional Trade in Iraq, Arabia, and the Gulf, 1745–1900* (Albany: State University of New York Press, 1997), 159–83.

A great number of the Arabian and Persian horses transported to India were eventually traded to China. See, for example, the following account of the trade in horses and elephants from Sri Lanka to China in the thirteenth century: Rugua Zhao, *Chau Ju-Kua: His Work on the Chinese and Arab Trade in the Twelfth and Thirteenth Centuries, entitled Chu-Fan-Chï*, trans. Friedrich Hirth and W. W. Rockhill, 2 vols. (New York: Paragon Book Reprint Corp., 1966), 73. See also the following record of horses moving from Samarkand to Ming China: Li Gouxiang, Yang Chang, Wang Yude, et al., eds., *Ming Shi Lu Lei Zuan: She Wai Shi Liao Juan* (Wuhan Shi: Wuhan Chu Ban She: Xin Hua Shu Dian Jing Xiao, 1991), 1063–87. Generally on the trade in horses to China in various periods, see Yokkaichi Yasuhiro, "Horses in the East-West Trade between China and Iran under Mongol Rule," in *Pferde in Asien: Geschichte, Handel Und Kultur*, ed. Bert G. Fragner, Ralph Kauz, Roderich Ptak, and Angela Schottenhammer (Vienna: Österreichischen Akademie der Wissenschaften, 2009), 87–97; S. Jagchid and C. R. Bawden, "Some Notes on the Horse Policy of the Yüan Dynasty," *Central Asiatic Journal* 10 (1965): 246–68; Morris Rossabi, "The Tea and Horse Trade with Inner Asia During the Ming," *Journal of Asian History* 4 (1970): 136–68; Christopher I. Beckwith, "The Impact of the Horse and Silk Trade on the Economies of T'ang China and the Uighur Empire: On the Importance of International Commerce in the Early Middle Ages," *Journal of the Economic and Social History of the Orient* 34 (1991): 183–98; Roderich Ptak, "Pferde auf See: Ein vergessener Aspekt des maritimen chinesischen Handels im frühen 15.

Jahrhundert," *Journal of the Economic and Social History of the Orient* 34 (1991): 199–233; Ranabir Chakravarti, "Early Medieval Bengal and the Trade in Horses: A Note," *Journal of the Economic and Social History of the Orient* 42 (1999): 194–211; Bin Yang, "Horses, Silver, and Cowries: Yunnan in Global Perspective," *Journal of World History* 15 (2004), 294–300.

35. On the Egyptian demand for Indian products other than animals, see al-Jabartī, *ʿAjāʾib al-Āthār* (1994), 1: 50, 52, 553; 2: 120, 329–30; 4: 233.
36. Ibid., 3: 67, 94, 255.
37. Ibid., 4: 361.
38. Ibid., 4: 389.
39. Ibid., 4: 393.
40. On the place of monkeys in Islamic intellectual history, see Remke Kruk, "Traditional Islamic Views of Apes and Monkeys," in *Ape, Man, Apeman: Changing Views since 1600*, ed Raymond Corbey and Bert Theunissen (Leiden: Leiden University, 1995), 29–42; Michael Cook, "Ibn Qutayba and the Monkeys," *Studia Islamica* 89 (1999): 43–74. For an eleventh-century description of monkeys' physical attributes, see Ibn Bakhtīshūʿ, *Ṭabāʾiʿ al-Ḥayawān wa al-Insān*, 36v.
41. Among non-animal goods, one of the primary products to buck the usual unidirectionality of Indian Ocean trade—from India to Egypt—was gold. Thanks to its connections to the West African gold trade, Egypt had ample supplies to export east, but of course only when the gold-rich South Asian market allowed for this. Humphreys, "Egypt in the World System of the later Middle Ages," 454–55.
42. Michael Gorgas, "Animal Trade Between India and Western Eurasia in the Sixteenth Century—The Role of the Fuggers in Animal Trading," in *Indo-Portuguese Trade and the Fuggers of Germany, Sixteenth Century*, ed. K. S. Mathew (New Delhi: Manohar, 1997), 195–96.
43. Humphreys, "Egypt in the World System of the later Middle Ages," 449; Janet L. Abu-Lughod, *Before European Hegemony: The World System A.D. 1250–1350* (New York: Oxford University Press, 1989).
44. Casale, *Ottoman Age of Exploration*, 91–92.
45. China was also sometimes linked to the Ottoman Empire through the early modern trade in charismatic megafauna. In the 1520s, for example, Ming emperor Jiajing received in his court a group of rhinoceroses, lions, and horses. The envoys who brought these animals as gifts were identified as being Rumis, a suggestion that they were perhaps from the Ottoman Empire, though this is far from certain. That they brought lions with them suggests that they were perhaps from Iran or somewhere in Central Asia. This uncertainty aside, the Chinese court seems to have refused the animals on several grounds. Not knowing how to care for them, the Chinese would have had to bring in "barbarian" non-Chinese to look after the animals. Salaries for these handlers and the costs of feeding and maintaining the animals would be high, and it was determined to be of little use to keep the animals since they seemed to serve no productive function. Moreover, refusing such gifts was said to show great virtue. Gouxiang, Chang, Yude, et al., *Ming Shi Lu Lei Zuan*, 1060–62. On Chinese knowledge of the Turks, see Ayşe Onat, *Çin Kaynaklarında Türkler: Han Hanedanı Tarihinde "Batı Bölgeleri"* (Ankara: Türk Tarih Kurumu Basımevi, 2012). On Ottoman knowledge of and interest in China, see Mevhibe Pınar Emiralioğlu, "Cognizance of the Ottoman World: Visual and Textual Representations in the Sixteenth-Century Ottoman Empire (1514–1596)" (Ph.D. diss., University of Chicago, 2006), 181–222; Pınar Emiralioğlu, "Relocating the Center of the Universe: China and the Ottoman Imperial Project in the Sixteenth Century," *Journal of Ottoman Studies* 39 (2012): 161–87. The active horse and elephant trade in Southeast Asia was also often connected to Ottoman, Safavid, and Mughal economic networks. Clarence-Smith, "Elephants, Horses, and the Coming of Islam."
46. As the fourteenth-century Persian writer Qazvīnī makes clear, interest in elephants was largely a legacy of the three empires' shared Mongol past. In explaining the importance of elephants, Qazvīnī offers the following snapshot of some of the zoological and medicinal understandings of the animal current in fourteenth-century Iran. "It is bigger than most other animals, and more sagacious. And notwithstanding the bulk of its body, and that it has no joints except at the shoulder, it moves actively. And since its height is very great, and its

neck extremely short, God most high has given it a trunk, fleshy and without bone, in which he has planted a faculty such that it can do everything with it… There are many elephants in Qandahār, and the tallest elephants are in Aghbāb. And the elephant and the lion are enemies; and the snake is the enemy of the young elephant. And when the elephant is sick, it eats a snake, and recovers. Its age reaches 300 and 400 years. It never sleeps on its side, because having no joints, if it were to sleep on its side it could not get up, and would perish; hence it sleeps standing… More than twenty men can sit on the back of one; and they reckon one elephant with the men on its back as equal to a thousand horsemen…

If its earwax is given in drink to anybody with his food, he will not go to sleep for a week. If its bile be used as an ointment for three days on a (maculo-anaesthetic) leprous patch, this will disappear. Smelling its fat causes tubercular leprosy. Its bones are ivory, which commands a high price; and the teeth are the best of its bones; and if ivory be tied round a child's neck, it will be safe against epilepsy. The smoke of its bones will make sweet the fruit of a sour tree, and will keep off worms and [other] pests from the land, and will drive mosquitos out of a house; scrapings of ivory scattered on a wound or a burn will bring about a cure. To sleep on an elephants' skin will cure convulsions; and the smoke of its skin will take away piles. If its urine be sprinkled in a house, rats will leave it. The smoke from its dung will cure fever and colic." Qazvīnī, *Zoological Section of the Nuzhatu-l-Qulūb*, 23–24. See also the following eleventh-century account of elephants: Ibn Bakhtīshū', *Ṭabā'i' al-Ḥayawān wa al-Insān*, 14v–15v. For an example of the gifting of elephant tusks in seventeenth-century Ottoman Egypt, see Evliyâ Çelebi, *Evliyâ Çelebi Seyahatnâmesi*, 10: 441.

47. Elephants were involved in diplomacy in earlier periods and different places as well. At the end of the tenth century, for example, the clearly very impressed court of the Chinese Emperor T'ai-tsung asked a visiting Arab embassy how it was that it captured elephants. One of the ambassadors replied, "To capture elephants, we use decoy elephants to get so near them that we can catch them with a big lasso." Rugua Zhao, *Chinese and Arab Trade*, 117–18.
48. Hedda Reindl-Kiel, "Dogs, Elephants, Lions, a Ram and a Rhino on Diplomatic Mission: Animals as Gifts to the Ottoman Court," in *Animals and People in the Ottoman Empire*, ed. Suraiya Faroqhi (Istanbul: Eren, 2010), 280. This elephant came to Istanbul during a period of intense administrative correspondence and military maneuvering between the Ottoman Empire and Iran. Animals were integral to this back and forth between the two imperial states as they aided in the movement of military supplies, emissaries, and various envoys from Ottoman lands to Iran. In 1731, for example, five hundred mules were collected to move men and equipment toward the Iranian border. BOA, Cevdet Askeriye 16011 (3 M 1144/8 July 1731). And in 1735 and 1741, arrangements were made to move barley and other foodstuffs for the horses and other animals transporting the sultan's troops and emissaries to the Ottomans' eastern front. BOA, Cevdet Askeriye 28573 (29 C 1148/16 Nov. 1735); BOA, Cevdet Hariciye 3089 (2 Ra 1154/18 May 1741).
49. On the gifting of elephants in early modern South India, see Gorgas, "Animal Trade Between India and Western Eurasia," 206–11.
50. On the material and symbolic importance of elephants in pre-Mughal South Asia, see Thomas R. Trautmann, "Elephants and the Mauryas," in *India, History and Thought: Essays in Honour of A.L. Basham*, ed. S. N. Mukherjee (Calcutta: Subarnarekha, 1982), 254–81; Simon Digby, *War-horse and Elephant in the Delhi Sultanate: A Study of Military Supplies* (Oxford: Orient Monographs, 1971).
51. See, for example, the relevant sections of Babur, *The Baburnama: Memoirs of Babur, Prince and Emperor*, trans. and ed. Wheeler M. Thackston, intro. Salman Rushdie (New York: Modern Library, 2002); Jahangir, *The Jahangirnama: Memoirs of Jahangir, Emperor of India*, trans. and ed. Wheeler M. Thackston (Washington, DC: Freer Gallery of Art; New York: Oxford University Press, 1999); Abul Fazl 'Allami, *Ain i Akbari*.
52. On the role of elephants and other animals in the Mughal court, see Annemarie Schimmel, *The Empire of the Great Mughals: History, Art and Culture*, trans. Corinne Attwood, ed. Burzine K. Waghmar (London: Reaktion Books, 2004), 213–23. See also Asok Kumar Das, "The Elephant in Mughal Painting," in *Flora and Fauna in Mughal Art*, ed. Som Prakash Verma (Mumbai: Marg Publications, 1999), 36–54; Sujit Sivasundaram, "Trading Knowledge: The

East India Company's Elephants in India and Britain," *The Historical Journal* 48 (2005): 27–63. Like the Ottomans and Safavids, the rulers of Aceh in the early modern period also learned a great deal from the Mughals about the royal uses of elephants. Clarence-Smith, "Elephants, Horses, and the Coming of Islam," 274 and 279.

53. For an account of Seydi ʿAli Reis's visit to the Mughal court, see Alam and Subrahmanyam, *Indo-Persian Travels*, 111–20. See also Palmira Brummett, "What Sidi Ali Saw," *Portuguese Studies Review* 9 (2002): 232–53; Casale, *Ottoman Age of Exploration*, 120–23.
54. On the uses of horses (and other animals) in Safavid royal pageantry, see Willem Floor, "The Talar-i Tavila or Hall of Stables, a Forgotten Safavid Palace," *Muqarnas* 19 (2002): 19–63.
55. Reindl-Kiel, "Animals as Gifts to the Ottoman Court," 279–80.
56. Karl Saurer and Elena Hinshaw-Fischli, "They Called Him Suleyman: The Adventurous Journey of an Indian Elephant from the Forests of Kerala to the Capital of Vienna in the Middle of the Sixteenth Century," in *Maritime Malabar and the Europeans, 1500–1962*, ed. K. S. Mathew (Gurgaon: Hope India Publications, 2003), 153–64; Annemarie Jordan Gschwend, *The Story of Süleyman: Celebrity Elephants and Other Exotica in Renaissance Portugal* (Philadelphia: Pachyderm, 2010).
57. Saurer and Hinshaw-Fischli, "They Called Him Suleyman," 154–55.
58. Both during his life and after his death in 1553, Süleyman was beloved in Vienna. Homes decorated their facades with elephant motifs; "to the elephant" became a popular name for coffeehouses, guesthouses, and other businesses; stories, verse, and legend developed around Süleyman, his life, and his journey to Vienna; artists depicted him; and several of the city's leading scientists began to write about elephants. Ibid., 159–60.
59. Not surprisingly, the Ottomans were not the first dynasts to employ elephants to display royal power in the city. The sixth-century Byzantine Hebdomon Palace had a large cistern—which the Ottomans called *fil damı* (elephant stable)—that housed a group of elephants who were brought to the palace on special occasions. There were other elephant stables in Byzantine Constantinople as well, some of which were eventually converted into mosques. The mosque of the Tekke of Hekim Çelebi in Koska, for example, used to be an elephant stable. Hafiz Hüseyin Ayvansarayî, *The Garden of the Mosques: Hafiz Hüseyin Ayvansarayî's Guide to the Muslim Monuments of Ottoman Istanbul*, trans. Howard Crane (Leiden: Brill, 2000), 101.
60. MacLean, "The Sultan's Beasts," 151.
61. Reindl-Kiel, "Animals as Gifts to the Ottoman Court," 280.
62. About the tiger, Qazvīnī had this to say: "It is a tyrannical and arrogant animal, very strong, swift, a [good] jumper, and of handsome appearance. Its back is extremely weak, and gives way if it feels but a slight degree of pain. It and the snake are close friends, but it is at enmity with all other animals. When it kills its prey it sleeps three days and nights, and on the fourth day it goes forth to hunt again." Qazvīnī, *Zoological Section of the Nuzhatu-l-Qulūb*, 35. For an illuminating study of the human-tiger relationship in parts of Southeast Asia, see Peter Boomgaard, *Frontiers of Fear: Tigers and People in the Malay World, 1600–1950* (New Haven, CT: Yale University Press, 2001).
63. Parts of the next few pages lean heavily on Suraiya Faroqhi, "Exotic Animals at the Sultan's Court," in *Another Mirror for Princes: The Public Image of the Ottoman Sultans and Its Reception* (Piscataway: Gorgias Press; Istanbul: Isis Press, 2009), 87–101.
64. Ibid., 96–97.
65. As part of the celebration of the entry of Süleyman the elephant to Trent while on his way to Vienna in late 1551, there were likewise wooden sculptures of the animal constructed in which fireworks were shot out of the fake pachyderms' nostrils. Saurer and Hinshaw-Fischli, "They Called Him Suleyman," 156.
66. For a useful comparative analysis of the role of elephants in Britain in roughly the same period, see Christopher Plumb, "'Strange and Wonderful': Encountering the Elephant in Britain, 1675–1830," *Journal for Eighteenth-Century Studies* 33 (2010): 525–43.
67. For a discussion of Mughal elephant paintings, see Das, "The Elephant in Mughal Painting." Another example of the very clear connections the Ottomans drew between elephants and India is found in Deli Birader's sixteenth-century account of bestiality. "It is said that in India, people fuck elephants. When they desire an elephant, one man climbs on the shoulders of

another. Then they put not only their cocks into the elephant's vagina, but their balls." Kuru, "Deli Birader," 254.

68. O. Kurz, *European Clocks and Watches in the Near East* (Leiden: Brill, 1975), 28–29; Faroqhi, "Exotic Animals," 97. For further discussion of the ubiquity of elephant imagery in Renaissance Europe, see Robert Elena Hinschaw-Fischli, "Some Italian and German Representations of India in the Early Renaissance—with a Predilection for Elephants," in *Maritime Malabar and the Europeans, 1500–1962*, ed. K. S. Mathew (Gurgaon: Hope India Publications, 2003), 165–96.
69. Kurz, *European Clocks and Watches*, 29.
70. Abī al-ʿIzz Ismāʿīl al-Jazarī, *al-Jāmiʿ bayna al-ʿIlm wa al-ʿAmal al-Nāfiʿ fī Ṣināʿat al-Ḥiyal*, ed. Aḥmad Yūsuf al-Ḥasan, with ʿImād Ghānim, Mālik al-Mallūḥī, and Muṣṭafā Taʿmurī (Aleppo: Jāmiʿat Ḥalab, Maʿhad al-Turāth al-ʿIlmī al-ʿArabī, 1979). For an English translation of this text, see Ismāʿīl ibn al-Razzāz al-Jazarī, *The Book of Knowledge of Ingenious Mechanical Devices: Fī Maʿrifat al-Ḥiyal al-Handasiyya*, trans. Donald R. Hill (Dordrecht: Reidel, 1974). See also Kurz, *European Clocks and Watches*, 8–9. The operation of an elephant clock is described as follows. When the half-hour strikes, "the bird on top begins to sing, the cupola turns, the man at the top left moves his arm, the falcon drops a metal pellet from his beak into the mouth of the upper dragon, the two dragons move, and the pellet finally lands in the small vase, the mahout moves his ankus [elephant goad] and beats the drum on the elephant's head, while the scribe turns and indicates the hour on a scale." Ibid., 9.
71. Faroqhi, "Exotic Animals," 98–99. An elephant was acquired by the sultan's menagerie at some point in the middle of 1740s, but nothing is known about how it came to Istanbul.
72. BOA, CS 6683 (29 Za 1151/10 Mar. 1739).
73. There is some discrepancy in the citation of the amount to be spent on the elephant and his handlers. The sum of 15,300 akçes assumes a thirty-day month, but this case specifically states that for accounting purposes a month is to be considered as having only twenty-nine days.
74. BOA, CS 7256 (2 C 1152/6 Sept. 1739); BOA, CS 4499 (29 Z 1152/28 Mar. 1740).
75. For comparative purposes, see the following account of the amounts and kinds of foods elephants were given in Vijayanagara: Gorgas, "Animal Trade Between India and Western Eurasia," 206.
76. BOA, CS 4301 (2 Z 1159/16 Dec. 1746). Also cited in Faroqhi, "Exotic Animals," 98.
77. BOA, CS 6410 (29 N 1154/8 Dec. 1741); BOA, CS 2067 (26 L 1154/4 Jan. 1742); BOA, CS 2274 (4 Ca 1155/7 July 1742).
78. BOA, CS 6410 (29 N 1154/8 Dec. 1741); BOA, CS 2067 (26 L 1154/4 Jan. 1742).
79. BOA, CS 5842 (22 Ş 1155/22 Oct. 1742).
80. BOA, CS 3559 (11 L 1155/9 Dec. 1742).
81. BOA, CS 6410 (29 N 1154/8 Dec. 1741); BOA, CS 2067 (26 L 1154/4 Jan. 1742). See also: Faroqhi, "Exotic Animals," 98.
82. BOA, CS 6410 (29 N 1154/8 Dec. 1741); BOA, CS 2067 (26 L 1154/4 Jan. 1742); BOA, CS 5842 (22 Ş 1155/22 Oct. 1742); BOA, CS 3559 (11 L 1155/9 Dec. 1742).
83. BOA, CS 6410 (29 N 1154/8 Dec. 1741); BOA, CS 2067 (26 L 1154/4 Jan. 1742); BOA, CS 2274 (4 Ca 1155/7 July 1742).
84. Faroqhi, "Exotic Animals," 98.
85. Intimating that elephant handlers were often Muslims, Suraiya Faroqhi suggests that the prospect of being able to perform the pilgrimage as part of the journey from South Asia to the Ottoman Empire was good incentive for a mahout to travel such a long distance with his charge. Ibid.
86. It was also the case with the elephant Süleyman whose mahouts brought him from South Asia to Vienna in 1552. Saurer and Hinshaw-Fischli, "They Called Him Suleyman," 157–59 and 162.
87. BOA, CS 8605 (5 Ra 1191/13 Apr. 1777). This case concerns the problem of dealing with the personal effects (which totaled 36.5 guruş) of an elephant handler on his way to Istanbul.
88. BOA, CS 6016 (22 Za 1192/12 Dec. 1778). Suraiya Faroqhi identifies Jalbār as an "Indian principality." Faroqhi, "Exotic Animals," 99. The name may refer to the town of Chalbar in modern-day Afghanistan.

89. For a later case about transporting an elephant from Egypt to Edirne, see BOA, Cevdet Dahiliye 10256 (28 Ca 1235/13 Mar. 1820).
90. In discussing this case, Suraiya Faroqhi observes that the assignment of such a minor official to accompany the elephant evidences Sultan Abdülhamid I's indifference toward the animal. Faroqhi, "Exotic Animals," 99.
91. BOA, CS 6483 (27 Za 1192/17 Dec. 1778).
92. BOA, CS 6778 (19 C 1233/26 Apr. 1818). Each of these handlers was paid thirty guruş a month.
93. Faroqhi, "Exotic Animals," 99.
94. For an example of the gifting of an elephant and other animals to the Mamluk sultan from the ruler of the Dahlak Archipelago in the Red Sea at the end of the fourteenth century, see Shehada, *Mamluks and Animals*, 72.
95. Reindl-Kiel, "Animals as Gifts to the Ottoman Court," 279; Celia J. Kerslake, "The Correspondence between Selīm I and Ḳānṣūh al-Ğawrī," *Prilozi za Orijentalnu Filologiju* 30 (1980), 227–28.
96. Muḥammad ibn Aḥmad ibn Iyās al-Ḥanafī, *Badāʾiʿ al-Zuhūr fī Waqāʾiʿ al-Duhūr*, ed. Muḥammad Muṣṭafā, 5 vols. (Cairo: al-Hayʾa al-Miṣriyya al-ʿĀmma lil-Kitāb, 1982), 4: 284.
97. For useful accounts of Egypt in this period, see Carl F. Petry, *Twilight of Majesty: The Reigns of the Mamlūk Sultans al-Ashrāf Qāytbāy and Qanṣūh al-Ghawrī in Egypt* (Seattle: University of Washington Press, 1993); idem., *Protectors or Praetorians?: The Last Mamlūk Sultans and Egypt's Waning as a Great Power* (Albany: State University of New York Press, 1994).
98. On Ottoman-Mamluk relations, see Emire Cihan Muslu, "Ottoman-Mamluk Relations: Diplomacy and Perceptions" (Ph.D. diss., Harvard University, 2007); Kerslake, "Correspondence between Selīm I and Ḳānṣūh al-Ğawrī," 219–34.
99. Generally on the role of gift giving in Ottoman-Mamluk relations, see Muslu, "Ottoman-Mamluk Relations," 182–209; Elias I. Muhanna, "The Sultan's New Clothes: Ottoman-Mamluk Gift Exchange in the Fifteenth Century," *Muqarnas* 27 (2010): 189–207.
100. Sending a fethname with the head of one of an enemy sovereign's allies had many precedents in the Muslim world and, importantly, was not always a symbol of aggression or confrontation. On these points and for a discussion of an earlier fethname sent to Mamluk Egypt with a severed head, see Matthew Melvin-Koushki, "The Delicate Art of Aggression: Uzun Hasan's *Fathnama* to Qaytbay of 1469," *Iranian Studies* 44 (2011): 193–214.
101. For later examples of the dispatch of elephants from Egypt to Istanbul, see BOA, CS 2152 (28 R 1233/7 Mar. 1818); BOA, Cevdet Dahiliye 10256 (28 Ca 1235/13 Mar. 1820); BOA, Cevdet Maarif 3798 (16 C 1235/31 Mar. 1820). In all three of these cases, the elephants likely came to Egypt from India, but only the final two cases state this explicitly.
102. For an argument that the Ottoman conquest of Egypt was largely motivated by Ottoman desires to benefit from Egypt's established Indian Ocean trade links, see Casale, *Ottoman Age of Exploration*, 25–29.
103. al-Jabartī, *ʿAjāʾib al-Āthār* (1994), 2: 2.
104. Another example of Indians accompanying animals to Egypt was the embassy of Tipu, Sultan of Mysore, to the Ottoman Empire at the end of the eighteenth century. One of Tipu's chief objectives in dispatching this mission was to secure the support of Ottoman Sultan Abdülhamid I for his military efforts against the British in India. The delegation left Tipu's court for Istanbul in 1785 and stopped in Egypt on their way back home in the summer of 1788. According to al-Jabartī, Tipu's men brought many gifts to Istanbul. Along with a pulpit made of aloe wood, gold, and silver and a large couch that could seat six, they arrived to the Ottoman court with two birds who spoke Hindi. For al-Jabartī's accounts of the delegation's visit to Cairo and Istanbul and his description of the talking birds, see ibid., 2: 263–64; 3: 107–08. On the history and significance of Tipu's delegation to the Ottoman Empire, see Yusuf Hikmet Bayur, "Maysor Sultanı Tipu ile Osmanlı Pâdişahlarından I. Abdülhamid ve III. Selim Arasındaki Mektuplaşma," *Belleten* 12/47 (1948): 617–54; Iqbal Husain, "The Diplomatic Vision of Tipu Sultan: Briefs for Embassies to Turkey and France, 1785-86," in *State and Diplomacy under Tipu Sultan: Documents and Essays*, ed. Irfan Habib (New Delhi: Tulika, 2001), 19–65; Alam and Subrahmanyam, *Indo-Persian Travels*, 314–27.

When Napoleon invaded Egypt, he had designs for an alliance with Tipu against the British. In early 1798 there were reportedly French plans to send a force of over ten thousand troops from Suez to India to help Tipu fight the British. Two events in May 1799, however, ensured that this alliance never materialized—the defeat of Napoleon's armies at Acre and Tipu's death during a battle against British forces. Generally on Tipu's reign, see Irfan Habib, ed., *Confronting Colonialism: Resistance and Modernization under Haidar Ali and Tipu Sultan* (London: Anthem, 2002); Aniruddha Ray, ed., *Tipu Sultan and His Age: A Collection of Seminar Papers* (Kolkata: Asiatic Society, 2001); Kate Brittlebank, *Tipu Sultan's Search for Legitimacy: Islam and Kingship in a Hindu Domain* (Delhi: Oxford University Press, 1997). For the account of an East India Company army officer who fought against Tipu's army at Trichinopoly in 1799, see Letter, 18 March 1799, from Thomas Fiott De Havilland to His Father, Thomas Fiott De Havilland Correspondence (1799–1844), OSB MSS 136, Box 1, Folder 1, Beinecke Rare Book and Manuscript Library, Yale University.

Also noteworthy is Tipu's use of tiger symbolism to project his sovereignty. Many of his kingdom's banners, coins, weapons, and seals incorporated tiger motifs, and he is of course well-known as the Tiger of Mysore. He maintained a menagerie of tigers in his palace at Seringapatam and, quite famously, commissioned a moving and sounding mechanical sculpture of a tiger atop a British soldier taking a bite out of the man's white flesh. On Tipu's tiger symbolism, see Kate Brittlebank, "Sakti and Barakat: The Power of Tipu's Tiger, An Examination of the Tiger Emblem of Tipu Sultan of Mysore," *Modern Asian Studies* 29 (1995): 257–69; Maya Jasanoff, *Edge of Empire: Lives, Culture, and Conquest in the East, 1750–1850* (New York: Vintage, 2006), 177–80; Lawrence James, *Raj: The Making and Unmaking of British India* (New York: St. Martin's Press, 1997), 68.

105. For another example of the maritime commerce between India and Ottoman Egypt in this period, see BOA, MM, 9: 71 (Evail Z 1189/23 Jan.–1 Feb 1776).

106. For a discussion of Indian mahouts in Mamluk Egypt, see Shehada, *Mamluks and Animals*, 174–75 and 192–93.

107. For an earlier seventeenth-century example of how Egyptians similarly amused themselves with a young elephant calf, see Evliyâ Çelebi, *Evliyâ Çelebi Seyahatnâmesi*, 10: 442.

Nineteenth- and early twentieth-century Americans engaged in very different kinds of elephant spectacles. Perhaps the most prominent of these was the circus. Another less savory but still common one was the public execution—usually through electrocution, shooting, or hanging—of elephants who had killed circus workers or patrons. On the circus elephant in America, see Susan Nance, *Entertaining Elephants: Animal Agency and the Business of the American Circus* (Baltimore: Johns Hopkins University Press, 2013); Jennifer L. Mosier, "The Big Attraction: The Circus Elephant and American Culture," *Journal of American Culture* 22 (1999): 7–18. For a discussion of American elephant executions, see Amy Louise Wood, "'Killing the Elephant': Murderous Beasts and the Thrill of Retribution, 1885–1930," *Journal of the Gilded Age and Progressive Era* 11 (2012): 405–44.

108. A Briton resident in Istanbul at the end of the sixteenth century wrote of a giraffe who was similarly made to kneel for amusement by his handlers. MacLean, "The Sultan's Beasts," 152.

109. In the early nineteenth century, a few decades after the previous account from 1776, another group of elephants was brought to Cairo from India. Rumor preceded their actual arrival in the city. On June 16, 1817, word got out that the ship carrying these animals from the subcontinent was in Jidda and would soon leave this port for its final destination in Suez. With a buzz of anticipation in the air, the three elephants entered Cairo on Thursday, July 10, 1817. al-Jabartī describes the scene: "The three elephants arrived from Suez, one of which was bigger than the other two though it was the middle in age. They were brought into Cairo through Bāb al-Naṣr, passed through the center of the city, and went out through Bāb Zuwayla by al-Aḥmar Street on the way to Kara Maydān. Adults and children alike along with soldiers and Dalāt, mounted and on foot, rushed to look at the elephants, trailing behind them and crowding the market. The large elephant bore a wooden howdah on its back." al-Jabartī, *'Ajā'ib al-Āthār* (1994), 4: 390–91. Although this account gives no information about the identities of those who brought the elephants to Cairo from India, the mention of the animals' ages clearly suggests that those handling them knew a great deal about them. Like the elephant in 1776, these three also clearly elicited a degree of awe, fantasy, and wonder from

onlookers as they were paraded through Cairo. Moreover, in both cases, the animals were displayed in a palace or other royal structure.

110. Specifically, political leaders were to strive toward cultivating the heart of the lion. For a discussion of this and other political animal symbolism, see Dankoff, "Animal Traits."

111. The fourteenth-century Persian writer Qazvīnī offers the following description of the lion: "He is the most powerful and terrible and majestic of the beasts of prey...When the lion kills his prey he eats its heart and some of its other parts, and leaves the rest, and does not return to his half-eaten meal; he does no injury to a menstruous woman; and any animal that approaches him with submission he neither pursues nor injures; he is jealous of the female. And these qualities are those of kings; therefore he is called the King of Beasts. In spite of his majesty, he is continually attacked by fever; from other diseases he is freed by eating an ape. The lion dreads a white cock, and a peacock, and fire; he is an enemy of the elephant; from fear of ants he betakes himself to salt-pans, since if an ant gets into his paw he cannot free himself from it, and perishes...

Its brain allays tremors. Its bile makes a man brave and courageous, and cures epilepsy and ringworm of the scalp; used as an eyewash it cures bloodshot eyes; as an ointment it resolves tuberculous glands. Its fat is beneficial for piles, severe inflammations, and boils...Its blood is beneficial in cancer." Qazvīnī, *Zoological Section of the Nuzhatu-l-Qulūb*, 27–28.

On the history of the famous Persian cat, see Jean-Pierre Digard, "Chah des Chats, Chat de Chah? Sur les traces du chat persan," in *Hommes et terres d'Islam: Mélanges offerts à Xavier de Planhol*, ed. Daniel Balland, 2 vols. (Tehran: Institut de recherche en Iran, 2000), 1: 321–38; Willem Floor, "A Note on Persian Cats," *Iranian Studies* 36 (2003): 27–42; *Encyclopaedia Iranica*, s.v. "Cat II. Persian Cat" (Jean-Pierre Digard): http://www.iranicaonline.org/articles/cat-ii-persian-cat (accessed February 19, 2013). For a useful discussion of the animal entries in *Encyclopaedia Iranica*, see Steven C. Anderson, "Fauna," *Iranian Studies* 31 (1998): 399–405. On cats in Selçuk and Ottoman art, see Yaşar Çoruhlu, "Türk Sanatı ve Sembolizminde Kedi: Selçuklu ve Osmanlı Sanatında Kedi," *Toplumsal Tarih* 124 (2004): 40–45. And more generally on cats in Islamic history and literature, see Annemarie Schimmel, *Die Orientalische Katze* (Köln: Diederichs, 1983).

112. Reindl-Kiel, "Animals as Gifts to the Ottoman Court," 271.

113. Ibid., 280–83.

114. BOA, CS 2051 (9 Ra 1156/3 May 1743).

115. BOA, CS 328 (26 R 1175/24 Nov. 1761). This case simply instructs that four cats are to be sent to Istanbul; it does not delineate the breakdown between numbers of lions and tigers.

116. BOA, CS 977 (25 B 1221/8 Oct. 1806).

117. BOA, CS 6712 (11 N 1231/5 Aug. 1816).

118. Reindl-Kiel, "Animals as Gifts to the Ottoman Court," 283.

119. On the lion house, see BOA, Ali Emiri Üçüncü Ahmed 10344 (19 Ra 1120/8 June 1708); BOA, Ali Emiri Üçüncü Ahmed 10345 (19 Ra 1120/8 June 1708); BOA, CS 328 (26 R 1175/24 Nov. 1761); BOA, HAT 1451/17 (10 B 1203/6 Apr. 1789); BOA, HAT 246/13857 (29 Z 1213/3 June 1799); BOA, HAT 119/4849 (29 Z 1218/10 Apr. 1804); BOA, Cevdet Askeriye 38353 (29 M 1219/10 May 1804). For reference to the Ankara lion house, see BOA, Cevdet Evkaf 1902 (18 L 1209/8 May 1795).

120. Faroqhi, "Exotic Animals," 92–96.

121. Ibid., 93. For references both to the lion house's former status as a church and the use of materials from other Christian structures to repair it, see BOA, CS 6712 (11 N 1231/5 Aug. 1816); BOA, Cevdet Belediye 3526 (6 Ca 1247/13 Oct. 1831).

122. Faroqhi, "Exotic Animals," 92.

123. Ibid. It is indeed noteworthy that cages, chains, and other means of controlling animals are not mentioned more regularly in accounts of the lion house. This would seem to suggest that the animals were allowed some amount of limited autonomy while inside. That said, of course the animals were subdued when the need arose. For a specific request from the chief lion keeper for chains and ropes to tie up lions and tigers, see BOA, CS 328 (26 R 1175/24 Nov. 1761). For correspondence about the need for halters (*yularlar*) to control lions and tigers, see BOA, Ali Emiri Üçüncü Ahmed 10344 (19 Ra 1120/8 June 1708). For other requests and orders for materials to control the animals, see BOA, HAT 246/13857

(29 Z 1213/3 June 1799); BOA, Ali Emiri Üçüncü Ahmed 10345 (19 Ra 1120/8 June 1708); BOA, CS 6460 (27 Ş 1217/23 Dec. 1802); BOA, HAT 119/4849 (29 Z 1218/10 Apr. 1804).

124. Svetlana Kirillina, "Representing the Animal World of the Ottoman Empire: The Accounts of Russian Orthodox Pilgrims (Sixteenth—Eighteenth Centuries)," in *Animals and People in the Ottoman Empire*, ed. Suraiya Faroqhi (Istanbul: Eren, 2010), 83.
125. Ibid.
126. On the early modern cabinet of curiosities, see Paula Findlen, "Inventing Nature: Commerce, Art, and Science in the Early Modern Cabinet of Curiosities," in *Merchants and Marvels: Commerce, Science, and Art in Early Modern Europe*, ed. Pamela H. Smith and Paula Findlen (New York: Routledge, 2002), 297–323.
127. Faroqhi, "Exotic Animals," 93–94.
128. Kirillina, "Representing the Animal World," 82.
129. Faroqhi, "Exotic Animals," 94–95.
130. On the replacement of portions of the *arslanhane* with the *cebehane*, see BOA, HAT 119/4849 (29 Z 1218/10 Apr. 1804); BOA, Cevdet Askeriye 38353 (29 M 1219/10 May 1804).
131. BOA, CS 6460 (27 Ş 1217/23 Dec. 1802); BOA, CS 4817 (4 R 1218/24 July 1803).
132. BOA, CS 4817 (4 R 1218/24 July 1803). This case includes a detailed expense report for repairs to the lion house during the first six months of 1803.
133. BOA, CS 977 (25 B 1221/8 Oct. 1806); BOA, Cevdet Belediye 5177 (29 N 1221/10 Dec. 1806).
134. BOA, CS 6712 (11 N 1231/5 Aug. 1816).
135. Ibid. Also cited in Faroqhi, "Exotic Animals," 89 and 95.
136. Such carts had also been constructed several years earlier just after the 1802 fire. BOA, HAT 119/4849 (29 Z 1218/10 Apr. 1804).
137. MacLean, "The Sultan's Beasts," 153–54.
138. Tezcan, *Second Ottoman Empire*, 205.
139. BOA, CS 2051 (9 Ra 1156/3 May 1743).
140. For cases to, from, or about lion keepers, see BOA, CS 3660 (13 Ca 1113/16 Oct. 1701); BOA, CS 328 (26 R 1175/24 Nov. 1761); BOA, CS 2999 (25 Za 1186/17 Feb. 1773); BOA, CS 4640 (2 Ca 1216/10 Sept. 1801); BOA, CS 6460 (12 Ş 1217/8 Dec. 1802).
141. Evliyâ Çelebi, *Evliyâ Çelebi Seyahatnâmesi*, 1: 321. Also cited in Faroqhi, "Exotic Animals," 95. With the reference to ʿAli, Evliya means to imply that these men were Shiites.
142. BOA, CS 3660 (13 Ca 1113/16 Oct. 1701).
143. Faroqhi, "Exotic Animals," 95–96.
144. BOA, CS 2999 (25 Za 1186/17 Feb. 1773).
145. BOA, CS 4640 (2 Ca 1216/10 Sept. 1801).
146. For reasons unstated, Ismail's stipend was only two akçes. BOA, Maliyeden Müdevver 9989, p. 45 (1171/1757–58), cited in Faroqhi, "Exotic Animals," 96.
147. The abolition of this office did not mean that leopards no longer came to Istanbul. For the case of a leopard sent by the Vali of Syria Mehmed Necib from Gaza to Istanbul in 1842, see BOA, İradeler Dahiliye 2574 (30 Z 1257/12 Feb. 1842).
148. Marc David Baer, *Honored by the Glory of Islam: Conversion and Conquest in Ottoman Europe* (New York: Oxford University Press, 2008), 231–44.
149. Reindl-Kiel, "Animals as Gifts to the Ottoman Court," 283.
150. This association with the Safavid Empire is made clearer by the text that accompanies this image which specifically states that the lion came from Baghdad. The city was in an important border zone between the two empires and regularly changed hands. Thus, as was the case with Maximilian's elephant Süleyman in Vienna, Ottoman political and military rivalries with bordering early modern states are expressed in this image through the representation of a lion.
151. Reindl-Kiel, "Animals as Gifts to the Ottoman Court," 283.
152. The empire's chief lion keeper was clearly an important part of Ottoman efforts to tame the animal's wild nature. In his sixteenth-century account of sex in the Ottoman Empire, Deli Birader puts forth his interpretation of how the chief lion keeper came to control the animals

in his charge. "One day, one of the Sultan's lions broke free of its chains and ran away into the market place. It leaped on a man and mounted him. The head of the lion-tamers was informed and he arrived. As soon as he saw the lion, he understood that it was so frightened and nervous that it could not be approached. He stood in front of the lioness and, untying his belt, took out his penis and pointed it at it. The lioness saw it and calmly approached to the lion-tamer and the tamer put the chain around its neck. The onlookers were amazed and understood that from time to time he fucked the lion." Kuru, "Deli Birader," 255–56. According to Deli Birader, "Lion-tamers claim that if you fuck a lioness, it turns calm and friendly." Ibid., 255.

153. Faroqhi, "Exotic Animals," 95 and 99–100.
154. Sumathi Ramaswamy, "Conceit of the Globe in Mughal Visual Practice," *Comparative Studies in Society and History* 49 (2007), 777–78. See also Ebba Koch, *Mughal Art and Imperial Ideology: Collected Essays* (New Delhi: Oxford University Press, 2001).
155. Faroqhi, "Exotic Animals," 100.
156. Generally on the Ottoman hunt, see Artan, "A *Book of Kings* as a Treatise on Hunting"; idem., "Ahmed I and 'Tuhfet'ül-Mülûk ve's-Selâtîn': A Period Manuscript on Horses, Horsemanship and Hunting," in *Animals and People in the Ottoman Empire*, ed. Suraiya Faroqhi (Istanbul: Eren, 2010), 235–69; idem., "Ahmed I's Hunting Parties: Feasting in Adversity, Enhancing the Ordinary," in *Starting with Food: Culinary Approaches to Ottoman History*, ed. Amy Singer (Princeton, NJ: Markus Wiener Publishers, 2011), 93–138; Borromeo, "The Ottomans and Hunting"; Baer, *Honored by the Glory of Islam*, 179–203 and 231–44. See also the relevant sections of Karin Ådahl, ed., *The Sultan's Procession: The Swedish Embassy to Sultan Mehmed IV in 1657–1658 and the Rålamb Paintings* (Istanbul: Swedish Research Institute in Istanbul, 2006). For a wider perspective, see Allsen, *Royal Hunt in Eurasian History*.
157. This understanding of the mortal danger posed by powerful animals is famously depicted in a drawing of a tiger attacking a man from seventeenth-century Safavid Iran. Massumeh Farhad, "An Artist's Impression: Muʿin Musavvir's *Tiger Attacking a Youth*," *Muqarnas* 9 (1992): 116–23. On hunting, danger, and interspecies competition, see also the following analysis of the French hunting treatise *La Vénerie* written by Jacques du Fouilloux and first published in Poitiers in 1561: Suzanne J. Walker, "Making and Breaking the Stag: The Construction of the Animal in the Early Modern Hunting Treatise," in *Early Modern Zoology: The Construction of Animals in Science, Literature and the Visual Arts*, ed. Karl A. E. Enenkel and Paul J. Smith, 2 vols. (Leiden: Brill, 2007), 2: 317–37.
158. Artan, "A *Book of Kings* as a Treatise on Hunting," 300.
159. For a general account of the role of hunting in Turkish and Ottoman history, see Emine Gürsoy Naskali and Hilal Oytun Altun, eds., *Av ve Avcılık Kitabı* (Istanbul: Kitabevi Yayınları, 2008).
160. Allsen, *Royal Hunt in Eurasian History*; Dankoff, "Animal Traits."
161. On the relationship between *ghaza* and hunting, see Baer, *Honored by the Glory of Islam*, 165.
162. Artan, "A *Book of Kings* as a Treatise on Hunting," 301.
163. Ibid., 300–01; Baer, *Honored by the Glory of Islam*, 165 and 181. Many of Süleyman's favorite hunting spots were on the Anatolian side of the Bosphorus, in the garden of Sultaniye in Beykoz and the garden of Fenerbahçe in Üsküdar, for example. Shirine Hamadeh, *The City's Pleasures: Istanbul in the Eighteenth Century* (Seattle: University of Washington Press, 2008), 37. Later sultans also regularly undertook royal hunts on the Anatolian side of the Bosphorus, especially in Üsküdar. In 1618, for example, Osman II, only months into his reign, fell off his horse during a hunt in Üsküdar, badly injuring his head. Tezcan, *Second Ottoman Empire*, 119.
164. This observation was made by the Venetian *bailo* Bernardo Navagero in 1553. Gülrü Necipoğlu, "The Suburban Landscape of Sixteenth-Century Istanbul as a Mirror of Classical Ottoman Garden Culture," in *Gardens in the Time of the Great Muslim Empires: Theory and Design*, ed. Attilio Petruccioli (Leiden: Brill, 1997), 33. Some chroniclers criticized Süleyman for, among other things, hunting too much to the detriment of his other responsibilities. Artan, "A *Book of Kings* as a Treatise on Hunting," 301. For a discussion and analysis

of some of the other critiques of Süleyman's reign, see Cemal Kafadar, "The Myth of the Golden Age: Ottoman Historical Consciousness in the Post-Süleymanic Era," in *Süleymân the Second and His Time*, ed. Halil İnalcık and Cemal Kafadar (Istanbul: Isis Press, 1993), 37–48; Cornell H. Fleischer, *Bureaucrat and Intellectual in the Ottoman Empire: The Historian Mustafa Âli (1541–1600)* (Princeton, NJ: Princeton University Press, 1986), 191–231.

165. Baer, *Honored by the Glory of Islam*, 181.
166. For a discussion of various classes of hunting animals and the administration built around them in the early seventeenth century, see Borromeo, "The Ottomans and Hunting," 219–33.
167. On the use of falcons in Ottoman hunting, see Gilles Veinstein, "Falconry in the Ottoman Empire of the Mid-Sixteenth Century," in *Animals and People in the Ottoman Empire*, ed. Suraiya Faroqhi (Istanbul: Eren, 2010), 205–18; idem., "La fauconnerie dans l'Empire ottoman au milieu du XVIe siècle: une institution en péril," in *Hommes et terres d'Islam: Mélanges offerts à Xavier de Planhol*, ed. Daniel Balland, 2 vols. (Tehran: Institut de recherche en Iran, 2000), 1: 343–59; Bistra Cvetkova, "La fauconnerie dans les sancaks de Nicopol et de Vidin aux XVe et XVIe siècles," *İstanbul Üniversitesi Edebiyat Fakültesi Tarih Dergisi* 32 (1979): 795–818; *Encyclopaedia of Islam*, 2nd ed. (Leiden: Brill Online, 2013), s.v. "Doghandjı" (Halil İnalcık). See also François Viré, "Essai du détermination des oiseaux-de-vol mentionnés dans les principaux manuscrits arabes médiévaux sur la fauconnerie," *Arabica* 24 (1977): 138–49. On Mughal emperor Jahangir's use of hunting falcons, see Mutribi al-Asamm Samarqandi, *Conversations with Emperor Jahangir*, 40. On the history of Mamluk hunting falcons, see Shehada, *Mamluks and Animals*, 68–70 and 297–99. For further comparative perspective, see Robin S. Oggins, *The Kings and Their Hawks: Falconry in Medieval England* (New Haven, CT: Yale University Press, 2004); John Cummins, *The Hound and the Hawk: The Art of Medieval Hunting* (London: Weidenfeld and Nicolson, 1988); Richard Grassby, "The Decline of Falconry in Early Modern England," *Past and Present* 157 (1997): 37–62.
168. Artan, "A *Book of Kings* as a Treatise on Hunting," 301.
169. Ibid., 300–02; idem., "Ahmed I's Hunting Parties," 101–10; Baer, *Honored by the Glory of Islam*, 145.
170. On Ottoman hunting parties in the seventeenth century, see Artan, "Ahmed I's Hunting Parties"; Baer, *Honored by the Glory of Islam*, 182–83. The Mughals were also fond of using hunting trips in this way. Schimmel, *Empire of the Great Mughals*, 201–03.
171. Baer, *Honored by the Glory of Islam*, 181.
172. For a discussion of the French embassy's use of hunting as a means of diplomacy in the early seventeenth-century Ottoman Empire, see Borromeo, "The Ottomans and Hunting," 229–31.
173. Artan, "A *Book of Kings* as a Treatise on Hunting," 306.
174. Thus hunting plays a key role in the familiar tension between pleasure and active rule that runs throughout Ottoman literature on the topic of sovereignty. Idem., "Ahmed I's Hunting Parties," 101–02.
175. On the perceived positive aspects of hunting for sultans, see Baer, *Honored by the Glory of Islam*, 180–82.
176. This was part of Muṣṭafā ʿĀlī's critique of Sultan Murad III at the end of the sixteenth century, for example. Fleischer, *Bureaucrat and Intellectual*, 293–307.
177. Gabriel Piterberg, *An Ottoman Tragedy: History and Historiography at Play* (Berkeley: University of California Press, 2003), 196–98; Leslie P. Peirce, *The Imperial Harem: Women and Sovereignty in the Ottoman Empire* (Oxford: Oxford University Press, 1993), 9; Walter G. Andrews, "Singing the Alienated 'I': Guattari, Deleuze and Lyrical Decodings of *the Subject* in Ottoman *Divan* Poetry," *Yale Journal of Criticism* 6 (1993), 199; Alan Mikhail, "The Heart's Desire: Gender, Urban Space and the Ottoman Coffee House," in *Ottoman Tulips, Ottoman Coffee: Leisure and Lifestyle in the Eighteenth Century*, ed. Dana Sajdi (London: I. B. Tauris, 2007), 134. For comparative perspective, see Allsen, *Royal Hunt in Eurasian History*, 34–51.
178. Artan, "A *Book of Kings* as a Treatise on Hunting," 307–08.

179. For a useful comparative perspective from early modern German horsemanship literature, see Pia F. Cuneo, "(Un)Stable Identities: Hippology and the Professionalization of Scholarship and Horsemanship in Early Modern Germany," in *Early Modern Zoology: The Construction of Animals in Science, Literature and the Visual Arts*, ed. Karl A. E. Enenkel and Paul J. Smith, 2 vols. (Leiden: Brill, 2007), 2: 339–59.
180. On Osman II, see Tezcan, *Second Ottoman Empire*, 119. On Süleyman Pasha, see Artan, "A *Book of Kings* as a Treatise on Hunting," 300.
181. Artan, "A *Book of Kings* as a Treatise on Hunting," 308. A similar sort of logic undergirded the prescription that if the sultan needed to leave the palace quickly on horseback to join a military campaign or for any other reason, he should do so completely surrounded by his advisors and closest military men so that onlookers would not see him riding. Ibid.
182. In this regard, see also François Viré, "La chasse au guépard d'après sources arabes et les oeuvres d'art musulman par Ahmad Abd ar-Raziq," *Arabica* 21 (1974): 85–88.
183. Artan, "Ahmed I and 'Tuhfet'ül-Mülûk ve's-Selâtîn,'" 235–69. For a discussion of Mamluk precedents, see Shehada, *Mamluks and Animals*, 24–30 and 169–72; Howard R. Turner, *Science in Medieval Islam: An Illustrated Introduction* (Austin: University of Texas Press, 1995), 158–59. Further examples of this lineage in the Muslim world include G. Rex Smith, *Medieval Muslim Horsemanship: A Fourteenth-Century Arabic Cavalry Manual* (London: British Library, 1979). More generally on the importance of horses in world history, see David W. Anthony, *The Horse, the Wheel, and Language: How Bronze-Age Riders from the Eurasian Steppes Shaped the Modern World* (Princeton, NJ: Princeton University Press, 2007); Kelekna, *The Horse in Human History*.
184. Artan, "A *Book of Kings* as a Treatise on Hunting," 306–07.
185. On the connection between these virtues of the hunt and notions of Ottoman masculinity, see Baer, *Honored by the Glory of Islam*, 180–81.
186. Artan, "A *Book of Kings* as a Treatise on Hunting," 308.
187. For comparative examples of the didactic potential of animals for human sovereigns, see Allsen, *Royal Hunt in Eurasian History*, 141–59.
188. On this point, see also Baer, *Honored by the Glory of Islam*, 181.
189. Ibid., 180 and 231–44. He went on over fifty hunting expeditions during his nearly forty years as sultan. Ibid., 182.
190. Artan, "A *Book of Kings* as a Treatise on Hunting," 309. Some accounts of Ahmet I's reign thus go out of their way to state explicitly that he always made time for prayer and other religious duties while on his hunts. Idem., "Ahmed I's Hunting Parties," 101 and 121. On Ahmet I's hunting, see also Borromeo, "The Ottomans and Hunting," 223–29.
191. For a comparative analysis of the exorbitant costs of hunting in the ʿAbbasid court, see Muhammad Manazir Ahsan, "A Note on Hunting in the Early ʿAbbasid Period: Some Evidence on Expenditure and Prices," *Journal of the Economic and Social History of the Orient* 19 (1976): 101–05.
192. Baer, *Honored by the Glory of Islam*, 232–33.
193. Artan, "A *Book of Kings* as a Treatise on Hunting," 309.
194. Ibid., 303.
195. This claim obviously somewhat contradicts the previously discussed assertion that one of the benefits of hunting was that it allowed sultans to escape the eyes of their imperial subjects. One can of course imagine different scenarios in which the hunt met both ideals.
196. Baer, *Honored by the Glory of Islam*, 185–203.
197. Artan, "A *Book of Kings* as a Treatise on Hunting," 308–09; idem., "Ahmed I's Hunting Parties."
198. Idem., "A *Book of Kings* as a Treatise on Hunting," 305. Ottoman subjects were, however, sometimes employed during hunting expeditions to drive animals into designated hunting areas. Baer, *Honored by the Glory of Islam*, 184–85.
199. Baer, *Honored by the Glory of Islam*, 180–83. For useful comparative examples, see Allsen, *Royal Hunt in Eurasian History*, 119–40.
200. Artan, "A *Book of Kings* as a Treatise on Hunting," 310.

6. Encagement

1. Alain Silvera, "The First Egyptian Student Mission to France under Muhammad Ali," *Middle Eastern Studies* 16 (1980): 1–22. For a recent analysis of the impact of Egyptians and other Arabs on France in the early nineteenth century, see Ian Coller, *Arab France: Islam and the Making of Modern Europe, 1798–1831* (Berkeley: University of California Press, 2011). Most of these individuals made their way to France in the aftermath of Napoleon's occupation of Egypt; some returned with the French army in 1801 and others arrived later.
2. Hourani, *Arabic Thought in the Liberal Age*, 67–102; al-Rāfʿī, *ʿAṣr Muḥammad ʿAlī*, 426–64. al-Ṭahṭāwī's account of his time in Paris is Rifāʿa Badawī Rāfiʿ al-Ṭahṭāwī, *Takhlīṣ al-Ibrīz fī Talkhīṣ Bārīz* (Cairo: Dār al-Kutub wa al-Wathāʾiq al-Qawmiyya, 2005).
3. For an argument that al-Ṭahṭāwī had already developed his reformist ideas before going to France, see Peter Gran, "Tahtawi in Paris," *al-Ahram Weekly*, January 10–16, 2002.
4. The story of this giraffe and her journey from Egypt to Paris is entertainingly related in Michael Allin, *Zarafa: A Giraffe's True Story, from Deep in Africa to the Heart of Paris* (New York: Delta, 1998).
5. Giraffes were, by contrast, well-known in the Muslim world. For example, the fourteenth-century Persian scholar Qazvīnī had this to say about the animal: "Its flesh may be eaten, since it is begotten by the camel and the mountain cow. Its neck is like that of the camel, and its skin like that of the leopard, and its four extremities like those of the cow; its fore-legs are longer than its hind-legs." Qazvīnī, *Zoological Section of the Nuzhatu-l-Qulūb*, 16. For a discussion of earlier giraffe gifts to and from Egypt, see Shehada, *Mamluks and Animals*, 71–74; Belozerskaya, *The Medici Giraffe*, 107–129. For a more general account of giraffes as diplomatic gifts, see Thierry Buquet, "*La belle captive*. La girafe dans les ménageries princières au Moyen Âge," in *La bête captive au Moyen Âge et à l'époque moderne*, ed. Corinne Beck and Fabrice Guizard (Amiens: Encrage, 2012), 65–90.
6. For al-Ṭahṭāwī's thoughts on human-animal difference, see Rifāʿa Rāfiʿ al-Ṭahṭāwī, *al-Siyāsa wa al-Waṭaniyya wa al-Tarbiyya*, vol. 2 of *al-Aʿmāl al-Kāmila li-Rifāʿa Rāfiʿ al-Ṭahṭāwī*, ed. Muḥammad ʿImāra (Beirut: al-Muʾassasa al-ʿArabiyya lil-Dirāsāt wa al-Nashr, 1973), 299–305. His main argument is that while animal power derives from physical strength and ability, human power is constituted through rationality, language, and the capacity for discernment. For further discussion of al-Ṭahṭāwī's ideas about species difference, see Samera Esmeir, *Juridical Humanity: A Colonial History* (Stanford, CA: Stanford University Press, 2012), 134–36.
7. For later cases about the transport of animals from Egypt to France, see BOA, İradeler Mısır 16.528 (15 M 1269/29 Oct. 1852). Like France's first giraffe, London's first hippo also came from Egypt. For an account of this animal's journey in 1850 and his new life in London, see Nina J. Root, "Victorian England's Hippomania," *Natural History* 102 (1993): 34–39.
8. It was not just living animals that were sought in Egypt. Beginning with Napoleon's invasion, many European zoologists and other scientists took specimens of dead creatures from Egypt back to their labs in Europe. See, for example, the following letters about a British zoologist's needs for firearms and preserving alcohol during a collecting expedition in Egypt and Abyssinia in 1868: TNA, MT 23/15 (18 Jan. 1868). European interest in specimens of both living and dead animals and humans was, of course, part of the larger phenomenon of nineteenth-century Egyptomania. For histories of European Egyptomania see, for example, James Stevens Curl, *Egyptomania: The Egyptian Revival, a Recurring Theme in the History of Taste* (Manchester: Manchester University Press, 1994); Elliott Colla, *Conflicted Antiquities: Egyptology, Egyptomania, Egyptian Modernity* (Durham, NC: Duke University Press, 2007). Of course, even before the nineteenth century, Europeans went to Egypt in search of what one document terms unspecified "objects of natural history" (*objets d'histoire naturelle*). ACM, LC 12, 94r–94v (27 May 1783).
9. On Hagenbeck's life and career, see Lothar Dittrich and Annelore Rieke-Müller, *Carl Hagenbeck (1844–1913): Tierhandel und Schaustellungen im Deutschen Kaiserreich* (Frankfurt: Peter Lang, 1998); Rothfels, *Savages and Beasts*; Eric Ames, *Carl Hagenbeck's Empire of Entertainments* (Seattle: University of Washington Press, 2008); Herman Reichenbach, "A Tale of Two Zoos: The Hamburg Zoological Garden and Carl Hagenbeck's

Tierpark," in *New Worlds, New Animals: From Menagerie to Zoological Park in the Nineteenth Century*, ed. R. J. Hoage and William A. Deiss (Baltimore: Johns Hopkins University Press, 1996), 51–62.

10. Rothfels, *Savages and Beasts*, 44–80.
11. Quoted in ibid., 50.
12. Hagenbeck paid £2,750 for the 128 animals he took from Suez to Alexandria. Ibid., 214, n. 16.
13. Ibid., 50. Collector that he was, Hagenbeck enumerated the total combined animal transport that left Alexandria in 1870. It included "a rhinoceros, five elephants, two warthogs, four aardvarks, fourteen giraffes, twelve antelopes and gazelles, four wild Nubian buffalo, sixty large and small beasts of prey, including thirty spotted and striped hyenas, seven young lions, eight leopards and cheetahs as well as several other wilds cats, etc. Beyond all of those, came twenty-six ostriches, among which were sixteen full-grown birds . . . The transport was completed by twenty large crates with monkeys and birds, as well as seventy-two Nubian milk-goats, a wandering dairy which provided milk for our young animals."
14. For examples of Egypt's trade in animals to Crimea, Antalya, Serbia, Malta, Crete, Salonica, Benghazi, Italy, Belgium, and elsewhere, see BOA, İradeler Mısır 17.608 (29 C 1271/19 Mar. 1855); BOA, Sadaret Âmedî Kalemi 62.40 (1271/1854 and 1855); BOA, Sadaret Âmedî Kalemi 62.41 (1271/1854 and 1855); BOA, Meclis-i Vâlâ Riyâseti 1038.10 (14 Za 1283/20 Mar. 1867); BOA, Yıldız Perakende Evrakı Mabeyn Başkitabeti 6.39 (24 N 1299/9 Aug. 1882); BOA, BEO 2997.224730 (6 M 1325/19 Feb. 1907); BOA, BEO 2918.218845 (12 Ş 1324/1 Oct. 1906); BOA, DMK 714.8 (27 S 1321/25 May 1903); BOA, DMK 597.29 (14 B 1320/17 Oct. 1902); BOA, DMK 489.56 (19 M 1320/28 Apr. 1902); BOA, BEO 1844.138239 (29 M 1320/8 May 1902); BOA, DMK 1475.91 (24 R 1305/9 Jan. 1888); BOA, BEO 1076.80636 (11 N 1315/3 Feb. 1898); BOA, Dahiliye Islahat 14.4 (22 Ra 1315/21 Aug. 1897); BOA, DMK 1616.30 (14 Ş 1306/15 Apr. 1889); BOA, DMK 1471.105 (10 R 1305/26 Dec. 1887); BOA, Yıldız Perakende Evrakı Askerî Maruzat 29.84 (9 S 1303/17 Nov. 1885). More examples of Egypt's involvement in the global animal trade are given in subsequent citations.
15. For examples of the increasingly global dimensions and demands of the nineteenth-century animal trade, see the following cases about the movement of animals from Iran to Trieste via Trabzon: BOA, Sadaret Âmedî Kalemi 78.36 (1273/1856 and 1857); BOA, Hariciye Nezareti Mektubî Kalemi 161 (26 M 1273/26 Sept. 1856). Much of the bureaucratic work needed to organize these animal deliveries took place in Istanbul and Beirut.
16. As with the giraffe sent to Paris in 1826, Mehmet ʿAli often used Egypt's animal wealth to great diplomatic effect as a means of currying favor with rival sovereigns. Although perhaps seemingly similar to early modern gift-giving practices, these animals were not used as signs of Mehmet ʿAli's abilities to harmonize disparate parts of nature. Rather—much like Egypt's antiquities, which Mehmet ʿAli also happily used as political capital—these animals were considered the personal property of the Egyptian Pasha that he could dole out as exotic trinkets whenever he saw fit. He, for example, sent various animal gifts to Istanbul over the course of his tense relationship with the Ottoman capital. BOA, CS 2152 (28 R 1233/7 Mar. 1818); BOA, Cevdet Maarif 3798 (16 C 1235/31 Mar. 1820). For a case about a group of elephants Mehmet ʿAli sent to Edirne, see BOA, Cevdet Dahiliye 10256 (28 Ca 1235/13 Mar. 1820). Some of these elephants died before arriving at their intended destination. Mehmet ʿAli also sent lions and tigers to Istanbul in the winter of 1817–18, along with eighteen lion and tiger keepers. BOA, Cevdet Belediye 3083 (29 C 1233/6 May 1818); BOA, CS 8909 (25 S 1233/4 Jan. 1818). For a comparative perspective on Mehmet ʿAli's gifting habits, see the following report about the transport of a sarcophagus from the third pyramid at Giza that he gave to the British Museum: TNA, FO 926/8 (20 April 1838).
17. For example, whereas tigers historically came from South Asia or Iran, at the end of the nineteenth century Istanbul brought four tigers from Yemen for the captial's newly established zoological garden (*hayvanat bahçesi*). BOA, Yıldız Perakende Evrakı Umum Vilayetler Tahriratı 31.17 (26 S 1307/10 Oct. 1889). On the establishment of this animal collection, see BOA, Maarif Nezareti Evrakı 65.99 (22 Ş 1297/30 July 1880); BOA, Yıldız Maarif Nezareti Maruzâtı 1.11 (22 Ş 1297/30 July 1880). For an order from roughly the same period to send animals from Egypt to Yemen, see BOA, İradeler Dahiliye 643.44739 (14 L 1288/27

Dec. 1871). For examples of Egypt's involvement in the nineteenth-century Mediterranean animal trade, see BOA, İradeler Mısır 11.276 (8 Z 1258/10 Jan. 1843); BOA, Meclis-i Vâlâ Riyâseti 761.43 (6 N 1278/7 Mar. 1862); BOA, Yıldız Sadaret Hususî Maruzat Evrakı 171.54 (7 L 1299/22 Aug. 1882); BOA, Meclis-i Vâlâ Riyâseti 1038.10 (14 Za 1283/20 Mar. 1867); BOA, Hariciye Nezareti Tercüme Odası 260.83 (22 Mar. 1883); BOA, Şûra-yı Devlet 594.25 (3 S 1325/18 Mar. 1907); BOA, BEO 2918.218845 (12 Ş 1324/1 Oct. 1906); BOA, Meclis-i Vükelâ Mazbataları 111.88 (15 C 1323/17 Aug. 1905); BOA, İradeler Hususi 96.1320/M–106 (29 M 1320/8 May 1902); BOA, Yıldız Sadaret Hususî Maruzat Evrakı 261.149 (26 Za 1309/22 June 1892); BOA, DMK 1699.6 (24 C 1307/15 Feb. 1890); BOA, DMK 1613.94 (7 Ş 1306/8 Apr. 1889).

18. For an early twentieth-century example of Egyptian authorities sending animals from the Sudan to Istanbul, see BOA, Yıldız Perakende Evrakı Mabeyn Başkitabeti 81.63 (29 Z 1327/11 Jan. 1910). In addition to the Sudan, Egypt was also vital to British imperial projects in Ethiopia, coastal Somalia, Zanzibar, and other part of Africa at the end of the nineteenth century. These imperial interests manifested themselves in the growing fields of colonial botany and zoology. For correspondence between British imperial officials and Egyptian authorities about the transfer of botanical and zoological specimens from various parts of Africa to Britain, see TNA, FO 143/10 (1873–78); TNA, 141/70 (10 Apr. 1869). Generally on Egyptian-Sudanese relations in the nineteenth and early twentieth centuries, see Powell, *A Different Shade of Colonialism*; Sharkey, *Living with Colonialism*; Walz, *Trade between Egypt and Bilād as-Sūdān*.

19. For an Ottoman order from 1883 to acquire animals from Anatolia for use in Ottoman-Egyptian military operations in Sudan, see BOA, Yıldız Sadaret Hususî Maruzat Evrakı 173.49 (6 C 1300/14 Apr. 1883).

20. For various examples of Egypt's purchase and import of animals, see BOA, İradeler Mısır 16.537 (18 Ca 1269/27 Feb. 1853); BOA, Cevdet İktisat 33.1640 (11 N 1260/24 Sept. 1844); BOA, Sadaret Mühimme Kalemi Evrakı 260.65 (25 L 1279/15 Apr. 1863); BOA, Sadaret Mektubî Kalemi Nezaret ve Devâir Yazışmaları 140.99 (13 B 1271/1 Apr. 1865); BOA, Meclis-i Vâlâ Riyâseti 771.19 (10 Za 1280/17 Apr. 1864); BOA, Sadaret Mühimme Kalemi Evrakı 296.36 (28 L 1280/6 Apr. 1864); BOA, Meclis-i Vâlâ Riyâseti 529.107 (26 Za 1283/1 Apr. 1867); BOA, Meclis-i Vâlâ Riyâseti 780.53 (21 B 1283/29 Nov. 1866); BOA, Hariciye Nezareti Tercüme Odası 260.83 (22 Mar. 1883); BOA, DMK 1609.56 (23 B 1306/25 Mar. 1889); BOA, DMK 911.30 (15 N 1322/23 Nov. 1904); BOA, BEO 3179.238384 (4 L 1325/10 Nov. 1907); BOA, Dahiliye İdare 101.4 (1 Z 1328/4 Dec. 1910).

21. For useful insights from the literature on colonial veterinary medicine, see Diana K. Davis, "Prescribing Progress: French Veterinary Medicine in the Service of Empire," *Veterinary Heritage* 29 (2006): 1–7; idem., "Brutes, Beasts, and Empire"; Brown and Gilfoyle, *Healing the Herds*; Shaun Milton, "Western Veterinary Medicine in Colonial Africa: A Survey, 1902–1963," *Argos* 18 (1998): 313–22; Richard Waller, " 'Clean' and 'Dirty': Cattle Disease and Control Policy in Colonial Kenya, 1900–40," *Journal of African History* 45 (2004): 45–80.

22. Although I do not focus on it here, the animal body itself was of course also variously understood as having its own inherent medicinal properties. For a discussion of the historic uses of various animals' body parts as *materia medica* in Yemen, see Schönig, "Animal Drugs in Yemen."

23. For studies of Arab and Islamic zoology, see Eisenstein, *Einführung in die arabische Zoographie*; Ullmann, *Die Natur- und Geheimwissenschaften*, 5–61.

24. For accounts of Geoffroy Saint-Hilaire's life and work, see Isidore Geoffroy Saint-Hilaire, *Vie, travaux et doctrine scientifique d'Étienne Geoffroy Saint-Hilaire* (Paris: P. Bertrand, 1847); Théophile Cahn, *La vie et l'oeuvre d'Étienne Geoffroy Saint-Hilaire* (Paris: Presses universitaires de France, 1962); Hervé Le Guyader, *Étienne Geoffroy Saint-Hilaire, 1772–1844: A Visionary Naturalist*, trans. Marjorie Grene (Chicago: University of Chicago Press, 2004); Michael A. Osborne, "Zoos in the Family: The Geoffroy Saint-Hilaire Clan and the Three Zoos of Paris," in *New Worlds, New Animals: From Menagerie to Zoological Park in the Nineteenth Century*, ed. R. J. Hoage and William A. Deiss (Baltimore: Johns Hopkins University Press, 1996), 33–42.

25. For accounts of Geoffroy Saint-Hilaire's time in Egypt, see Geoffroy Saint-Hilaire, *Vie, travaux et doctrine scientifique*, 74–123; Nina Burleigh, *Mirage: Napoleon's Scientists and the Unveiling of Egypt* (New York: HarperCollins, 2007), 195–207. For some of his writings about Egypt, see Étienne Geoffroy Saint-Hilaire, "Miscellaneous Notes in Egypt (1798–1801 and 1830s)," Étienne Geoffroy Saint Hilaire Collection, Mss.B.G287p, American Philosophical Society, Philadelphia.
26. Toby A. Appel, *The Cuvier-Geoffroy Debate: French Biology in the Decades before Darwin* (New York: Oxford University Press, 1987).
27. One biographer goes so far as to liken Geoffroy Saint-Hilaire's time in Egypt to Darwin's voyage on the *Beagle*. Le Guyader, *Étienne Geoffroy Saint-Hilaire*, 21–22.
28. Charles Darwin, *On the Origin of Species by Means of Natural Selection, or the Preservation of Favoured Races in the Struggle for Life* (London: John Murray, 1859), 434.
29. For a collection of studies of early modern (mostly European) zoology, see Karl A. E. Enenkel and Paul J. Smith, eds., *Early Modern Zoology: The Construction of Animals in Science, Literature and the Visual Arts*, 2 vols. (Leiden: Brill, 2007).
30. ʿAbd al-Karīm, *Tārīkh al-Taʿlīm*, 309.
31. Antoine Barthélemy Clot-Bey, *Aperçu general sur l'Égypte*, 2 vols. (Bruxelles: Meline, cans, 1840), 2: 440. Of the two, Hamont would play the more important role in the history of veterinary medicine in Egypt since Pretot quickly fell ill upon his arrival in Egypt and soon returned to France where he died less than a year after first coming to Egypt. ʿAbd al-Karīm, *Tārīkh al-Taʿlīm*, 309. James Heyworth-Dunne writes that Pretot died in Smyrna. James Heyworth-Dunne, *An Introduction to the History of Education in Modern Egypt* (London: Luzac, 1939), 133. For Hamont's general comments on Egypt's animals, see P. N. Hamont, *L-Egypt sous Méhémet-Ali*, 2 vols. (Paris: Leautey et Lecointe, 1843), 1: 523–61. Hamont knew Étienne Geoffroy Saint-Hilaire and consulted him on various veterinary matters in Egypt. See, for example, Pierre Nicolas Hamont, "Observations communiqué par M. P.N. Hamont, medecin vétérinaire d'Alfot professeur de medecine veterinaire en Egypte au Service se S.A. Mehamet Ali Pacha (27 April 1829)," Étienne Geoffroy Saint Hilaire Collection, Mss.B.G287p, American Philosophical Society, Philadelphia.
32. ʿAbd al-Karīm, *Tārīkh al-Taʿlīm*, 310. On these points, ʿAbd al-Karīm cites Hamont, *L-Egypt sous Méhémet-Ali*, 2: 122–24.
33. Generally on the history of the school of veterinary medicine, see Clot-Bey, *Aperçu general*, 2: 437–48.
34. Interestingly, the only book on science or technology to be published in both Turkish and Arabic by Mehmet ʿAli's new Būlāq Press (founded in 1822) was a textbook of veterinary medicine. Richard N. Verdery, "The Publications of the Būlāq Press under Muḥammad ʿAlī of Egypt," *Journal of the American Oriental Society* 91 (1971), 131. For a complete list of the books available from the Būlāq Press in 1845, including nine on veterinary medicine, two of which were the Arabic and Turkish versions of the text just mentioned (*Ṭibb Bayṭarī Qānūnnāmasī*), see the following press catalog: *Mısır'da Bir Kütüphaneye ait Fihrist*, SK, Hüsrev Paşa 828. For a general history of the Būlāq Press and its publications, see Abū al-Futūḥ Raḍwān, *Tārīkh Maṭbaʿat Būlāq wa Lamḥa fī Tārīkh al-Ṭibāʿa fī Buldān al-Sharq al-Awsaṭ* (Cairo: al-Maṭbaʿa al-Amīriyya, 1953); J. Heyworth-Dunne, "Printing and Translations under Muḥammad ʿAlī of Egypt: The Foundation of Modern Arabic," *Journal of the Royal Asiatic Society* 72 (1940): 325–49.
35. ʿAbd al-Karīm, *Tārīkh al-Taʿlīm*, 313; Heyworth-Dunne, *History of Education in Modern Egypt*, 133.
36. ʿAbd al-Karīm, *Tārīkh al-Taʿlīm*, 313.
37. For his part, Clot-Bey was also very interested in Egypt's animals. Clot-Bey, *Aperçu general*, 1: 203–40.
38. ʿAbd al-Karīm, *Tārīkh al-Taʿlīm*, 313.
39. There is some discrepancy here. Amira el-Azhary Sonbol puts the school's highest enrollment at one hundred, while ʿAbd al-Laṭīf Muḥammad al-Ṣabbāgh claims it was only fifty. Amira el-Azhary Sonbol, *The Creation of a Medical Profession in Egypt, 1800–1922* (Syracuse, NY: Syracuse University Press, 1991), 43; ʿAbd al-Laṭīf

Muḥammad al-Ṣabbāgh, "al-Ṭibb al-Bayṭarī fī Miṣr (1828–1849): Dirāsa Wathā'iqiyya," *Miṣr al-Ḥadītha* 8 (2009), 24.

40. Heyworth-Dunne, *History of Education in Modern Egypt*, 136.
41. About this military confrontation, see Fahmy, *All the Pasha's Men*; Kutluoğlu, *The Egyptian Question*.
42. Hamont, *L-Egypt sous Méhémet-Ali*, 1: 523–39 and 562–73.
43. Hamont's rival Clot-Bey also made special note of the very poor state of Egypt's horses. Clot-Bey, *Aperçu general*, 2: 439–40.
44. ʿAbd al-Karīm, *Tārīkh al-Taʿlīm*, 319. In 1836, the Agriculture School was also transferred to Shubra. Hamont would come to oversee both institutions. Heyworth-Dunne, *History of Education in Modern Egypt*, 134 and 152.
45. Clot-Bey, *Aperçu general*, 2: 445–48.
46. al-Ṣabbāgh, "al-Ṭibb al-Bayṭarī," 30–31.
47. Ibid., 31.
48. ʿAbd al-Munʿim Ibrāhīm al-Jumayʿī, ed., *Wathā'iq al-Taʿlīm al-ʿĀlī fī Miṣr khilāl al-Qarn al-Tāsiʿ ʿAshar* (Cairo: Dār al-Kutub wa al-Wathā'iq al-Qawmiyya, 2004), 290.
49. al-Ṣabbāgh, "al-Ṭibb al-Bayṭarī," 43. Given the distances involved and the organizational time needed, veterinarians in the countryside were instructed to stock up on medicines to make sure they did not run out.
50. Ibid., 39–40. While, as Chapter Two shows, domestic animals played a far less central role in the labor regime of the countryside in the early nineteenth century, they obviously did not disappear entirely and were still often used for various purposes.
51. Ibid., 40.
52. Ibid.
53. Another example of extreme temperatures leading to problems for animal populations is the following case from the late nineteenth century about very cold weather preventing the transport of dogs, chickens, and pigeons to the imperial palace in Istanbul: BOA, Yıldız Perakende Evrakı Mabeyn Başkitabeti 38.67 (2 C 1312/1 Dec. 1894).
54. al-Ṣabbāgh, "al-Ṭibb al-Bayṭarī," 40–41.
55. Egyptian veterinarians were generally very concerned about what animals ate, and there thus developed a whole section of the veterinary establishment charged with overseeing dietary health. Ibid., 41.
56. Ibid.
57. Ammonia in the form of sal ammoniac was made in Giza through the crystallization of the gas produced by the slow burning of a mixture of animal dung and straw. Kerr, "Letters," 10.
58. For a comparison with human quarantines in this period, see Mikhail, *Nature and Empire*, 230–41.
59. al-Ṣabbāgh, "al-Ṭibb al-Bayṭarī," 44.
60. Such breeding efforts would continue throughout the nineteenth century. For example, on behalf of the Egyptian Khedive, British viceroy of Egypt Lord Cromer sent a letter from Cairo in 1891 to one of his fellow colonial officers in the Arabian Peninsula stating that "the Egyptian Government wish to buy a good donkey stallion for breeding purposes—one of the large Hedjaz breed." TNA, FO 633/5 (22 Apr. 1891). For a useful analysis of colonial breeding attempts aimed as species "improvement," see Greg Bankoff, "A Question of Breeding: Zootechny and Colonial Attitudes toward the Tropical Environment in the Late Nineteenth-Century Philippines," *Journal of Asian Studies* 60 (2001): 413–37.
61. al-Ṣabbāgh, "al-Ṭibb al-Bayṭarī," 44.
62. For a discussion of other Ottoman attempts to undertake merino breeding in the nineteenth century, see M. Erdem Kabadayı, "The Introduction of Merino Sheep Breeding in the Ottoman Empire: Successes and Failures," in *Animals and People in the Ottoman Empire*, ed. Suraiya Faroqhi (Istanbul: Eren, 2010), 153–69.
63. al-Ṣabbāgh, "al-Ṭibb al-Bayṭarī," 44.
64. Ibid., 31–32 and 45–47.
65. From the time of its transfer to Shubra in 1837 to its closing twelve years later, the school's enrollment steadily dropped: from 122 in 1837 to 79 in 1839, to 64 in 1841, to 50 in 1842, to 11 in 1843. ʿAbd al-Karīm, *Tārīkh al-Taʿlīm*, 319.

66. On Egyptians' efforts to hide sick relatives from medical officials, see Kuhnke, *Lives at Risk*, 91; Mikhail, *Nature and Empire*, 230–36.
67. ʿAbd al-Karīm, *Tārīkh al-Taʿlīm*, 320–22.
68. *Ẓāhir ẓuhūr al-shams.*
69. BOA, Sadaret Mektubî Kalemi 160.94 (29 Z 1264/26 Nov. 1848). For another example of the transfer of animals from Teke and Hamid to Egypt, see BOA, İradeler Mısır 11.295 (21 Za 1260/2 Dec. 1844). For later examples of the importation of animals from Anatolia and Greater Syria to Egypt during epizootics, see BOA, Yıldız Esas ve Sadrazam Kâmil Paşa 129.96 (18 Ra 1317/27 July 1899); BOA, Meclis-i Vükelâ Mazbataları 93.3 (11 Ra 1315/10 Aug. 1897).
70. BOA, Sadaret Mühimme Kalemi Evrakı 270.86 (6 S 1280/23 July 1863); TNA, FO 83/349 (1856–66); TNA, FO 143/10 (1873–78); TNA, FO 143/11 (1880–82).
71. DWQ, ʿAI, Agriculture, 12/2 (15 Feb. 1866).
72. DWQ, ʿAI, Agriculture, 12/2 (22 May 1871).
73. al-Ṣabbāgh, "al-Ṭibb al-Bayṭarī," 47.
74. Khedive Ismail tried to overcome these problems by rehiring some of the best of the veterinary school's former staff, including a teacher named ʿAbd al-Hadi Effendi Ismail who had taught in the school in the 1830s and 1840s.
75. DWQ, ʿAI, Agriculture, 12/2 (22 May 1871).
76. DWQ, ʿAI, Agriculture, 12/2 (7 Dec. 1871).
77. DWQ, ʿAI, Agriculture, 12/2 (15 Feb. 1866).
78. DWQ, ʿAI, Agriculture, 12/2 (23 Dec. 1871).
79. DWQ, ʿAI, Agriculture, 12/2 (17 Feb. 1866). The Russian steppe was also often thought to be the origin of many human diseases. See, for example, the following case from the nineteenth century about the quarantine of Tatar *ḥajj* pilgrims in Alexandria: TSMA, E. 5026/242 (n.d.).
80. BOA, Sadaret Mühimme Kalemi Evrakı 283.60 (22 Ca 1280/4 Nov. 1863); BOA, Sadaret Mühimme Kalemi Evrakı 282.63 (22 Ca 1280/4 Nov. 1863). For British concerns from roughly the same period about the spread of cattle plague from Odessa, see TNA, FO 83/349 (14 Mar. 1866).
81. BOA, Sadaret Mühimme Kalemi Evrakı 270.86 (21 M 1280/8 July 1863).
82. BOA, Sadaret Mühimme Kalemi Evrakı 296.37 (29 N 1279/20 Mar. 1863).
83. Even animals in cities at some distance from major ports were affected by epizootics in this period. For a case about the spread of cattle plague in landlocked Kayseri in 1849, see BOA, Sadaret Mektubî Kalemi Deavî Yazışmaları 13.60 (21 S 1265/16 Jan. 1849).
84. DWQ, ʿAI, Agriculture, 12/2 (15 Feb. 1866).
85. TNA, FO 83/349 (1856–66); TNA, FO 83/350 (1869); TNA, FO 83/351 (1870); TNA, FO 83/352 (1870); TNA, FO 83/353 (1870).
86. See, for example, the following letter dated May 30, 1856, written by a British official named J. A. Blackwell stationed in the Hanseatic port of Lübeck on the Baltic Sea: TNA, FO 83/349 (1856–66).
87. On the history of one of these ports—Qusayr—see Daniel Crecelius, "The Importance of Qusayr in the Late Eighteenth Century," *Journal of the American Research Center in Egypt* 24 (1987): 53–60.
88. On the history of Arabian horse breeding, see Derry, *Bred for Perfection*, 103–55; Charles Gladitz, *Horse Breeding in the Medieval World* (Dublin: Four Courts Press, 1997). On Egyptian Arabians in particular, see Philippe Paraskevas, *Breeding the Arabian Horse*, vol. 1 of *The Egyptian Alternative* (Exeter: Obelisque Publications, 2010); idem, *In Search of the Identity of the Egyptian Arabian Bloodlines*, vol. 2 of *The Egyptian Alternative* (Exeter: Obelisque Publications, 2012); Nasr Marei, *The Arabian Horse of Egypt* (Cairo: American University in Cairo Press, 2010).
89. For an example of the quarantine of animals sent from Egypt, see the following case about quarantine in the Anatolian port of Mersin: BOA, Cevdet Sıhhiye 611 (24 Ş 1259/19 Sept. 1843). See also the following case from the end of the nineteenth century about preventing the movement of animals, leather, and wool from Egypt during an epizootic: BOA, Sadaret Mühimme Kalemi Evrakı 578.19 (19 S 1317/29 June 1899).

90. For a later attempt by the Ottoman imperial administration to prevent the spread of animal diseases by outlawing the importation of animals to Egypt from Anatolia and Greater Syria, see BOA, DMK 2306.116 (15 L 1317/16 Feb. 1900).
91. TNA, FO 143/10 (1873–78).
92. See, for instance, TNA, FO 141/70 (19 Nov. 1869).
93. TNA, FO 143/10 (1873–78); TNA, FO 143/11 (1880–82).
94. TNA, FO 143/11 (1880–82).
95. For some of the many examples of efforts to quarantine humans in this period, see TNA, FO 141/131 (1879); TNA, FO 633/3 (1880–83). For a case about the quarantine of Tatar *ḥajj* pilgrims in Alexandria in the nineteenth century, see TSMA, E. 5026/242 (n.d.). For a general discussion of quarantine in nineteenth-century Egypt, see Kuhnke, *Lives at Risk*, 89–110 and 154–57.
96. DWQ, ʿAI, Agriculture, 12/2 (15 Feb. 1866).
97. For examples of early modern quarantine efforts aimed at human plague, see Daniel Panzac, *Quarantaines et lazarets: l'Europe et la peste d'Orient (XVIIe–XXe siècles)* (Aix-en-Provence: Édisud, 1986); Charles Carrière, Marcel Courdurié, and Ferréol Rebuffat, *Marseille ville morte: la peste de 1720* (Marseille: M. Garçon, 1968); ACM, LC 12, 15r–15v (30 May 1765); ACM, LC 12, 92r–93r (27 Aug. 1782).
98. Suraiya Faroqhi, "Horses Owned by Ottoman Officials and Notables: Means of Transportation but also Sources of Pride and Joy," in *Animals and People in the Ottoman Empire*, ed. Suraiya Faroqhi (Istanbul: Eren, 2010), 293–311.
99. On confrontations between Bedouin horsemen and Ottoman cavalry soldiers in Egypt, see Tietze, *Muṣṭafā ʿĀlī's Description of Cairo*, 34–35. On the competition for horses between soldiers, see ibid., 54.
100. For a broad account of the role of horses in Ottoman and Turkish history, see Kudret Emiroğlu and Ahmet Yüksel, *Yoldaşımız At* (Istanbul: Yapı Kredi Yayınları, 2009).
101. On some of the history of the crucial military roles of horses, see Pekka Hämäläinen, *The Comanche Empire* (New Haven, CT: Yale University Press, 2008); Digby, *War-Horse and Elephant in the Delhi Sultanate*; Jos Gommans, "The Horse Trade in Eighteenth-Century South Asia," *Journal of the Economic and Social History of the Orient* 37 (1994): 228–50; idem., "Warhorse and Post-Nomadic Empire in Asia, c. 1000–1800," *Journal of Global History* 2 (2007): 1–21; Arash Khazeni, "Through an Ocean of Sand: Pastoralism and the Equestrian Culture of the Eurasian Steppe," in *Water on Sand: Environmental Histories of the Middle East and North Africa*, ed. Alan Mikhail (New York: Oxford University Press, 2013), 133–58.
102. Clot-Bey, *Aperçu general*, 2: 219. Generally on the history of the cavalry horse, see Louis A. DiMarco, *War Horse: A History of the Military Horse and Rider* (Yardley: Westholme Publishing, 2008); Robert Drews, *Early Riders: The Beginnings of Mounted Warfare in Asia and Europe* (New York: Routledge, 2004).
103. Muḥammad Fuʾād Shukrī, ʿAbd al-Maqṣūd al-ʿInānī, and Sayyid Muḥammad Khalīl, *Bināʾ Dawlat Miṣr Muḥammad ʿAlī*, 2 vols. (Cairo: Dār al-Kutub wa al-Wathāʾiq al-Qawmiyya, 2009), 165. On the importation to Egypt from Greater Syria of horses and other animals for military purposes, see BOA, İradeler Dahiliye 1306.1311/S–04 (2 S 1311/15 Aug. 1893); BOA, Yıldız Esas ve Sadrazam Kâmil Paşa 84.50 (20 B 1318/13 Nov. 1900); BOA, İradeler Mısır 30.1731 (15 B 1318/8 Nov. 1900); BOA, DMK 2547.27 (9 B 1319/22 Oct. 1901); BOA, DMK 499.30 (29 M 1320/8 May 1902); BOA, BEO 1837.137725 (21 M 1320/30 Apr. 1902); BOA, DMK 733.13 (9 R 1321/5 July 1903); BOA, Yıldız Perakende Evrakı Umum Vilayetler Tahriratı 65.41 (9 R 1321/5 July 1903); BOA, DMK 555.67 (2 Ca 1320/7 Aug. 1902); BOA, BEO 40.2934 (2 M 1310/27 July 1892); BOA, Yıldız Perakende Evrakı Umum Vilayetler Tahriratı 64.25 (6 S 1321/4 May 1903); BOA, DMK 2906.46 (5 Ş 1327/22 Aug. 1909); BOA, DMK 2913.38 (12 Ş 1327/29 Aug. 1909). On the export of animals from Egypt for military uses elsewhere, see BOA, BEO 1726.129409 (17 C 1319/1 Oct. 1901); BOA, DMK 2537.21 (15 C 1319/29 Sept. 1901); BOA, DMK 2493.27 (16 S 1319/4 June 1901); BOA, Yıldız Perakende Evrakı Umum Vilayetler Tahriratı 56.10 (8 C 1319/22 Sept. 1901); BOA, DMK 455.32 (11 Z 1319/21 Mar. 1902); BOA, DMK 487.36 (15 M 1320/24 Apr. 1902). More generally on

the military uses of animals in the middle of the nineteenth century and later, see BOA, Sadaret Mektubî Kalemi Nezaret ve Devâir Yazışmaları 140.42 (10 B 1271/29 Mar. 1855); BOA, Sadaret Mektubî Kalemi Nezaret ve Devâir Yazışmaları 138.47 (29 C 1271/19 Mar. 1855); BOA, Sadaret Âmedî Kalemi 57.22 (1271/1854 and 1855); BOA, Meclis-i Vükelâ Mazbataları 195.15 (6 M 1333/24 Nov. 1914).

104. Clot-Bey, *Aperçu general*, 2: 219–20; Shukrī, al-ʿInānī, and Khalīl, *Bināʾ Dawlat Miṣr Muḥammad ʿAlī*, 166.
105. Fahmy, *All the Pasha's Men*, 65.
106. Ibid., 165.
107. For a useful comparative discussion of the military uses of Arabian horses by Ottoman and British armies in the nineteenth century, see Fattah, *Politics of Regional Trade*, 160–61 and 171–77.
108. For a discussion of how nineteenth-century developments in dressage and cavalry impacted the study of the physiology of locomotion in France, see Andreas Mayer, "The Physiological Circus: Knowing, Representing, and Training Horses in Motion in Nineteenth-Century France," *Representations* 111 (2010): 88–120.
109. Fahmy, *All the Pasha's Men*, 144–45; Mikhail, *Nature and Empire*, 170–200.
110. For an account of Noyes's stay in Egypt, see the relevant sections of his travel journal from May 1854: George F. Noyes Travel Journal (1850–1854), BANC MSS C–F 48, The Bancroft Library, University of California, Berkeley.
111. DWQ, ʿAI, al-Isṭablāt, 39/10 (5 Oct. 1865).
112. DWQ, ʿAI, al-Isṭablāt, 39/10 (30 July 1869).
113. DWQ, ʿAI, al-Isṭablāt, 39/10 (30 July 1869); DWQ, ʿAI, al-Isṭablāt, 39/10 (1 Sept. 1869).
114. Letter, 30 March 1881, from Arthur Rodgers to His Father, Arthur Rodgers Correspondence (1865–1902).
115. For a fourteenth-century account of the zoological and medicinal properties of silkworms, see Qazvīnī, *Zoological Section of the Nuzhatu-l-Qulūb*, 41–42.
116. For an exegesis of the development of the concept "animal capital," see Nicole Shukin, *Animal Capital: Rendering Life in Biopolitical Times* (Minneapolis: University of Minnesota Press, 2009).
117. For earlier examples of some small-scale silkworm cultivation in 1816 and 1817, see al-Jabartī, *ʿAjāʾib al-Āthār* (1994), 4: 360 and 396.
118. DWQ, ʿAI, Tarbiyyat Dūdat al-Qazz, 12/10 (n.d.).
119. I have been unable to locate any further information about this man.
120. DWQ, ʿAI, Tarbiyyat Dūdat al-Qazz, 12/10 (26 Apr. 1871). For more on these Italian imports, see B. Francis Cobb, "Report on Raw Silk," *The Journal of the Society of Arts and Institutions in Union, and Official Record of Annual International Exhibitions* 21 (1872–73), 729. The exact name and location of this town are unclear.
121. On Anketell's work with silk in Egypt, see also *The Entomologist's Annual for 1871* (London: John Van Voorst, 1871), 103–04; Cobb, "Report on Raw Silk," 729; *The London and China Telegraph* (4 Oct. 1869), 499.
122. DWQ, ʿAI, Tarbiyyat Dūdat al-Qazz, 12/10 (24 May 1871).
123. For a useful discussion of mulberry trees, silkworms, and the development of sericulture, see Charles E. Hatch, Jr., "Mulberry Trees and Silkworms: Sericulture in Early Virginia," *Virginia Magazine of History and Biography* 65 (1957): 3–61.
124. DWQ, ʿAI, Promenades et Plantations, Jardin Zoologique, 62/4 (23 Mar. 1871). Sauvadon was part of the community of zoologists associated with the Jardin Zoologique d'Acclimatation du Bois de Boulogne directed by Isidore Geoffroy Saint-Hilaire, son of Étienne Geoffroy Saint-Hilaire. He likely learned a great about Egypt's fauna from his work with the Geoffroy Saint-Hilaires and was an active participant in discussions about acclimatization in mid-nineteenth-century France. On the history of acclimatization, see Michael A. Osborne, *Nature, the Exotic, and the Science of French Colonialism* (Bloomington: Indiana University Press, 1994); Warwick Anderson, "Climates of Opinion: Acclimatization in Nineteenth-Century France and England," *Victorian Studies* 35 (1992): 135–57; Harriet Ritvo, "Going Forth and Multiplying: Animal Acclimatization and Invasion," *Environmental History* 17 (2012): 404–14.

125. On the life and work of Barillet-Deschamps, see Luisa Limido, *L'art des jardins sous le Second Empire: Jean-Pierre Barillet-Deschamps, 1824–1873* (Seyssel: Champ Vallon, 2002). Specifically on his work in Egypt, see Abu-Lughod, *Cairo*, 105–08, 112, 141–42.
126. DWQ, ʿAI, Tarbiyyat Dūdat al-Qazz, 12/10 (24 May 1871).
127. *Reports from Her Majesty's Consuls on the Manufactures, Commerce, &c., of their Consular Districts*, vol. 22 (London: Harrison and Sons, 1872), 380.
128. This was only the beginning according to Anketell. He projected that production could increase by at least twentyfold. Ibid.
129. DWQ, ʿAI, Tarbiyyat Dūdat al-Qazz, 12/10 (4 Dec. 1876).
130. DWQ, ʿAI, Tarbiyyat Dūdat al-Qazz, 12/10 (4 June 1878). This institution was founded in 1854 by Isidore Geoffroy Saint-Hilaire, again the son of Étienne Geoffroy Saint-Hilaire. On the life and work of Isidore Geoffroy Saint-Hilaire, see Émile Blanchard, *Notice sur la vie et les travaux d'Isidore Geoffroy Saint-Hilaire* (Paris, 1862); M. Dumas, *Eloge historique de Isidore Geoffroy Saint-Hilaire* (Paris: Typ. de F. Didot frères, fils et Cie, 1872); Osborne, "Zoos in the Family," 33–42.
131. DWQ, ʿAI, Tarbiyyat Dūdat al-Qazz, 12/10 (26 May 1879).
132. The ultimate raw material export from nineteenth-century Egypt was cotton. For an analysis of the cotton economy, see E. R. J. Owen, *Cotton and the Egyptian Economy, 1820–1914: A Study in Trade and Development* (Oxford: Clarendon Press, 1969).
133. Quoted in Peter Mansfield, *The British in Egypt* (London: Weidenfeld and Nicolson, 1971), 11.
134. These are again Anketell's words. DWQ, ʿAI, Tarbiyyat Dūdat al-Qazz, 12/10 (24 May 1871).
135. Berger, *About Looking*, 19.
136. Ibid.
137. For a recent useful discussion in this regard, see the following study of the modern zoo: Irus Braverman, *Zooland: The Institution of Captivity* (Stanford, CA: Stanford University Press, 2013).
138. On the politics of multiple forms of species rendering, see Shukin, *Animal Capital*.
139. Mitchell, *Colonising Egypt*, 4.
140. Zeynep Çelik, *Displaying the Orient: Architecture of Islam at Nineteenth-Century World's Fairs* (Berkeley: University of California Press, 1992), 145–51; Raymond, *Cairo*, 311–18; Doris Behrens-Abouseif, *Azbakiyya and Its Environs: From Azbak to Ismāʿīl, 1476–1879* (Cairo: Institut français d'archéologie orientale, 1985), 90; Abu-Lughod, *Cairo*, 105; Mitchell, *Colonising Egypt*, 17–18. One of the most esteemed of these visitors was Empress Eugenie of France, who in her own, albeit unwittingly, small way contributed to the history of the Cairo zoo. During her stay in Cairo to celebrate the opening of the Canal, her beloved pet tortoise ran away. Try as they may, she and her entourage could not find the reptile and returned to France without him. Sometime later the creature was found on the road to the Pyramids and was eventually taken to the Cairo zoo, where he died in 1949 at the age of 90. "Empress Eugenie's Turtle Dies in Cairo Zoo at 90," *New York Times*, February 12, 1949, 2.

    As part of their tours of Egypt, foreign dignitaries regularly visited the zoo in the early twentieth century. For example, after former U.S. president Theodore Roosevelt visited the Pyramids with his son in the spring of 1910, he made his way to the zoo before dinner. "Mr. Roosevelt at Cairo," *Manchester Guardian*, March 26, 1910, 6.
141. On this competition for animals, see Rothfels, *Savages and Beasts*.
142. On this point, see Thomas Veltre, "Menageries, Metaphors, and Meanings," in *New Worlds, New Animals: From Menagerie to Zoological Park in the Nineteenth Century*, ed. R. J. Hoage and William A. Deiss (Baltimore: Johns Hopkins University Press, 1996), 27.
143. DWQ, ʿAI, Promenades et Plantations, Jardin Zoologique, 62/4 (5 Dec. 1868). Animals were brought from as far away as Atmour in western Sudan near the modern Sudanese-Chadian border.
144. For later Anglo-Egyptian colonial efforts to deal with animal diseases in the Sudan, see R. T. Wilson, "The Incidence and Control of Livestock Diseases in Darfur, Anglo-Egyptian Sudan, during the Period of the Condominium, 1916–56," *International Journal of African Historical Studies* 12 (1979): 62–82.

145. For other examples of Egypt's role in the seizure of animals, plants, and physical objects from the Sudan, Abyssinia, Somalia, and other parts of Africa, see TNA, FO 141/70 (10 Apr. 1869); TNA, FO 143/10 (1876). On the Egyptian slave trade in these areas, see Terence Walz and Kenneth M. Cuno, eds., *Race and Slavery in the Middle East: Histories of Trans-Saharan Africans in Nineteenth-Century Egypt, Sudan, and the Ottoman Mediterranean* (Cairo: American University in Cairo Press, 2010). See also Ehud R. Toledano, *Slavery and Abolition in the Ottoman Middle East* (Seattle: University of Washington Press, 1998). For Maltese evidence of the slave trade in eighteenth-century Egypt, see ACM, LC 12, 148r–49r (30 Apr. 1791); ACM, LC 12, 180r–83r (22 Apr. 1763); ACM, LC 12, 199r–200r (21 June 1774).

    Ottoman commentators on slavery in the nineteenth century drew a sharp distinction between the practice in the empire and the United States. Taking for granted the far more humane treatment of human slaves in the Ottoman Empire than in the United States, the Tanzimat-era writer Ahmet Midhat, for example, described American slavery using animal imagery. In contrast to the Ottomans, Americans treated their slaves like beasts (*behaim*) and animals (*hayvanat*). Ehud R. Toledano, "Late Ottoman Concepts of Slavery (1830s–1880s)," *Poetics Today* 14 (1993), 492.

146. For Barillet-Deschamps's letter to Riaz Pacha making Sauvadon Chef de la partie zoologique du Jardin du Palais de Ghezireh, see DWQ, ʿAI, Promenades et Plantations, Jardin Zoologique, 62/4 (23 Mar. 1871).
147. DWQ, ʿAI, Promenades et Plantations, Jardin Zoologique, 62/4 A (n.d.).
148. Rothfels, *Savages and Beasts*, 13–43 and 143–88.
149. DWQ, ʿAI, Promenades et Plantations, Jardin Zoologique, 62/4 (22 Apr. 1871); DWQ, ʿAI, Promenades et Plantations, Jardin Zoologique, 62/4 B (n.d.).
150. This determination of the number of animals in the Khedive's garden was also picked up by the Egyptian press. Mona L. Russell, *Creating the New Egyptian Woman: Consumerism, Education, and National Identity, 1863–1922* (New York: Palgrave Macmillan, 2004), 32–33.
151. On the relationship between legibility and government, see Scott, *Seeing Like a State.*
152. Another manifestation of these interlocking impulses of collection, organization, and classification in the Ottoman Empire at the turn of the twentieth century was the identification of dog breeds. The imperial administration in this period began to compile and translate lists of dog breeds and their classic characteristics. See, for example, BOA, Yıldız Perakende Evrakı Elçilik, Şehbenderlik, ve Ataşemiliterlik 47.79 (3 R 1323/7 June 1905); BOA, Yıldız Mabeyn Erkanı ve Saray Görevlileri Maruzâtı 11.120 (29 Ra 1330/18 Mar. 1912). The sale of a Pomeranian in 1896 was evidently important enough to be recognized and recorded by the empire, complete with copies and translations of a certificate of registration and pedigree from the London Kennel Club. BOA, Yıldız Müteferrik 3.74 (26 B 1314/31 Dec. 1896).

    Sultan Abdülhamid II was apparently a great dog lover. He for example took much delight in an eight-month-old English mastiff given to him in 1885 by the U.S. minister to the Ottoman Empire Lew Wallace. The sultan liked to keep the dog close by his side, despite the fact that many of his advisors feared the animal because it purportedly resembled a lion. C. Max Kortepeter, "The Life and Times of General Lew Wallace, Minister Extraordinary to the Ottoman Court, 1881–1885," in *Cultural Horizons: A Festschrift in Honor of Talat S. Halman*, ed. Jayne L. Warner, 2 vols. (Syracuse, NY: Syracuse University Press; Istanbul: Yapı Kredi Yayınları, 2001), 1: 127. Further evidence of the pride he took in acquiring breeds from as many different locations as possible, Abdülhamid also enjoyed a dog given to him by his German counterpart in 1898. BOA, Yıldız Perakende Evrakı Hariciye Nezareti Maruzâtı 25.40 (4 Za 1315/27 Mar. 1898).

153. A comparative example of the logic and practice of collecting geographically disparate animals in a single institution was Louis XIV's menagerie in which animals were selected, in Peter Sahlins's words, "for their perceived characteristics of peacefulness, grace, and beauty—a certain model of *civilité*—marked by an exoticism of geographic distance more than rarity itself." Sahlins, "Royal Menageries of Louis XIV," 250.
154. On the history of the zoological specimen, see: Samuel J. M. M. Alberti, ed., *The Afterlives of Animals: A Museum Menagerie* (Charlottesville: University of Virginia Press, 2011).
155. DWQ, ʿAI, Agriculture, 12/2 (2 Feb. 1878).

156. Sauvadon had in fact studied fertilized chicken eggs and indeed gained a worldwide reputation for his scientific work. He patented a procedure which used a ten-inch tube of India rubber to test whether or not incubated eggs had been successfully fertilized. By putting eggs in these tubes and then holding them up to sunlight, one could see into the eggs to judge whether or not chicken fibers were developing properly. If by the fifth day of incubation nothing was seen inside the egg, it most likely had not been fertilized. *Appletons' Journal of Literature, Science, and Art* 12 (28 Nov. 1874), 703.
157. For a discussion of the central role birds played in embodying and modeling notions of aristocratic civility in the menagerie of Louis XIV, see Sahlins, "Royal Menageries of Louis XIV." Generally on early modern European ornithology, see Paul J. Smith, "On Toucans and Hornbills: Readings in Early Modern Ornithology from Belon to Buffon," in *Early Modern Zoology: The Construction of Animals in Science, Literature and the Visual Arts*, ed. Karl A. E. Enenkel and Paul J. Smith, 2 vols. (Leiden: Brill, 2007), 1: 75–119. See also Marcy Norton, "Going to the Birds: Animals as Things and Beings in Early Modernity," in *Early Modern Things: Objects and their Histories, 1500–1800*, ed. Paula Findlen (London: Routledge, 2013), 53–83. For a compelling analysis of the modern American companionate relationship with parrots, see Patricia K. Anderson, "A Bird in the House: An Anthropological Perspective on Companion Parrots," *Society and Animals* 11 (2003): 393–418. More broadly on the history of human-parrot relations, see Bruce Thomas Boehrer, *Parrot Culture: Our 2,500-Year-Long Fascination with the World's Most Talkative Bird* (Philadelphia: University of Pennsylvania Press, 2004).
158. The native habitats of African elephants were too far south to make their acquisition desirable or practical for Egyptian officials. Native North African elephants most likely went extinct in the sixth century CE. Ronald M. Nowak, *Walker's Mammals of the World*, 6th ed., 2 vols. (Baltimore: Johns Hopkins University Press, 1999), 2: 1002. For evidence of elephants in pre-Pharaonic Egypt, see Béatrix Midant-Reynes, *The Prehistory of Egypt: From the First Egyptians to the First Pharaohs*, trans. Ian Shaw (Oxford: Blackwell, 2000), 30, 97, 103, 149.
159. TNA, FO 143/10 (1873–78).
160. TNA, FO 141/131 (1879); TNA, FO 143/10 (1873–78).
161. TNA, FO 141/131 (12 Dec. 1879).
162. TNA, FO 141/131 (27 Jan. 1879). The men returned to India in August 1878.
163. See, for example, TNA, FO 141/131 (27 Jan. 1879); TNA, FO 141/131 (20 June 1879); TNA, FO 141/131 (15 July 1879); TNA, FO 141/131 (12 Dec. 1879).
164. DWQ, ʿAI, Agriculture, 12/2 (1 Dec. 1877).
165. In coming to this recommendation, the zoologists invoked Étienne Geoffroy Saint-Hilaire's earlier work on Egyptian animals.
166. DWQ, ʿAI, Agriculture, 12/2 (2 Feb. 1878).
167. In a later letter, Sauvadon cites the "rusticness" of Bazadaise cattle as evidence that they might also do well in Egypt. DWQ, ʿAI, Agriculture, 12/2 (4 Feb. 1878).
168. For a discussion of how the German animal supplier Carl Hagenbeck addressed both the zoological and the public relations challenges of rendering living animals to Europe, see Rothfels, *Savages and Beasts*, 50–80.
169. DWQ, ʿAI, Promenades et Plantations, Jardin Zoologique, 62/4 (15 Oct. 1878).
170. DWQ, ʿAI, Promenades et Plantations, Jardin Zoologique, 62/4 (17 Oct. 1878).
171. The following very useful modern technical guide to zoo management makes all of this very clear: Paul A. Rees, *An Introduction to Zoo Biology and Management* (Chichester: Wiley-Blackwell, 2011).
172. Never mind that descriptions of hippopotami in Egypt date back to the ancient period, with the animal serving as inspiration for Taurt, the ancient Egyptian goddess of protection in pregnancy and childbirth. Angela P. Thomas, *Egyptian Gods and Myths* (Aylesbury: Shire Publications, 1986), 13 and 60; Manfred Lurker, *The Gods and Symbols of Ancient Egypt: An Illustrated Dictionary* (London: Thames and Hudson, 1980), 119. For depictions of Taurt, see Roberto Pagliero, ed., *Aegyptica Animalia: Il bestiario del Nilo* (Torino: Museo di Antropologia ed Etnografia dell'Università di Torino, 2000), 31 and 91; David Paton, *Animals of Ancient Egypt* (Princeton, NJ: Princeton University Press, 1925), 33; Lurker, *Gods and Symbols*, 120. The hippopotamus also figured in the image of Ammit, a part-hippopotamus, part-lion,

part-crocodile monster that ate the souls of those who failed the judgment test of the dead. On Ammit, see Robert A. Armour, *Gods and Myths of Ancient Egypt* (Cairo: American University in Cairo Press, 2001), 161. More generally on the various roles of hippopotami in ancient Egyptian religious beliefs, see ibid., 37, 78–80, 83, 93, 141; Paton, *Animals of Ancient Egypt*, 32–33; Lurker, *Gods and Symbols*, 63–64; Szpakowska, *Daily Life in Ancient Egypt*, 126.

173. Vincent Stochove, "Voyage en Égypte," in *Voyage en Egypte: Vincent Stochove, Gilles Fermanel, Robert Fauvel, 1631*, ed. Baudouin van de Walle (Cairo: Institut français d'archéologie orientale, 1975), 8.

174. Anthoine Morison, *Voyage en Égypte d'Anthoine Morison, 1697*, ed. Georges Goyon (Cairo: Institut français d'archéologie orientale, 1976), 119.

175. Gonzales, *Voyage en Égypte*, 2: 653–54. Gonzales draws parallels between the medicinal lessons gained by observing hippos and those garnered from watching dogs. Ibid., 2: 614.

176. al-Jabartī, *'Ajā'ib al-Āthār* (1994), 4: 411.

177. For an illuminating comparative analysis of the development, evolution, and social role of such stories of mythic beasts (both real and imagined), see Jay M. Smith, *Monsters of the Gévaudan: The Making of a Beast* (Cambridge, MA: Harvard University Press, 2011).

178. For helpful insights from the cultural history of taxidermy, see Rachel Poliquin, *The Breathless Zoo: Taxidermy and the Cultures of Longing* (University Park: Pennsylvania State University Press, 2012); Stephen Christopher Quinn, *Windows on Nature: The Great Habitat Dioramas of the American Museum of Natural History* (New York: Harry N. Abrams, 2006).

179. Rivlin, *Agricultural Policy of Muḥammad 'Alī*, 248.

180. Some dramatic examples of how irrigation projects on the Nile hurt hippo populations come from attempts at the end of the nineteenth century to break through the Sudanese Sudd—the over-four-hundred-mile swampland that made the river impassable to ships and contributed to the evaporation of over half of the water trapped in that section of the river. During attempts to cut through these acres of swamp vegetation, large chunks of land would often break free, releasing enormous amounts of land, water, and silt. The release of these Sudd plugs, as they were known, would destroy anything in their path as they rushed downstream—people, animals, machines, ships, and buildings. In 1873, an officer of the Egyptian state witnessed one such dramatic Sudd plug explosion and wrote that "the hippopotamuses were carried down, screaming and snorting; crocodiles were whirled round and round, and the river was covered with dead and dying hippopotamuses, crocodiles, and fish who had [been] crushed by the mass." Collins, *Nile*, 60.

181. Hippos were not completely extinct in Egypt at the end of the nineteenth century. Take, for instance, the account of the British adventure traveler and hunter Samuel White Baker who published a book in 1868 concerned largely with his hunting of hippopotami and other animals in Egypt. He relates an incident in which he tracks and kills one of a group of six hippopotami with his Ceylon No. 10 double rifle. This animal's meat served as his and his wife's daily breakfast for the rest of their hunting expedition. Samuel White Baker, *Exploration of the Nile Tributaries of Abyssinia: The Sources, Supply, and Overflow of the Nile; the Country, People, Customs, Etc. Interspersed with Highly Exciting Adventures of the Author among Elephants, Lions, Buffaloes, Hippopotami, Rhinoceros, Etc. Accompanied by Expert Nature Sword Hunters* (Hartford: O. D. Case and Co., 1868), 59–66. Elsewhere Baker refers to following "a herd of hippopotami." Ibid., 354. He also expressed surprise at the fact that Egyptians did not seem to use or value hippopotamus tusks. He noted that in Europe they "are far more valuable than elephant ivory," since dentists used them to make false human teeth, a task to which they were especially suited since they did not easily change color. Ibid., 361. For other of Baker's hunting adventures with hippopotami, see ibid., 271–74, 352, 354–61, 412–18.

182. This section on ostriches relies heavily on Nicolas Manlius, "The Ostrich in Egypt: Past and Present," *Journal of Biogeography* 28 (2001): 945–53.

183. Ibid., 947–48. On the use of ostrich eggshells in ancient Egypt, see Jacke Phillips, "Ostrich Eggshells," in *Ancient Egyptian Materials and Technology*, ed. Paul T. Nicholson and Ian Shaw (Cambridge: Cambridge University Press, 2000), 332–33.

184. Borromeo, "The Ottomans and Hunting," 228–29.

185. Stochove, "Voyage en Égypte," 106; Gilles Fermanel, "Voyage en Égypte," in *Voyage en Egypte: Vincent Stochove, Gilles Fermanel, Robert Fauvel, 1631*, ed. Baudouin van de Walle (Cairo: Institut français d'archéologie orientale, 1975), 107.
186. Morison, *Voyage en Égypte*, 109; Veryard, *An Account of Divers Choice Remarks*, 300–01; Manlius, "Ostrich in Egypt," 948.
187. Gonzales, *Voyage en Égypte*, 2: 606–07.
188. Manlius, "Ostrich in Egypt," 948.
189. Ibid.
190. Ibid.
191. Ned H. Greenwood, *The Sinai: A Physical Geography* (Austin: University of Texas Press, 1997), 107.
192. Manlius, "Ostrich in Egypt," 948.
193. There was, for example, a massive rainstorm in the Delta in 1860 that wreaked havoc on ships in Alexandria and Suez destined for China, Austria, Greece, and France. TNA, BJ 7/31 (11 Feb. 1860).
194. On drought and famine in the Ottoman Empire at the end of the nineteenth century, see Mehmet Yavuz Erler, *Osmanlı Devleti'nde Kuraklık ve Kıtlık Olayları, 1800–1880* (Istanbul: Libra Kitap, 2010).
195. For general discussions of these and similar hunting reserves elsewhere, see Stephen Mosley, *The Environment in World History* (New York: Routledge, 2010), 24; Jonathan S. Adams and Thomas O. McShane, *The Myth of Wild Africa: Conservation without Illusion* (Berkeley: University of California Press, 1996), 46.
196. Greenwood, *Sinai*, 107.
197. For a long view on the history of firearms and other weapons, see Alfred W. Crosby, *Throwing Fire: Projectile Technology Through History* (Cambridge: Cambridge University Press, 2002).
198. Manlius, "Ostrich in Egypt," 950.
199. Ross Gordon Cooper and Jaroslaw Olav Horbańczuk, "Anatomical and Physiological Characteristics of Ostrich (*Struthio camelus var. domesticus*) Meat Determine Its Nutritional Importance for Man," *Animal Science Journal* 73 (2002): 167–73.
200. Manlius, "Ostrich in Egypt," 950. For other examples of the use of various animals' body parts for medicinal purposes, see Schönig, "Animal Drugs in Yemen."
201. Manlius, "Ostrich in Egypt," 950.
202. On this trade, see Sarah Abrevaya Stein, *Plumes: Ostrich Feathers, Jews, and a Lost World of Global Commerce* (New Haven, CT: Yale University Press, 2008).
203. For mention of ostrich feathers in Cairo's markets, see Gonzales, *Voyage en Égypte*, 1: 109. Later Gonzales discusses the immense danger and great difficulty involved in acquiring ostrich feathers of different colors and varieties. Ibid., 2: 606. In this regard, see also Veryard, *An Account of Divers Choice Remarks*, 304; Stochove, "Voyage en Égypte," 106.
204. al-Jabartī, *ʿAjāʾib al-Āthār* (1994), 4: 442. For more on ostriches in the Sudan, see Baker, *Exploration of the Nile Tributaries*, 574.
205. Stein, *Plumes*, 84–93.
206. Demand for ostrich feathers was itself likely also a contributing factor to population reductions. Although the most sustainable and profitable means of harvesting feathers was to keep ostriches alive for as long as possible, the black market in feathers and the lure of quick profits surely led to the deaths of many birds. In ancient Egypt, ostriches were regularly killed for their feathers. Manlius, "Ostrich in Egypt," 950.
207. The National Zoo in Washington, DC, opened two years earlier. For a brief celebratory account of the official opening of the Cairo zoo, see the following pamphlet published on the occasion of its centennial: *al-ʿĪd al-Miʾawī li-Ḥadāʾiq al-Ḥayawān, 1891–1991* (Cairo: al-Idāra al-Markaziyya li-Ḥadāʾiq al-Ḥayawān wa al-Ḥifāẓ ʿalā al-Ḥayāh al-Barriyya, 1991). See also Abu-Lughod, *Cairo*, 142.

Conclusion: The Human Ends

1. Mitchell, *Colonising Egypt*, 34.
2. On prisons, see Khaled Fahmy, "Medical Conditions in Egyptian Prisons in the Nineteenth Century," in *Marginal Voices in Literature and Society: Individual and Society in the Mediterranean Muslim World*, ed. Robin Ostle (Strasbourg: European Science Foundation in collaboration

with Maison méditerranéenne des sciences de l'homme d'Aix-en-Provence, 2000), 135–55; Rudolph Peters, "Controlled Suffering: Mortality and Living Conditions in 19th-Century Egyptian Prisons," *International Journal of Middle East Studies* 36 (2004): 387–407; idem., "Egypt and the Age of the Triumphant Prison: Legal Punishment in Nineteenth Century Egypt," *Annales Islamologiques* 36 (2002): 253–85; idem., "Prisons and Marginalisation in Nineteenth-century Egypt," in *Outside In: On the Margins of the Modern Middle East*, ed. Eugene Rogan (London: I. B. Tauris, 2002), 31–52. For a useful recent account of the global history of the emergence of prisons, see Mary Gibson, "Global Perspective on the Birth of the Prison," *American Historical Review* 116 (2011): 1040–63. On conscription, see Fahmy, *All the Pasha's Men*. For a discussion of quarantine, see Kuhnke, *Lives at Risk*; Mikhail, *Nature and Empire*, 230–41. On hospitals, see Kuhnke, *Lives at Risk*; Sonbol, *Creation of a Medical Profession in Egypt*; Khaled Fahmy, "Women, Medicine, and Power in Nineteenth-Century Egypt," in *Remaking Women: Feminism and Modernity in the Middle East*, ed. Lila Abu-Lughod (Princeton, NJ: Princeton University Press, 1998), 35–72. About schools, see Mitchell, *Colonising Egypt*, 63–94; ʿAbd al-Karīm, *Tārīkh al-Taʿlīm*; Heyworth-Dunne, *History of Education in Modern Egypt*. On asylums in Egypt, see Eugene Rogan, "Madness and Marginality: The Advent of the Psychiatric Asylum in Egypt and Lebanon," in *Outside In: On the Margins of the Modern Middle East*, ed. Eugene Rogan (London: I. B. Tauris, 2002), 104–25; Marilyn Anne Mayers, "A Century of Psychiatry: The Egyptian Mental Hospitals" (Ph.D. diss., Princeton University, 1984). On the police, see Khaled Fahmy, "The Police and the People in Nineteenth-Century Egypt," *Die Welt des Islams* 39 (1999): 340–77; Mirfat Aḥmad al-Sayyid, "Idārat al-Shurṭa fī Miṣr fī al-ʿAṣr al-ʿUthmānī," *Annales Islamologiques* 40 (2006): 51–70; ʿAbd al-Wahhāb Bakr, *al-Būlīs al-Miṣrī, 1922–1952* (Cairo: Maktabat Madbūlī, 1988).

3. On this point see, for example, the following discussion of the relationships between animality and human insanity: Michel Foucault, *History of Madness*, ed. Jean Khalfa, trans. Jonathan Murphy and Jean Khalfa (London: Routledge, 2006), 132–59. For excerpted portions of this and other of Foucault's relevant work about animals and for Clare Palmer's helpful article "Madness and Animality in Michel Foucault's *Madness and Civilization*," see Matthew Calarco and Peter Atterton, eds., *Animal Philosophy: Essential Readings in Continental Thought* (London: Continuum, 2004), 63–84. On Foucault and animals, see also Clare Palmer, " 'Taming the Wild Profusion of Existing Things'? A Study of Foucault, Power, and Human/Animal Relationships," *Environmental Ethics* 23 (2001): 339–58; Sahlins, "Royal Menageries of Louis XIV," 248–50; Ralph Acampora, "Zoos and Eyes: Contesting Captivity and Seeking Successor Practices," *Society and Animals* 13 (2005), 77–80.
4. The history of barbed wire as a technology of enclosure and form of politics makes this point as well. Barbed wire began as a way of corralling and keeping cattle and then became a means of imprisoning humans. Its history can thus only be properly understood through a human-animal analysis. Reviel Netz, *Barbed Wire: An Ecology of Modernity* (Middletown, CT: Wesleyan University Press, 2004).
5. For a discussion of Thomas Aquinas on these points, see Alves, *The Animals of Spain*, 5–6; Erica Fudge, *Brutal Reasoning*, 159. For a summary of some of the modern literature from psychology and social work about how and why violence toward animals leads to violence against humans, see Randall Lockwood and Frank R. Ascione, eds., *Cruelty to Animals and Interpersonal Violence: Readings in Research and Application* (West Lafayette, IN: Purdue University Press, 1998); Frank R. Ascione and Phil Arkow, eds., *Child Abuse, Domestic Violence, and Animal Abuse: Linking the Circles of Compassion for Prevention and Intervention* (West Lafayette, IN: Purdue University Press, 1999). See also Clifton P. Flynn, "Animal Abuse in Childhood and Later Support for Interpersonal Violence in Families," *Society and Animals* 7 (1999): 161–72; Maya Gupta, "Functional Links Between Intimate Partner Violence and Animal Abuse: Personality Features and Representations of Aggression," *Society and Animals* 16 (2008): 223–42.
6. Piers Beirne, "From Animal Abuse to Interhuman Violence? A Critical Review of the Progression Thesis," *Society and Animals* 12 (2004): 39–65.
7. In the case of Egypt, this dominant narrative is embodied in Hourani, *Arabic Thought in the Liberal Age*; al-Rāfʿī, *ʿAṣr Muḥammad ʿAlī*; Marsot, *Egypt in the Reign of Muhammad Ali*; ʿAbd al-Karīm, *Tārīkh al-Taʿlīm*; Dodwell, *Founder of Modern Egypt*.

8. I borrow this language from Horowitz, *Inside of a Dog*. The original quote is from Groucho Marx: "Outside of a dog, a book is a man's best friend. Inside of a dog, it's too dark to read."
9. For several non-Egyptian examples, see Scott, *Seeing Like a State*.
10. Adrian Franklin, *Animals and Modern Cultures: A Sociology of Human-Animal Relations in Modernity* (London: Sage, 1999); Kete, *Beast in the Boudoir*; Tuan, *Dominance and Affection*; Erica Fudge, *Pets*.
11. Bulliet, *Hunters, Herders, and Hamburgers*, 25–27 and 189–94. For a discussion of the use of dog imagery in early twentieth-century Ottoman political cartoons, see Palmira Brummett, "Dogs, Women, Cholera, and Other Menaces in the Streets: Cartoon Satire in the Ottoman Revolutionary Press, 1908–11," *International Journal of Middle East Studies* 27 (1995), 438–43.
12. Tester, *Animals and Society*; LaCapra, *History and Its Limits*; Talal Asad, *Formations of the Secular: Christianity, Islam, Modernity* (Stanford, CA: Stanford University Press, 2003), 127–58; Julian H. Franklin, *Animal Rights and Moral Philosophy* (New York: Columbia University Press, 2004); Ted Benton, *Natural Relations: Ecology, Animal Rights, and Social Justice* (London: Verso, 1993).
13. Although it is not my goal here, one could trace a genealogy of discourses about animal rights in the Muslim world. For example, in both of the following texts—the first in Arabic from the seventeenth century and the second in Persian from the eighteenth—the authors invoke the rights (*ḥuqūq*, s. *ḥaqq*) of different animals: Muḥammad ibn ʿAbd al-Muʿṭī al-Isḥāqī, *Laṭā'if Akhbār al-Uwal fī man Taṣarrafa fī Miṣr min Arbāb al-Duwal*, Beinecke, Landberg MSS 690, 120v; Amīr Muḥammad Ḥusayn ibn Muḥammad Ṣāliḥ ibn ʿAbd al-Vāsiʿ Khātūnābādī, *Muḥāsin al-Ḥiṣān*, Caro Minasian Collection of Arabic and Persian Manuscripts, University of California, Los Angeles, Box 166, Mss. 1394, 43. My thanks to Arash Khazeni for bringing the latter manuscript to my attention. For different versions of the former Arabic text, see Muḥammad ibn ʿAbd al-Muʿṭī al-Isḥāqī, *Laṭā'if Akhbār al-Uwal fī man Taṣarrafa fī Miṣr min Arbāb al-Duwal*, Beinecke, Arabic MSS 497; Muḥammad ibn ʿAbd al-Muʿṭī al-Isḥāqī, *Kitāb Laṭā'if Akhbār al-Uwal fī man Taṣarrafa fī Miṣr min Arbāb al-Duwal*, ed. Muḥammad Raḍwān Muhannā (al-Manṣūra: Maktabat al-Īmān, 2000). For studies of ideas about animal rights in Islam, see Masri, *Animal Welfare in Islam*; Ḥasanayn Muḥammad Makhlūf, *al-Rifq bi-al-Ḥayawān fī Dīn al-Islām* (Cairo: Kashīda lil-Nashr wa al-Tawzīʿ, 2011); Nadeem Haque and Al-Hafiz Basheer Ahmad Masri, "The Principles of Animal Advocacy in Islam: Four Integrated Ecognitions," *Society and Animals* 19 (2011): 279–90; Muḥammad al-Zaybaq, *The Animal: Its Particulars and Its Rights within Islam*, trans. Ṣaḥeeḥ International (Jeddah: Abul-Qasim, 2001). For other premodern examples of animal rights discourse and animal protective policies, consider the case of the late seventeenth- and early eighteenth-century Tokogawa leader Tsunayoshi, whose stance against dog violence earned him the sobriquet Dog Shogun. Beatrice M. Bodart-Bailey, *The Dog Shogun: The Personality and Policies of Tokugawa Tsunayoshi* (Honolulu: University of Hawaiʻi Press, 2006). For an extremely useful analysis of how British colonial law came to enact anti-cruelty legislation in Egypt at the end of the nineteenth century to ostensibly "save" Egypt's animals, see Esmeir, *Juridical Humanity*, 109–47.
14. Foucault, *History of Sexuality*, 143.
15. Esmeir, *Juridical Humanity*, 109–47. On the development of anti-cruelty legislation in Britain, see Brian Harrison, "Animals and the State in Nineteenth-Century England," *English Historical Review* 88 (1973): 786–820. See also James Turner, *Reckoning with the Beast: Animals, Pain, and Humanity in the Victorian Mind* (Baltimore: Johns Hopkins University Press, 1980).
16. TNA, FO 633/5, p. 389, "Lord Cromer to Mrs. Adlam" (1 Dec. 1892).
17. TNA, FO 633/5, pp. 407–08, "Lord Cromer to Tigrane Pasha" (28 Mar. 1894); TNA, FO 633/5, p. 393, "Lord Cromer to Mrs. Adlam" (2 Feb. 1893); TNA, FO 633/5, p. 389, "Lord Cromer to Mrs. Adlam" (1 Dec. 1892); TNA, FO 633/5, p. 408, "Lord Cromer to Mrs. Adlam" (5 Apr. 1894); TNA, FO 633/5, p. 414, "Lord Cromer to Mrs. Summers-Hutchinson" (20 Oct. 1894). In late 1895, *The Times* of London reported that "His Highness [Abbas Hilmi II] has also warmly supported the new law which recognizes societies for preventing cruelty to animals, inflicts fines for ill-treatment, and authorizes the destruction of incurable animals. The Cairo and Alexandria societies, which have been doing good work, will now find their efficiency greatly increased." *The Times* (November 29, 1895), 5. The Egyptian society was

established in 1883 through the efforts of Cromer and with the support of Khedive Tawfiq; other governmental officials; and the consuls general of Italy, the United States, and Holland. Samera Esmeir, *Juridical Humanity*, 126.

18. Roger Owen, *Lord Cromer: Victorian Imperialist, Edwardian Proconsul* (New York: Oxford University Press, 2004), 381; Esmeir, *Juridical Humanity*, 126 and 138.
19. John T. Chalcraft, *The Striking Cabbies of Cairo and Other Stories: Crafts and Guilds in Egypt, 1863–1914* (Albany: State University of New York Press, 2004), 165. For a wider discussion of anti-cruelty legal proceedings in colonial Egypt, see Esmeir, *Juridical Humanity*, 126–29.
20. J. E. Marshall, *The Egyptian Enigma, 1890–1928* (London: John Murray, 1928), 34–35, also cited in Chalcraft, *Striking Cabbies*, 165. Marshall goes on to describe other examples of Egyptian cruelty toward animals. He writes: "He [the Egyptian] is, through sheer ignorance, very cruel to his animals. He thinks nothing of tethering his cow or buffalo by the ear with a rope, which in course of time cuts right through it, or of leaving untended any saddle gall of his donkey, or his horse, or of putting the same saddle over a festering wound, which must cause intense agony to the poor brute." Marshall, *Egyptian Enigma*, 35.
21. For a discussion of some of these beatings, see Chalcraft, *Striking Cabbies*, 165–77.
22. In Samera Esmeir's words, "The story of rescuing Egyptians by colonization and the story of rescuing animals therefore ran along parallel lines." Esmeir, *Juridical Humanity*, 124. On the connections between human and animal rights, see E. S. Turner, "Animals and Humanitarianism," in *Animals and Man in Historical Perspective*, ed. Joseph and Barrie Klaits (New York: Harper and Row, 1974), 144–69; Franklin, *Animal Rights and Moral Philosophy*; Benton, *Natural Relations*; Tester, *Animals and Society*.
23. Foucault, *Order of Things*, 160.
24. Merchant, *Death of Nature*, xviii.
25. Netz, *Barbed Wire*, 16.

BIBLIOGRAPHY

Archival Sources

ARCHIVES OF THE CATHEDRAL OF MDINA, MDINA

Lettere Consolari 12

ARCHIVIO DI STATO, VENICE

Lettere e Scritture Turchesche 2, 3, 4
Miscellanea Documenti Algeri, Egitto, Marocco, Persia, Tripoli, Tunisi
  Documenti Algeri 1, 5
  Documenti Egitto 1, 4
  Documenti Marocco 2, 3, 4, 7, 9–10
Miscellanea Documenti Turchi 167, 169, 253, 254, 554, 714, 933, 1050, 1193, 1329
Senato, Deliberazioni, Costantinopoli 1

BAŞBAKANLIK OSMANLI ARŞİVİ, ISTANBUL

Ali Emiri Üçüncü Ahmed 10344, 10345
Bâb-ı Âlî Evrak Odası 40.2934, 1076.80636, 1726.129409, 1837.137725, 1844.138239, 2918.218845, 2997.224730, 3179.238384
Cevdet Adliye 3271
Cevdet Askeriye 16011, 28573, 38353
Cevdet Belediye 3083, 3526, 5177
Cevdet Dahiliye 1722, 10256
Cevdet Evkaf 1902
Cevdet Hariciye 3089
Cevdet İktisat 33.1640
Cevdet Maarif 3798
Cevdet Maliye 15566
Cevdet Nafia 120
Cevdet Saray 328, 977, 2051, 2067, 2152, 2274, 2999, 3559, 3660, 4301, 4499, 4640, 4817, 5088, 5842, 6016, 6410, 6460, 6483, 6637, 6683, 6712, 6778, 7256, 8605, 8909, 8981
Cevdet Sıhhiye 611
Dahiliye Islahat 14.4
Dahiliye İdare 55.5, 101.4
Dahiliye Mektubi Kalemi 455.32, 487.36, 489.56, 499.30, 555.67, 597.29, 714.8, 733.13, 911.30, 1471.105, 1475.91, 1609.56, 1613.94, 1616.30, 1699.6, 2306.116, 2493.27, 2537.21, 2547.27, 2906.46, 2913.38

Dahiliye Nezareti Kalem-i Mahsûs Müdüriyeti 43.36
Hariciye Nezareti İstanbul Murahhaslığı 82.42
Hariciye Nezareti Mektubî Kalemi 161
Hariciye Nezareti Tercüme Odası 260.83
Hatt-ı Hümayun 16/716A, 28/1354, 29/1361, 86/3520, 88/3601, 95/3856A, 119/4849, 209/11213, 240/13451, 245/13801A, 246/13857, 1399/56283, 1412/57500, 1446/34, 1451/17
İbnülemin Askeriye 7691
İradeler Dahiliye 643.44739, 1306.1311/S–04, 2574
İradeler Hususi 96.1320/M–106
İradeler Mısır 11.276, 11.295, 16.528, 16.537, 17.608, 30.1731
Maarif Nezareti Evrakı 65.99
Meclis-i Vâlâ Riyâseti 529.107, 761.43, 771.19, 780.53, 1038.10
Meclis-i Vükelâ Mazbataları 93.3, 111.88, 195.15
Mühimme-i Mısır 1, 3, 4, 5, 6, 8, 9, 10, 11, 12, 13
Sadaret Âmedî Kalemi 57.22, 62.40, 62.41, 78.36
Sadaret Mektubî Kalemi 160.94
Sadaret Mektubî Kalemi Deavî Yazışmaları 13.60
Sadaret Mektubî Kalemi Nezaret ve Devâir Yazışmaları 138.47, 140.42, 140.99
Sadaret Mühimme Kalemi Evrakı 260.65, 270.86, 282.63, 283.60, 296.36, 296.37, 578.19
Şûra-yı Devlet 594.25
Yıldız Esas ve Sadrazam Kâmil Paşa 84.50, 129.96
Yıldız Maarif Nezareti Maruzâtı 1.11
Yıldız Mabeyn Erkanı ve Saray Görevlileri Maruzâtı 11.120
Yıldız Müteferrik 3.74
Yıldız Perakende Evrakı Askerî Maruzat 29.84
Yıldız Perakende Evrakı Elçilik, Şehbenderlik, ve Ataşemiliterlik 47.79
Yıldız Perakende Evrakı Hariciye Nezareti Maruzâtı 25.40
Yıldız Perakende Evrakı Mabeyn Başkitabeti 6.39, 38.67, 81.63
Yıldız Perakende Evrakı Umum Vilayetler Tahriratı 31.17, 56.10, 64.25, 65.41
Yıldız Sadaret Hususî Maruzat Evrakı 171.54, 173.49, 261.149

DĀR AL-WATHĀʾIQ AL-QAWMIYYA, CAIRO

ʿAṣr Ismāʿīl (Documents concernant Le regime de Khédive Ismail)
Agriculture 12/2
al-Isṭablāt 39/10
Promenades et Plantations, Jardin Zoologique 62/4
Tarbiyyat Dūdat al-Qazz 12/10
al-Dīwān al-ʿĀlī 1, 2
al-Jusūr al-Sulṭāniyya 784
Maḍābiṭ al-Daqahliyya 19, 20
Maḥkamat Asyūṭ 1, 2, 4, 5, 7, 8, 9
Maḥkamat al-Baḥayra 5, 8, 10, 11, 16, 21, 22, 23, 24, 25, 26, 30, 37, 38
Maḥkamat Isnā 3, 6, 8
Maḥkamat Manfalūṭ 1, 2, 3
Maḥkamat al-Manṣūra 1, 3, 4, 7, 11, 12, 14, 16, 17, 18, 19, 24
Maḥkamat Miṣr al-Qadīma 90
Maḥkamat Rashīd 122, 123, 124, 125, 130, 132, 134, 137, 139, 142, 144, 145, 146, 148, 151, 154, 157
Taḥrīr Aṭyān al-Sharāqī wa al-Riyy 3019, 3020, 3021, 3022, 3023, 3024, 3025, 3026, 3027, 3028

KAIREIOS LIBRARY OF THE ISLAND OF ANDROS, CHORA

11, 82, 191, 195, 212, 277, 318, 354

THE NATIONAL ARCHIVES OF THE UNITED KINGDOM, KEW

Colonial Office 1069/855

Foreign Office 24/1, 83/349, 83/350, 83/351, 83/352, 83/353, 141/70, 141/131, 143/10, 143/11, 633/3, 633/5, 633/45, 926/8, 926/14, 926/15
Ministry of Transport 23/4, 23/13, 23/15, 23/18, 23/27
Privy Council 1/19/24
Records of the Meteorological Office 7/31
Treasury 1/12928

TOPKAPI SARAYI MÜZESİ ARŞİVİ, ISTANBUL

Evrak 510, 664/4, 664/7, 664/10, 664/40, 664/51, 664/52, 664/59, 664/60, 664/63, 664/64, 664/66, 1173/72, 3522, 4675/2, 4830, 5026/242, 5207/49, 5207/57, 5207/58, 7016/95, 9320/1

Manuscript Sources

AMERICAN PHILOSOPHICAL SOCIETY, PHILADELPHIA

Étienne Geoffroy Saint Hilaire Collection. Mss.B.G287p.

BANCROFT LIBRARY, UNIVERSITY OF CALIFORNIA, BERKELEY

Arthur Rodgers Correspondence (1865–1902). BANC MSS 2004/224 cz.
George F. Noyes Travel Journal (1850–1854). BANC MSS C–F 48.

BEINECKE RARE BOOK AND MANUSCRIPT LIBRARY, YALE UNIVERSITY, NEW HAVEN

al-ʿAwfī, Ibrāhīm ibn Abī Bakr. *Tarājim al-Ṣawāʿiq fī Wāqiʿat al-Ṣanājiq*. Landberg MSS 228.
al-Damīrī, Muḥammad ibn Mūsā. *Ḥayāt al-Ḥayawān al-Kubrā*. 2 vols. Salisbury MSS 78–79.
Dhikr Kalām al-Nās fī Manbaʿ al-Nīl wa Makhrajihi wa Ziyādatihi. Landberg MSS 365.
al-Ḥallāq, Muḥammad ibn Yūsuf. *Tuḥfat al-Aḥbāb bi-man Malaka Miṣr min al-Mulūk wa al-Nūwāb*. Landberg MSS 229.
Ibn ʿAbd al-Ghanī al-Ḥanafī al-Miṣrī, Aḥmad Shalabī. *Awḍaḥ al-Ishārāt fīman Tawallā Miṣr al-Qāhira min al-Wuzarāʾ wa al-Bāshāt*. Landberg MSS 3.
Ibn Abī al-Surūr al-Bakrī al-Ṣiddīqī, Muḥammad. *al-Nuzha al-Zahiyya fī Dhikr Wulāt Miṣr wa al-Qāhira al-Muʿizziyya*. Landberg MSS 231.
Ibn al-Marzubān, Muḥammad ibn Khalaf. *Faḍl al-Kilāb ʿalā Kathīr mimman Labisa al-Thiyāb*. Landberg MSS 350.
Ibn Ẓahīra, Muḥammad ibn Muḥammad. *al-Faḍāʾil al-Bāhira fī Maḥāsin Miṣr wa al-Qāhira*. Arabic MSS suppl. 395.
———. *al-Faḍāʾil al-Bāhira fī Maḥāsin Miṣr wa al-Qāhira*. Landberg MSS 105.
Ibn Zunbul, Aḥmad ibn ʿAlī. *Ghazwat al-Sulṭān Salīm Khān maʿa al-Sulṭān al-Ghūrī*. Landberg MSS 461.
al-Isḥāqī, Muḥammad ibn ʿAbd al-Muʿṭī. *Laṭāʾif Akhbār al-Uwal fī man Taṣarrafa fī Miṣr min Arbāb al-Duwal*. Arabic MSS 497.
———. *Laṭāʾif Akhbār al-Uwal fī man Taṣarrafa fī Miṣr min Arbāb al-Duwal*. Landberg MSS 690.
al-Jāḥiẓ, Abī ʿUthmān ʿAmr ibn Baḥr. *Kitāb al-Ḥayawān*. Pt. I. Landberg MSS 236.
al-Karmī, Marʿī ibn Yūsuf. *Nuzhat al-Nāẓirīn fī Taʾrīkh man Waliya Miṣr min al-Khulafāʾ wa al-Salāṭīn*. Arabic MSS suppl. 397.
———. *Nuzhat al-Nāẓirīn fī Taʾrīkh man Waliya Miṣr min al-Khulafāʾ wa al-Salāṭīn*. Landberg MSS 11a.
———. *Nuzhat al-Nāẓirīn fī Taʾrīkh man Waliya Miṣr min al-Khulafāʾ wa al-Salāṭīn*. Landberg MSS 232.
———. *Nuzhat al-Nāẓirīn fī Taʾrīkh man Waliya Miṣr min al-Khulafāʾ wa al-Salāṭīn*. Salisbury MSS 67.
al-Madābighī, Khalīl ibn Aḥmad. *Taʾrīkh*. Landberg MSS 630.
al-Maqrīzī, Aḥmad ibn ʿAlī. *al-Bayān wa al-Iʿrāb ʿammā bi-Arḍ Miṣr min al-Aʿrāb*. Landberg MSS 492.
Muḥibb al-Dīn al-Ḥamawī, Muḥammad ibn Abī Bakr. *al-Durra al-Muḍīʾa fī al-Riḥla al-Miṣriyya*. Landberg MSS 427.

Rūmī, Nūḥ ibn Muṣṭafā. *Kitāb Ta'rīkh Miṣr wa al-Nīl wa Khabar man Malakahā min Ibtidā' al-Zamān*. Landberg MSS 301.
al-Shirbīnī, Yūsuf ibn Muḥammad. *Hazz al-Quḥūf fī Sharḥ Qaṣīd Abī Shādūf*. Hartford Seminary Arabic MSS 56.
al-Suyūṭī, Jalāl al-Dīn ʿAbd al-Raḥmān. *Kawkab al-Rawḍa*. Landberg MSS 202.
———. *Kawkab al-Rawḍa*. Landberg MSS 566.
———. *Mufākhara bayna al-Rawḍa wa al-Miqyās wa Miṣr al-Qāhira*. Landberg MSS 525.
Ta'rīkh Miṣr wa ʿAjā'ibuhā. Landberg MSS 659.
Thomas Fiott De Havilland Correspondence (1799–1844). OSB MSS 136.

BIBLIOTECA APOSTOLICA VATICANA, VATICAN CITY

Ibn Qara Kamāl, Abū al-Fayḍ Aḥmad. *Jawāhir al-Bayān fī Dawlat Āl ʿUthmān*. Vatican Arabo 870.

BIBLIOTHÈQUE NATIONALE DE FRANCE, PARIS

al-Jawharī al-Khaṭīb, Nūr al-Dīn ʿAlī ibn Dāwud ibn Ibrāhīm. *al-Durr al-Thamīn al-Manẓūm fīmā Wurida fī Miṣr wa Aʿmālihā bi-al-Khuṣūṣ wa al-ʿUlūm*. Ms. Fonds Arabe 1812.

CARO MINASIAN COLLECTION OF ARABIC AND PERSIAN MANUSCRIPTS, UNIVERSITY OF CALIFORNIA, LOS ANGELES

Khātūnābādī, Amīr Muḥammad Ḥusayn ibn Muḥammad Ṣāliḥ ibn ʿAbd al-Vāsiʿ. *Muḥāsin al-Ḥiṣān*. Box 166, Mss. 1394.

DĀR AL-KUTUB AL-MIṢRIYYA, CAIRO

ʿAjā'ib al-Bilād wa al-Aqṭār wa al-Nīl wa al-Anhār wa al-Barārīy wa al-Biḥār. Jughrāfiyā Mīm 7.
ʿAwā'id al-Miṣriyyīn ʿand Izdiyād al-Nīl. Makhṭūṭāt al-Zakiyya 584.
Ibn al-ʿImād. *Kitāb al-Nīl al-Saʿīd al-Mubārak*. Jughrāfiyā Ḥalīm 5.
Ibn Riḍwān. *Risāla fī Ziyādat al-Nīl wa Naqṣihi ʿalā al-Dawām*. Majāmīʿ Mīm 213.
Ibn Shuʿayyib, Muḥammad. *Zahr al-Basāṭīn fī Faḍl al-Nīl wa Faḍl Miṣr wa al-Qāhira wa al-Qarāfa*. Buldān Taymūr 60.
Laqis al-Ḥanak fīmā Warada fī al-Samak. Makhṭūṭāt al-Zakiyya 373.
al-Maḥallī, Jalāl al-Dīn. *Kitāb Mabdāʿ al-Nīl*. Jughrāfiyā 381.
———. *Risāla fī Mabdāʿ al-Nīl*. Majāmīʿ Ṭalʿat 972.
al-Ṭahshawārī, Nūḥ Afandī. *Tārīkh Miṣr wa al-Nīl wa Khabar man Malakahā min Ibtidā' al-Zamān*. Tārīkh Ṭalʿat 2037.

AL-KHIZĀNA AL-ḤASANIYYA, RABAT

al-Ḥijāzī al-Anṣārī, Shihāb al-Dīn Aḥmad. *al-Nīl al-Rā'id min al-Nīl al-Zā'id*. Ms. 318.

KHUDA BAKHSH ORIENTAL PUBLIC LIBRARY, PATNA

al-Ḥijāzī al-Anṣārī, Shihāb al-Dīn Aḥmad ibn Muḥammad ʿAlī. *Nīl al-Rā'id fī al-Nīl al-Zā'id*. Ms. 1069.

MAʿHAD AL-MAKHṬŪṬĀT AL-ʿARABIYYA, CAIRO

Ibn ʿAbd al-Salām, Aḥmad ibn Muḥammad ibn Muḥammad. *al-Fayḍ al-Madīd fī Akhbār al-Nīl al-Saʿīd*. al-Jughrāfiyā wa al-Buldān 39.
Ibn Bakhtīshūʿ, ʿUbayd Allah ibn Jibrīl. *ʿIqd al-Jumān fī Ṭabā'iʿ al-Ḥayawān wa al-Insān*. al-Kīmiyā' wa al-Ṭabīʿiyāt 66.
al-Maqdisī, Marʿī al-Ḥanbalī. *Nuzhat al-Nāẓirīn fī Tārīkh man Waliya Miṣr min al-Khulafā' wa al-Salāṭīn*. al-Tārīkh 1283.

AL-MAKTABA AL-BALADIYYA BI-AL-ISKANDARIYYA, ALEXANDRIA

Ibn ʿAbd al-Salām, Aḥmad ibn Muḥammad ibn Muḥammad. *al-Fayḍ al-Madīd fī Akhbār al-Nīl al-Saʿīd*. Ms. 2591 Dāl.
Ibn al-ʿImād. *Kitāb al-Qaul al-Mufīd fī al-Nīl al-Saʿīd*. Ms. 4939 Dāl.
———. *Risāla fī Manbaʿ al-Nīl*. Ms. 1627 Bā'.

Risāla fī al-Nīl wa Ziyādatihi. Ms. 1707 Bāʾ (4).

AL-MAKTABA AL-WAṬANIYYA LIL-MAMLAKA AL-MAGHRIBIYYA, RABAT

al-Suyūṭī, Jalāl al-Dīn. *Kashf al-Ṣalṣala ʿan Waṣf al-Zalzala*. Ms. 1027 Kāf.

SÜLEYMANİYE KÜTÜPHANESİ, ISTANBUL

Abdülkerim ibn Abdurrahman. *Tarih-i Mısır*. Hekimoğlu 705.
al-Bulqīnī, Badr al-Dīn ʿAlī ibn Aḥmad ibn Muḥammad. *Kitāb al-Nīl al-Rāʾid fī al-Nīl al-Zāʾid*. Fatih 4181.
al-Damīrī, Muḥammad ibn Mūsā. *Ḥayāt al-Ḥayawān*. Damad İbrahim 859.
———. *Ḥayāt al-Ḥayawān*. Harput 372.
———. *Ḥayāt al-Ḥayawān*. Pertevniyal 763.
al-Ḥijāzī, Shihāb al-Dīn Aḥmad ibn Muḥammad ʿAlī al-Anṣārī. *al-Nīl al-Rāʾid fī al-Nīl al-Zāʾid*. Ayasofya 3528.
Hikaye-i Feth-i Mısır. Kemankeş 489/2.
Ibn al-ʿImād al-Aqfahsī, Shihāb al-Dīn. *Mabdāʾ Nīl Miṣr wa al-Ahrām wa Faḍīlat Miṣr wa al-Miqyās*. Veliyüddin Efendi 3182.
al-Jāḥiẓ, Abī ʿUthmān ʿAmr ibn Baḥr. *Kitāb al-Ḥayawān*. Reisülküttab 584.
al-Jawharī, Muḥammad ibn ʿAbd al-Muʾmin ibn Muḥammad. *Mūjaz fī Mabdāʾ al-Nīl wa Muntahāhu*. Lâleli 3752.
al-Maḥallī, Jalāl al-Dīn. *Muqaddima fī Dhikr Nīl Miṣr*. Kara Çelebi Zade 355/7.
———. *Muqaddima fī Mabdāʾ al-Nīl wa Muntahāhu*. Serez 3838/11.
———. *Muqaddima fī Nīl Miṣr al-Mubārak*. Ayasofya 3446.
Mısır'da Bir Kütüphaneye ait Fihrist. Hüsrev Paşa 828.
Ridvan Paşazade, ʿAbdullah Çelebi. *Mısır Tarihi*. Esad Efendi 2177.
———. *Tarih-i Mısır*. Fatih 4362.
———. *Tarih-i Mısır*. Reşid Efendi 624.
al-Suyūṭī, Jalāl al-Dīn ʿAbd al-Raḥmān. *Bulbul al-Rawḍa fī Waṣf Nīl Miṣr*. Reşid Efendi 865/7.
———. *Kitāb Dīwān al-Ḥayawān*. Fatih 4170.
Süheyli, Ahmed ibn Hemdem. *Tarih-i Mısır*. Mehmed Zeki Pakalın 99.
———. *Tarih-i Mısır ül-Cedid*. Hüsrev Paşa 353/2.
———. *Tarih-i Mısır ül-Cedid*. Reşid Efendi 631/3.
Yusuf Efendi, Çerkesler Katibi. *Tarih-i Fetih Mısır ve Hukmu*. Esad Efendi 2146.
———. *Tarih-i Mısır*. Esad Efendi 2148.

TOPKAPI SARAYI MÜZESİ KÜTÜPHANESİ, ISTANBUL

Ḳānūn-nāme-i Mıṣr. E. H. 2063.
Seyyid Lokman. *Hünername I*. H. 1523.
Sulṭān Selim'in İran ve Mıṣır Seferine dair Muḫāberāt. R. 1955.
Şehnâme-i Âli Osman. A. 3592.

VENERANDA BIBLIOTECA AMBROSIANA, MILAN

al-Ḥaymī, Sharaf al-Dīn al-Ḥasan ibn Aḥmad ibn Ṣāliḥ al-Yūsufī al-Jamālī. *Riḥlat al-Ḥaymī ilā Arḍ al-Ḥabasha*. Ms. 35 B.

YALE CENTER FOR BRITISH ART, YALE UNIVERSITY, NEW HAVEN

Kerr, Mary. Notes on Visits to Various Country Houses and Towns in Great Britain, 1789–1826. Rare Books and Manuscripts.

Published Primary Sources

Abul Fazl ʾAllami. *The Ain i Akbari*. Translated by H. Blochmann. 3 vols. Frankfurt am Main: Institute for the History of Arabic-Islamic Science at the Johann Wolfgang Goethe University, 1993.
al-Andalusī, Ṣāʿid. *Science in the Medieval World: "Book of the Categories of Nations."* Translated and edited by Semaʿan I. Salem and Alok Kumar. Austin: University of Texas Press, 1991.

al-ʿAwfī al-Ḥanbalī, Ibrāhīm ibn Abī Bakr al-Ṣawāliḥī. *Tarājim al-Ṣawāʿiq fī Wāqiʿat al-Ṣanājiq*. Edited by ʿAbd al-Raḥīm ʿAbd al-Raḥman ʿAbd al-Raḥīm. Cairo: Institut français d'archéologie orientale, 1986.

Babur. *The Baburnama: Memoirs of Babur, Prince and Emperor*. Translated and edited by Wheeler M. Thackston. Introduced by Salman Rushdie. New York: Modern Library, 2002.

Baker, Samuel White. *Exploration of the Nile Tributaries of Abyssinia: The Sources, Supply, and Overflow of the Nile; the Country, People, Customs, Etc. Interspersed with Highly Exciting Adventures of the Author among Elephants, Lions, Buffaloes, Hippopotami, Rhinoceros, Etc. Accompanied by Expert Nature Sword Hunters*. Hartford: O.D. Case and Co., 1868.

Barkan, Ömer Lûtfi, ed. *Kanunlar*. Vol. 1 of *XV ve XVIinci asırlarda Osmanlı İmparatorluğunda Ziraî Ekonominin Hukukî ve Malî Esasları*. İstanbul Üniversitesi Yayınlarından 256. Istanbul: Bürhaneddin Matbaası, 1943.

Bīdpāʾī. *Kitāb Kalīla wa Dimna: Taʾlīf Bīdpā al-Faylasūf al-Hindī, Tarjamahu ilā al-ʿArabiyya fī Ṣadr al-Dawla al-ʿAbbāsiyya ʿAbd Allah ibn al-Muqaffaʿ*. Bombay: Sharaf al-Dīn al-Kutubī, 1968.

Celâl-zâde Sâlih Çelebi. *Tarih-i Mısr-ı Cedîd: İnceleme—Metin*. Edited by Tuncay Bülbül. Ankara: Grafiker Yayınları, 2011.

Clot-Bey, Antoine Barthélemy. *Aperçu general sur l'Égypte*. 2 vols. Bruxelles: Meline, cans, 1840.

———. *De la peste observée en Égypte: recherches et considérations sur cette maladie*. Paris: Fortin, Masson, 1840.

Commission des sciences et arts d'Egypte. *Description de l'Égypte, ou, recueil de observations et des recherches qui ont été faites en Égypte pendant l'éxpédition de l'armée française, publié par les ordres de Sa Majesté l'empereur Napoléon le Grand*. 9 vols. in 3 pts. Paris: Imprimerie impériale, 1809–28.

al-Damardāshī, Muṣṭafā ibn al-Ḥājj Ibrāhīm tābiʿ al-Marḥūm Ḥasan Aghā ʿAzbān. *Tārīkh Waqāyiʿ Miṣr al-Qāhira al-Maḥrūsa: Kinānat Allāh fī Arḍihi*. Edited by Ṣalāḥ Aḥmad Harīdī ʿAlī. Cairo: Dār al-Kutub wa al-Wathāʾiq al-Qawmiyya, 2002.

al-Damīrī, Aḥmad ibn Aḥmad. *Quḍāt Miṣr fī al-Qarn al-ʿĀshir wa al-Rubʿ al-Awwal min al-Qarn al-Ḥādī ʿAshar al-Hijrī*. Edited by ʿAbd al-Rāziq ʿAbd al-Rāziq ʿĪsā and Yūsuf Muṣṭafā al-Maḥmūdī. Cairo: al-ʿArabī lil-Nashr wa al-Tawzīʿ, 2000.

al-Damīrī, Muḥammad ibn Mūsā. *Ad-Damîrî's Ḥayât al-Ḥayawân (A Zoological Lexicon)*. Translated by A. S. G. Jayakar. 2 vols. London: Luzac, 1906–08.

———. *Ḥayāt al-Ḥayawān al-Kubrā*. Edited by Ibrāhīm Ṣāliḥ. 4 vols. Damascus: Dār al-Bashāʾir lil-Ṭibāʿa wa al-Nashr wa al-Tawzīʿ, 2005.

al-Damurdāshī Katkhudā ʿAzabān, Aḥmad. *al-Damurdashi's Chronicle of Egypt, 1688–1755: al-Durra al-Musana fi Akhbar al-Kinana*. Translated by Daniel Crecelius and ʿAbd al-Wahhab Bakr. Leiden: Brill, 1991.

———. *Kitāb al-Durra al-Muṣāna fī Akhbār al-Kināna*. Edited by ʿAbd al-Raḥīm ʿAbd al-Raḥman ʿAbd al-Raḥīm. Cairo: Institut français d'archéologie orientale, 1989.

Dankoff, Robert, and Nuran Tezcan, eds. *Evliyâ Çelebi'nin Nil Haritası: "Dürr-i bî-misîl în Ahbâr-ı Nîl."* Istanbul: Yapı Kredi Yayınları, 2011.

Ebied, R. Y., and M. J. L. Young. "An Unpublished Legal Work on a Difference between the Shāfiʿites and Mālikites." *Orientalia Lovaniensia Periodica* 8 (1977): 251–62.

Evliyâ Çelebi bin Derviş Mehemmed Zıllî. *Evliyâ Çelebi Seyahatnâmesi*. Edited by Seyit Ali Kahraman, Yücel Dağlı, and Robert Dankoff. 10 vols. Istanbul: Yapı Kredi Yayınları, 2011.

———. *An Ottoman Traveller: Selections from the* Book of Travels *of Evliya Çelebi*. Translated by Robert Dankoff and Sooyong Kim. London: Eland, 2010.

Fermanel, Gilles. "Voyage en Égypte." In *Voyage en Egypte: Vincent Stochove, Gilles Fermanel, Robert Fauvel, 1631*. Edited by Baudouin van de Walle. Cairo: Institut français d'archéologie orientale, 1975.

Fleming, George. *Travels on Horseback in Mantchu Tartary: Being a Summer's Ride Beyond the Great Wall of China*. London: Hurst and Blackett, 1863.

Galloway, John Alexander. *Communication with India, China, &c.: Observations on the Proposed Improvements in the Overland Route via Egypt, with Remarks on the Ship Canal, the Boulac Canal, and the Suez Railroad*. London: John Weale, 1844.

Gonzales, Antonius. *Voyage en Égypte du Père Antonius Gonzales, 1665–1666*. Translated by Charles Libois. 2 vols. Cairo: Institut français d'archéologie orientale, 1977.

Gouxiang, Li, Yang Chang, Wang Yude, et al., eds. *Ming Shi Lu Lei Zuan: She Wai Shi Liao Juan.* Wuhan Shi: Wuhan Chu Ban She: Xin Hua Shu Dian Jing Xiao, 1991.

Hafız Hüseyin Ayvansarayî. *The Garden of the Mosques: Hafiz Hüseyin Ayvansarayî's Guide to the Muslim Monuments of Ottoman Istanbul.* Translated by Howard Crane. Leiden: Brill, 2000.

Hamont, P. N. *Destruction de la peste: lazarets et quarantaines.* Paris: Imprimerie de Bourgogne et Martinet, 1844.

———. *L-Egypt sous Méhémet-Ali.* 2 vols. Paris: Leautey et Lecointe, 1843.

Henniker, Frederick. *Notes during a Visit to Egypt, Nubia, the Oasis, Mount Sinai, and Jerusalem.* London: John Murray, 1823.

Herodotus. *The History.* Translated by David Grene. Chicago: University of Chicago Press, 1987.

Heywood, Colin. "A Red Sea Shipping Register of the 1670s for the Supply of Foodstuffs from Egyptian *Wakf* Sources to Mecca and Medina (Turkish Documents from the Archive of 'Abdurrahman ''Abdi' Pasha of Buda, I)." *Anatolia Moderna* 6 (1996): 111–74.

Ibn 'Abd al-Ghanī al-Ḥanafī al-Miṣrī, Aḥmad Shalabī. *Awḍaḥ al-Ishārāt fīman Tawallā Miṣr al-Qāhira min al-Wuzarā' wa al-Bāshāt: al-Mulaqqab bi-al-Tārīkh al-'Aynī.* Edited by 'Abd al-Raḥīm 'Abd al-Raḥmān 'Abd al-Raḥīm. Cairo: Maktabat al-Khānjī, 1978.

Ibn Abī al-Surūr al-Bakrī al-Ṣiddīqī, Muḥammad. *al-Minaḥ al-Raḥmāniyya fī al-Dawla al-'Uthmāniyya, wa Dhayluhu al-Laṭā'if al-Rabbāniyya 'alā al-Minaḥ al-Raḥmāniyya.* Edited by Laylā al-Ṣabbāgh. Damascus: Dār al-Bashā'ir, 1995.

———. *al-Nuzha al-Zahiyya fī Dhikr Wulāt Miṣr wa al-Qāhira al-Mu'izziyya.* Edited by 'Abd al-Rāziq 'Abd al-Rāziq 'Īsā. Cairo: al-'Arabī lil-Nashr wa al-Tawzī', 1998.

———. *al-Rawḍa al-Ma'nūsa fī Akhbār Miṣr al-Maḥrūsa.* Edited by 'Abd al-Rāziq 'Abd al-Rāziq 'Īsā. Cairo: Maktabat al-Thaqāfa al-Dīniyya, 1997.

———. *al-Tuḥfa al-Bahiyya fī Tamalluk Āl 'Uthmān al-Diyār al-Miṣriyya.* Edited by 'Abd al-Raḥīm 'Abd al-Raḥmān 'Abd al-Raḥīm. Cairo: Dār al-Kutub wa al-Wathā'iq al-Qawmiyya, 2005.

Ibn Iyās al-Ḥanafī, Muḥammad ibn Aḥmad. *Badā'i' al-Zuhūr fī Waqā'i' al-Duhūr.* Edited by Muḥammad Muṣṭafā. 5 vols. Cairo: al-Hay'a al-Miṣriyya al-'Āmma lil-Kitāb, 1982.

Ibn al-Marzubān, Muḥammad ibn Khalaf. *The Book of the Superiority of Dogs over Many of Those who Wear Clothes.* Translated and edited by G. R. Smith and M. A. S. Abdel Haleem. Warminster: Aris and Phillips, 1978.

Ibn Ṭufayl, Muḥammad ibn 'Abd al-Malik. *Ibn Tufayl's Hayy Ibn Yaqzān: A Philosophical Tale.* Translated by Lenn Evan Goodman. 5th ed. Los Angeles: Gee Tee Bee, 2003.

Ibn al-Zubayr, Aḥmad ibn al-Rashīd. *Book of Gifts and Rarities = Kitāb al-Hadāyā wa al-Tuḥaf.* Translated by Ghāda al-Ḥijjāwī al-Qaddūmī. Cambridge, MA: Center for Middle Eastern Studies of Harvard University, 1996.

Ibn Zunbul, Aḥmad ibn 'Alī. *Ākhirat al-Mamālīk: Wāqi'at al-Sulṭān al-Ghūrī ma'a Salīm al-'Uthmānī.* Edited by 'Abd al-Mun'im 'Āmir. Cairo: al-Dār al-Qawmiyya lil-Ṭibā'a wa al-Nashr, 1962.

al-Isḥāqī, Muḥammad ibn 'Abd al-Mu'ṭī. *Kitāb Laṭā'if Akhbār al-Uwal fī man Taṣarrafa fī Miṣr min Arbāb al-Duwal.* Edited by Muḥammad Raḍwān Muhannā. al-Manṣūra: Maktabat al-Īmān, 2000.

al-Jabartī, 'Abd al-Raḥman ibn Ḥasan. *'Abd al-Raḥman al-Jabartī's History of Egypt: 'Ajā'ib al-Āthār fī al-Tarājim wa al-Akhbār.* Edited by Thomas Philipp and Moshe Perlmann. 4 vols. Stuttgart: Franz Steiner Verlag, 1994.

———. *'Ajā'ib al-Āthār fī al-Tarājim wa al-Akhbār.* Edited by 'Abd al-Raḥīm 'Abd al-Raḥman 'Abd al-Raḥīm. 4 vols. Cairo: Maṭba'at Dār al-Kutub al-Miṣriyya, 1998.

Jahangir. *The Jahangirnama: Memoirs of Jahangir, Emperor of India.* Translated and edited by Wheeler M. Thackston. Washington, DC: Freer Gallery of Art; New York: Oxford University Press, 1999.

al-Jāḥiẓ, Abī 'Uthmān 'Amr ibn Baḥr. *Kitāb al-Ḥayawān.* Edited by Muḥammad Bāsil 'Uyūn al-Sūd. 4 vols. Beirut: Dār al-Kutub al-'Ilmiyya, 1998.

al-Jazarī, Ismā'īl ibn al-Razzāz. *The Book of Knowledge of Ingenious Mechanical Devices: Fī Ma'rifat al-Ḥiyal al-Handasiyya.* Translated by Donald R. Hill. Dordrecht: Reidel, 1974.

———. *al-Jāmi' bayna al-'Ilm wa al-'Amal al-Nāfi' fī Ṣinā'at al-Ḥiyal.* Edited by Aḥmad Yūsuf al-Ḥasan, with 'Imād Ghānim, Mālik al-Mallūḥī, and Muṣṭafā Ta'murī. Aleppo: Jāmi'at Ḥalab, Ma'had al-Turāth al-'Ilmī al-'Arabī, 1979.

al-Jumayʿī, ʿAbd al-Munʿim Ibrāhīm, ed. *Wathāʾiq al-Taʿlīm al-ʿĀlī fī Miṣr khilāl al-Qarn al-Tāsiʿ ʿAshar.* Cairo: Dār al-Kutub wa al-Wathāʾiq al-Qawmiyya, 2004.

al-Karmī al-Ḥanbalī, Marʿī. *Nuzhat al-Nāẓirīn fī Tārīkh man Waliya Miṣr min al-Khulafāʾ wa al-Salāṭīn.* Edited by ʿAbd Allāh Muḥammad al-Kandarī. Beirut: Dār al-Nawādir, 2012.

Kato, Hiroshi. "Egyptian Village Community under Muḥammad ʿAlī's Rule: An Annotation of 'Qānūn al-Filāḥa.'" *Orient* 16 (1980): 183–222.

al-Khashshāb, Ismāʿīl ibn Saʿd. *Akhbār Ahl al-Qarn al-Thānī ʿAshar: Tārīkh al-Mamālīk fī al-Qāhira.* Edited by ʿAbd al-ʿAzīz Jamāl al-Dīn and ʿImād Abū Ghāzī. Cairo: al-ʿArabī lil-Nashr wa al-Tawzīʿ, 1990.

———. *Khulāṣat mā Yurād min Akhbār al-Amīr Murād.* Edited and translated by Ḥamza ʿAbd al-ʿAzīz Badr and Daniel Crecelius. Cairo: al-ʿArabī lil-Nashr wa al-Tawzīʿ, 1992.

Lane, Edward William. *An Account of the Manners and Customs of the Modern Egyptians: The Definitive 1860 Edition.* Introduction by Jason Thompson. Cairo: American University in Cairo Press, 2003.

———. *Cairo Fifty Years Ago.* Edited by Stanley Lane-Poole. London: J. Murray, 1896.

al-Maqrīzī, Aḥmad ibn ʿAlī. *al-Bayān wa al-Iʿrāb ʿammā bi-Arḍ Miṣr min al-Aʿrāb lil-Maqrīzī, maʿa Dirāsāt fī Tārīkh al-ʿUrūba fī Wādī al-Nīl.* Edited by ʿAbd al-Majīd ʿĀbidīn. Cairo: ʿĀlam al-Kutub, 1961.

Morgan, James Morris. *Recollections of a Rebel Reefer.* Boston and New York: Houghton Mifflin, 1917.

Morison, Anthoine. *Voyage en Égypte d'Anthoine Morison, 1697.* Edited by Georges Goyon. Cairo: Institut français d'archéologie orientale, 1976.

Muḥibb al-Dīn al-Ḥamawī, Muḥammad ibn Abī Bakr. *Ḥādī al-Aẓʿān al-Najdiyya ilā al-Diyār al-Miṣriyya.* Edited by Muḥammad ʿAdnān al-Bakhīt. Muʾta: Jāmiʿat Muʾta ʿImādat al-Baḥth al-ʿIlmī wa al-Dirāsāt al-ʿUlyā, 1993.

Mutawallī, Aḥmad Fuʾād, trans. and intro. *Qānūn Nāmah Miṣr, alladhī Aṣdarahu al-Sulṭān al-Qānūnī li-Ḥukm Miṣr.* Cairo: Maktabat al-Anjlū al-Miṣriyya, 1986.

Mutribi al-Asamm Samarqandi. *Conversations with Emperor Jahangir.* Translated by Richard C. Foltz. Costa Mesa: Mazda Publishers, 1998.

Norden, Frederik Ludvig. *Voyage d'Égypte et de Nubie, par Frederic Louis Norden, ouvrage enrichi de cartes & de figures dessinées sur les lieux, par l'auteur même.* 2 vols. Copenhagen: Imprimerie de la Maison Royale, 1755.

Onat, Ayşe. *Çin Kaynaklarında Türkler: Han Hanedanı Tarihinde "Batı Bölgeleri."* Ankara: Türk Tarih Kurumu Basımevi, 2012.

Pellat, Charles. *The Life and Works of Jāḥiẓ, Translations of Selected Texts.* Translated by D. M. Hawke. Berkeley: University of California Press, 1969.

Polo, Marco. *The Travels of Marco Polo the Venetian.* Translated by W. Marsden. Revised by T. Wright. Edited by Peter Harris. New York: Alfred A. Knopf, 2008.

Qazvīnī, Ḥamd Allāh Mustawfī. *Nuzhat al-Qulūb.* Tehran: Dunyā-yi Kitāb, 1983/1984.

———. *The Zoological Section of the Nuzhatu-l-Qulūb of Ḥamdullāh al-Mustaufī al-Qazwīnī.* Edited and translated by J. Stephenson. London: Royal Asiatic Society, 1928.

Sāmī, Amīn. *Taqwīm al-Nīl.* 5 vols. in 3 pts. Cairo: Dār al-Kutub wa al-Wathāʾiq al-Qawmiyya, 2003.

Sanderson, John. *The Travels of John Sanderson in the Levant, 1584–1602: With his Autobiography and Selections from his Correspondence.* Edited by Sir William Foster. London: Hakluyt Society, 1931.

Sharqāwī, ʿAbd Allāh. *Tuḥfat al-Nāẓirīn fī man Waliya Miṣr min al-Mulūk wa al-Salāṭīn.* Edited by Riḥāb ʿAbd al-Ḥamīd al-Qārī. Cairo: Maktabat Madbūlī, 1996.

Shaw, Stanford J. *The Budget of Ottoman Egypt, 1005–1006/1596–1597.* The Hague: Mouton, 1968.

Shaw, Stanford J., ed. and trans. *Ottoman Egypt in the Eighteenth Century: The Nizâmnâme-i Mısır of Cezzâr Aḥmed Pasha.* Cambridge, MA: Center for Middle Eastern Studies of Harvard University, 1964.

al-Shirbīnī, Yūsuf ibn Muḥammad. *Kitāb Hazz al-Quḥūf bi-Sharḥ Qaṣīd Abī Shādūf.* Edited and translated by Humphrey Davies. 2 vols. Leuven: Peeters, 2005–07.

Smith, G. Rex. *Medieval Muslim Horsemanship: A Fourteenth-Century Arabic Cavalry Manual.* London: British Library, 1979.

St. John, James Augustus. *Egypt and Mohammed Ali; or, Travels in the Valley of the Nile*. 2 vols. London: Longman, Rees, Orme, Brown, Green, and Longman, 1834.

———. *Egypt and Nubia*. London: Chapman and Hall, 1845.

Stochove, Vincent. "Voyage en Égypte." In *Voyage en Egypte: Vincent Stochove, Gilles Fermanel, Robert Fauvel, 1631*. Edited by Baudouin van de Walle. Cairo: Institut français d'archéologie orientale, 1975.

al-Suyūṭī, Jalāl al-Dīn ʿAbd al-Raḥmān. *Bulbul al-Rawḍa, maʿa Dirāsa ʿan Jazīrat al-Rawḍa*. Edited by Nabīl Muḥammad ʿAbd al-Azīz Aḥmad. Cairo: Maktabat al-Anjilū al-Miṣriyya, 1981.

———. *Kashf al-Ṣalṣala ʿan Waṣf al-Zalzala*. Edited by ʿAbd al-Laṭīf al-Saʿdānī. Rabat: Wizārat al-Dawla al-Mukallafa bi-al-Shuʾūn al-Thaqāfiyya wa al-Taʿlīm al-Aṣlī, 1971.

———. *Kawkab al-Rawḍa*. Edited by Muḥammad al-Shishtāwī. Cairo: Dār al-Āfāq al-ʿArabiyya, 2002.

Taeschner, Franz. *Alt-Stambuler Hof-und Volksleben, ein türkisches Miniaturenalbum aus dem 17. Jahrhundert*. Hannover: Orient-Buchhandlung H. Lafaire, 1925.

al-Ṭahṭāwī, Rifāʿa Rāfiʿ. *al-Siyāsa wa al-Waṭaniyya wa al-Tarbiyya*. Vol. 2 of *al-Aʿmāl al-Kāmila li-Rifāʿa Rāfiʿ al-Ṭahṭāwī*. Edited by Muḥammad ʿImāra. Beirut: al-Muʾassasa al-ʿArabiyya lil-Dirāsāt wa al-Nashr, 1973.

———. *Takhlīṣ al-Ibrīz fī Talkhīṣ Bārīz*. Cairo: Dār al-Kutub wa al-Wathāʾiq al-Qawmiyya, 2005.

Tietze, Andreas. *Muṣṭafā ʿĀlī's Description of Cairo of 1599: Text, Transliteration, Translation, Notes*. Vienna: Verlag Der Österreichischen Akademie Der Wissenschaften, 1975.

Veryard, Ellis. *An Account of Divers Choice Remarks, as well Geographical, as Historical, Political, Mathematical, Physical, and Moral; Taken in a Journey through the Low-Countries, France, Italy, and Part of Spain; with the Isles of Sicily and Malta. As also, a Voyage to the Levant: A Description of Candia, Egypt, the Red Sea, the Desarts of Arabia, Mount-Horeb, and Mount-Sinai; the Coasts of Palestine, Syria, and Asia-Minor; the Hellespont, Propontis, and Constantinople; the Isles of the Carpathian, Egean, and Ionian Seas*. Exon: Sam. Farley, 1701.

Waghorn, Thomas. *Egypt as it is in 1837*. London: Smith, Elder, and Co., 1837.

———. *Truths Concerning Mahomet Ali, Egypt, Arabia, and Syria: Addressed to the Five Powers or to their Representatives in the Contemplated Congress*. London: Smith, Elder, and Co., 1840.

Warner, Nicholas. *The True Description of Cairo: A Sixteenth-Century Venetian View*. 3 vols. Oxford: Arcadian Library, in association with Oxford University Press, 2006.

Zhao, Rugua. *Chau Ju-Kua: His Work on the Chinese and Arab Trade in the Twelfth and Thirteenth Centuries, entitled Chu-Fan-Chï*. Translated by Friedrich Hirth and W. W. Rockhill. 2 vols. New York: Paragon Book Reprint Corp., 1966.

Secondary Sources

ʿAbd al-Karīm, Aḥmad ʿIzzat. *Tārīkh al-Taʿlīm fī ʿAṣr Muḥammad ʿAlī*. Cairo: Maktabat al-Nahḍa al-Miṣriyya, 1938.

ʿAbd al-Muʿṭī, Ḥusām Muḥammad. *al-ʿAlāqāt al-Miṣriyya al-Ḥijāziyya fī al-Qarn al-Thāmin ʿAshar*. Cairo: al-Hayʾa al-Miṣriyya al-ʿĀmma lil-Kitāb, 1999.

ʿAbd al-Raḥīm, ʿAbd al-Raḥīm ʿAbd al-Raḥman. *al-Rīf al-Miṣrī fī al-Qarn al-Thāmin ʿAshar*. Cairo: Maktabat Madbūlī, 1986.

Abir, M. "The 'Arab Rebellion' of Amir Ghālib of Mecca (1788–1813)." *Middle Eastern Studies* 7 (1971): 185–200.

———. "Modernisation, Reaction and Muhammad Ali's 'Empire.'" *Middle Eastern Studies* 13 (1977): 295–313.

Abou El Fadl, Khaled. "Dogs in the Islamic Tradition and Nature." *Encyclopedia of Religion and Nature*. London: Thoemmes Continuum, 2005.

Abou-El-Haj, Rifaʿat ʿAli. *Formation of the Modern State: The Ottoman Empire, Sixteenth to Eighteenth Centuries*. 2nd ed. Syracuse, NY: Syracuse University Press, 2005.

Abu-Lughod, Janet L. *Before European Hegemony: The World System A.D. 1250–1350*. New York: Oxford University Press, 1989.

———. *Cairo: 1001 Years of* The City Victorious. Princeton, NJ: Princeton University Press, 1971.

Acampora, Ralph. "Zoos and Eyes: Contesting Captivity and Seeking Successor Practices." *Society and Animals* 13 (2005): 69–88.

Acemoglu, Daron. "Technical Change, Inequality, and the Labor Market." *Journal of Economic Literature* 40 (2002): 7–72.
Ådahl, Karin, ed. *The Sultan's Procession: The Swedish Embassy to Sultan Mehmed IV in 1657–1658 and the Rålamb Paintings*. Istanbul: Swedish Research Institute in Istanbul, 2006.
Adams, Jonathan S., and Thomas O. McShane. *The Myth of Wild Africa: Conservation without Illusion*. Berkeley: University of California Press, 1996.
Ágoston, Gábor. *Guns for the Sultan: Military Power and the Weapons Industry in the Ottoman Empire*. Cambridge: Cambridge University Press, 2005.
Aguiar, Marian. *Tracking Modernity: India's Railway and the Culture of Mobility*. Minneapolis: University of Minnesota Press, 2011.
Aḥmad, Laylā ʿAbd al-Laṭīf. *al-Idāra fī Miṣr fī al-ʿAṣr al-ʿUthmānī*. Cairo: Maṭbaʿat Jāmiʿat ʿAyn Shams, 1978.
———. *al-Mujtamaʿ al-Miṣrī fī al-ʿAṣr al-ʿUthmānī*. Cairo: Dār al-Kitāb al-Jāmiʿī, 1987.
———. *Tārīkh wa Muʾarrikhī Miṣr wa al-Shām ibbāna al-ʿAṣr al-ʿUthmānī*. Cairo: Maktabat al-Khānjī, 1980.
Ahsan, Muhammad Manazir. "A Note on Hunting in the Early ʿAbbasid Period: Some Evidence on Expenditure and Prices." *Journal of the Economic and Social History of the Orient* 19 (1976): 101–05.
Alam, Muzaffar, and Sanjay Subrahmanyam. *Indo-Persian Travels in the Age of Discoveries, 1400–1800*. Cambridge: Cambridge University Press, 2007.
Alberti, Samuel J. M. M., ed. *The Afterlives of Animals: A Museum Menagerie*. Charlottesville: University of Virginia Press, 2011.
Alexander, Jennifer Karns. *The Mantra of Efficiency: From Waterwheel to Social Control*. Baltimore: Johns Hopkins University Press, 2008.
Alexander, John. "The Turks on the Middle Nile." *Archéologie du Nil Moyen* 7 (1996): 15–35.
Algazi, Gadi, Valentin Groebner, and Bernhard Jussen, eds. *Negotiating the Gift: Pre-Modern Figurations of Exchange*. Göttingen: Vandenhoeck and Ruprecht, 2003.
ʿAlī, Ṣalāḥ Aḥmad Harīdī. "al-Ḥayāh al-Iqtiṣādiyya wa al-Ijtimāʿiyya fī Madīnat Rashīd fī al-ʿAṣr al-ʿUthmānī, Dirāsa Wathāʾiqiyya." *Egyptian Historical Review* 30–31 (1983–84): 327–78.
Allin, Michael. *Zarafa: A Giraffe's True Story, from Deep in Africa to the Heart of Paris*. New York: Delta, 1998.
Allouche, Adel. *The Origins and Development of the Ottoman-Ṣafavid Conflict (906–962/1500–1555)*. Berlin: K. Schwarz Verlag, 1983.
Allsen, Thomas T. *The Royal Hunt in Eurasian History*. Philadelphia: University of Pennsylvania Press, 2006.
Altundağ, Şinasi. *Kavalalı Mehmet Ali Paşa Isyanı: Mısır Meselesi, 1831–1841*. Ankara: Türk Tarih Kurumu, 1988.
Alves, Abel A. *The Animals of Spain: An Introduction to Imperial Perceptions and Human Interaction with Other Animals, 1492–1826*. Leiden: Brill, 2011.
Ambraseys, Nicholas. *Earthquakes in the Eastern Mediterranean and Middle East: A Multidisciplinary Study of Seismicity up to 1900*. Cambridge: Cambridge University Press, 2009.
Ambraseys, N. N., and C. F. Finkel. *The Seismicity of Turkey and Adjacent Areas: A Historical Review, 1500–1800*. Istanbul: Eren, 1995.
Ambraseys, N. N., and C. P. Melville. *A History of Persian Earthquakes*. Cambridge: Cambridge University Press, 1982.
Ambraseys, N. N., C. P. Melville, and R. D. Adams. *The Seismicity of Egypt, Arabia and the Red Sea: A Historical Review*. Cambridge: Cambridge University Press, 1994.
Ames, Eric. *Carl Hagenbeck's Empire of Entertainments*. Seattle: University of Washington Press, 2008.
And, Metin. *16. Yüzyılda İstanbul: Kent—Saray—Günlük Yaşam*. Istanbul: Yapı Kredi Yayınları, 2011.
Anderson, Patricia K. "A Bird in the House: An Anthropological Perspective on Companion Parrots." *Society and Animals* 11 (2003): 393–418.
Anderson, Steven C. "Fauna." *Iranian Studies* 31 (1998): 399–405.
Anderson, Virginia DeJohn. *Creatures of Empire: How Domestic Animals Transformed Early America*. New York: Oxford University Press, 2004.

Anderson, Warwick. "Climates of Opinion: Acclimatization in Nineteenth-Century France and England." *Victorian Studies* 35 (1992): 135–57.

Andrews, Thomas G. "Contemplating Animal Histories: Pedagogy and Politics across Borders." *Radical History Review* 107 (2010): 139–65.

Andrews, Walter G. "Singing the Alienated 'I': Guattari, Deleuze and Lyrical Decodings of *the Subject* in Ottoman *Divan* Poetry." *Yale Journal of Criticism* 6 (1993): 191–219.

Anthony, David W. *The Horse, the Wheel, and Language: How Bronze-Age Riders from the Eurasian Steppes Shaped the Modern World*. Princeton, NJ: Princeton University Press, 2007.

Appel, Toby A. *The Cuvier-Geoffroy Debate: French Biology in the Decades before Darwin*. New York: Oxford University Press, 1987.

Appletons' Journal of Literature, Science, and Art 12. November 28, 1874.

Appuhn, Karl. "Ecologies of Beef: Eighteenth-Century Epizootics and the Environmental History of Early Modern Europe." *Environmental History* 15 (2010): 268–87.

———. *A Forest on the Sea: Environmental Expertise in Renaissance Venice*. Baltimore: Johns Hopkins University Press, 2009.

Arbel, Benjamin. "The Last Decades of Venice's Trade with the Mamluks: Importations into Egypt and Syria." *Mamlūk Studies Review* 8 (2004): 37–86.

Armour, Robert A. *Gods and Myths of Ancient Egypt*. Cairo: American University in Cairo Press, 2001.

Arnold, Dorothea. "An Egyptian Bestiary." *Metropolitan Museum of Art Bulletin* 52 (1995): 1 and 7–64.

Artan, Tülay. "Ahmed I's Hunting Parties: Feasting in Adversity, Enhancing the Ordinary." In *Starting with Food: Culinary Approaches to Ottoman History*. Edited by Amy Singer. Princeton, NJ: Markus Wiener Publishers, 2011.

———. "Ahmed I and 'Tuhfet'ül-Mülûk ve's-Selâtîn': A Period Manuscript on Horses, Horsemanship and Hunting." In *Animals and People in the Ottoman Empire*. Edited by Suraiya Faroqhi. Istanbul: Eren, 2010.

———. "Aspects of the Ottoman Elite's Food Consumption: Looking for 'Staples,' 'Luxuries,' and 'Delicacies' in a Changing Century." In *Consumption Studies and the History of the Ottoman Empire, 1550–1922: An Introduction*. Edited by Donald Quataert. Albany: State University of New York Press, 2000.

———. "A *Book of Kings* Produced and Presented as a Treatise on Hunting." *Muqarnas* 25 (2008): 299–330.

Asad, Talal. *Formations of the Secular: Christianity, Islam, Modernity*. Stanford, CA: Stanford University Press, 2003.

Ascione, Frank R., and Phil Arkow, eds. *Child Abuse, Domestic Violence, and Animal Abuse: Linking the Circles of Compassion for Prevention and Intervention*. West Lafayette, IN: Purdue University Press, 1999.

Asher, Catherine B., and Cynthia Talbot. *India before Europe*. Cambridge: Cambridge University Press, 2006.

Aslanian, Sebouh David. *From the Indian Ocean to the Mediterranean: The Global Trade Networks of Armenian Merchants from New Julfa*. Berkeley: University of California Press, 2011.

Aston, T. H., and C. H. E. Philpin, eds. *The Brenner Debate: Agrarian Class Structure and Economic Development in Pre-Industrial Europe*. Cambridge: Cambridge University Press, 1985.

Bacqué-Grammont, Jean-Louis. *Les Ottomans, les Safavides et leurs voisins: contribution à l'histoire des relations internationales dans l'Orient islamique de 1514 à 1524*. Istanbul: Nederlands Historisch-Archaeologisch Instituut te Istanbul, 1987.

Bacqué-Grammont, Jean-Louis, and Anne Kroell. *Mamlouks, Ottomans et Portugais en Mer Rouge: L'affaire de Djedda en 1517*. Cairo: Institut français d'archéologie orientale, 1988.

Bacqué-Grammont, Jean-Louis, Joséphine Lesur-Gebremariam, and Catherine Mayeur-Jaouen. "Quelques aspects de la faune nilotique dans la relation d'Evliyâ Çelebî, voyageur ottoman." *Journal Asiatique* 296 (2008): 331–74.

Badawi, El-Said, and Martin Hinds. *A Dictionary of Egyptian Arabic*. Beirut: Librairie du Liban, 1986.

Baer, Gabriel. "The Dissolution of the Egyptian Village Community." *Die Welt des Islams* 6 (1959): 56–70.

———. *A History of Landownership in Modern Egypt, 1800–1950*. London: Oxford University Press, 1962.

Baer, Marc David. *Honored by the Glory of Islam: Conversion and Conquest in Ottoman Europe*. New York: Oxford University Press, 2008.

Bagnall, Roger S. "The Camel, the Wagon, and the Donkey in Later Roman Egypt." *Bulletin of the American Society of Papyrologists* 22 (1985): 1–6.

Bailey, Michael R., ed. *Robert Stephenson—The Eminent Engineer*. Aldershot, UK: Ashgate, 2003.

Bakr, ʿAbd al-Wahhāb. *al-Būlīs al-Miṣrī, 1922–1952*. Cairo: Maktabat Madbūlī, 1988.

Bankoff, Greg. "A Question of Breeding: Zootechny and Colonial Attitudes toward the Tropical Environment in the Late Nineteenth-Century Philippines." *Journal of Asian Studies* 60 (2001): 413–37.

Barak, On. "Egyptian Times: Temporality, Personhood, and the Technopolitical Making of Modern Egypt, 1830–1930." Ph.D. diss., New York University, 2009.

Barkan, Ömer Lütfi. "İstanbul Saraylarına ait Muhasebe Defterleri." *Belgeler* IX/13 (1979): 1–380.

Barkey, Karen. *Bandits and Bureaucrats: The Ottoman Route to State Centralization*. Ithaca, NY: Cornell University Press, 1994.

Başaran, Betül. "Remaking the Gate of Felicity: Policing, Social Control, and Migration in Istanbul at the End of the Eighteenth Century, 1789–1793." Ph.D. diss., University of Chicago, 2006.

Bayly, C. A. *Imperial Meridian: The British Empire and the World, 1780–1830*. London: Longman, 1989.

Bayur, Yusuf Hikmet. "Maysor Sultanı Tipu ile Osmanlı Pâdişahlarından I. Abdülhamid ve III. Selim Arasındaki Mektuplaşma." *Belleten* 12/47 (1948): 617–54.

Beck, Alan M. *The Ecology of Stray Dogs: A Study of Free-Ranging Urban Animals*. West Lafayette, IN: NotaBell Books, 2002.

Beckett, Derrick. *Stephensons' Britain*. Newton Abbot, UK: David and Charles, 1984.

Beckwith, Christopher I. "The Impact of the Horse and Silk Trade on the Economies of T'ang China and the Uighur Empire: On the Importance of International Commerce in the Early Middle Ages." *Journal of the Economic and Social History of the Orient* 34 (1991): 183–98.

Bedini, Silvio A. *The Pope's Elephant*. Manchester: Carcanet Press, 1997.

Behrens-Abouseif, Doris. *Azbakiyya and Its Environs: From Azbak to Ismāʿīl, 1476–1879*. Cairo: Institut français d'archéologie orientale, 1985.

Beirne, Piers. "From Animal Abuse to Interhuman Violence? A Critical Review of the Progression Thesis." *Society and Animals* 12 (2004): 39–65.

Bekoff, Marc. *Minding Animals: Awareness, Emotions, and Heart*. Oxford: Oxford University Press, 2002.

———. *The Smile of a Dolphin: Remarkable Accounts of Animal Emotions*. New York: Discovery Books, 2000.

Belk, Russell W. "Metaphoric Relationships with Pets." *Society and Animals* 4 (1996): 121–45.

Belozerskaya, Marina. *The Medici Giraffe: And Other Tales of Exotic Animals and Power*. New York: Little, Brown and Company, 2006.

Belyaev, D. K. "Destabilizing Selection as a Factor in Domestication." *Journal of Heredity* 70 (1979): 301–08.

Belyaev, D. K., A. O. Ruvinsky, and L. N. Trut. "Inherited Activation-Inactivation of the Star Gene in Foxes: Its Bearing on the Problem of Domestication." *Journal of Heredity* 72 (1981): 267–74.

Bendiner, Kenneth. "Lewis, John Frederick (1804–1876)." *Oxford Dictionary of National Biography*. Oxford: Oxford University Press, 2004.

Benjamin, Andrew. *Of Jews and Animals*. Edinburgh: Edinburgh University Press, 2010.

Benkheira, Mohamed Hocine, Catherine Mayeur-Jaouen, and Jacqueline Sublet. *L'animal en islam*. Paris: Indes savantes, 2005.

Benton, Ted. "Marxism and the Moral Status of Animals." *Society and Animals* 11 (2003): 73–79.

———. *Natural Relations: Ecology, Animal Rights, and Social Justice*. London: Verso, 1993.

Ben-Zaken, Avner. *Reading Ḥayy Ibn-Yaqẓān: A Cross-Cultural History of Autodidacticism*. Baltimore: Johns Hopkins University Press, 2011.

Berger, John. *About Looking*. London: Writers and Readers Publishing Cooperative, 1980.

Berktay, Halil. "Three Empires and the Societies They Governed: Iran, India, and the Ottoman Empire." *Journal of Peasant Studies* 18 (1991): 242–63.

Bilge, Mustafa. "Suez Canal in the Ottoman Sources." In *Proceedings of the International Conference on Egypt during the Ottoman Era: 26–30 November 2007, Cairo, Egypt*. Edited by Research Centre for Islamic History, Art and Culture. Istanbul: IRCICA, 2010.

Bilgin, Arif. *Osmanlı Saray Mutfağı (1453–1650)*. Istanbul: Kitabevi Yayınları, 2004.

Blanchard, Émile. *Notice sur la vie et les travaux d'Isidore Geoffroy Saint-Hilaire*. Paris, 1862.

Bodart-Bailey, Beatrice M. *The Dog Shogun: The Personality and Policies of Tokugawa Tsunayoshi*. Honolulu: University of Hawai'i Press, 2006.

Bodson, Liliane, ed. *Les animaux exotiques dans les relations internationales: espèces, fonctions, significations*. Liège: Université de Liège, 1998.

Boehrer, Bruce Thomas. *Parrot Culture: Our 2,500-Year-Long Fascination with the World's Most Talkative Bird*. Philadelphia: University of Pennsylvania Press, 2004.

Bondeson, Jan. *Amazing Dogs: A Cabinet of Canine Curiosities*. Ithaca, NY: Cornell University Press, 2011.

Boogert, Maurits H. van den. *The Capitulations and the Ottoman Legal System: Qadis, Consuls and Beratlıs in the 18th Century*. Leiden: Brill, 2005.

Boogert, Maurits H. van den, and Kate Fleet, eds. *The Ottoman Capitulations: Text and Context*. Rome: Istituto per l'Oriente C.A. Nallino, 2004.

Boomgaard, Peter. *Frontiers of Fear: Tigers and People in the Malay World, 1600–1950*. New Haven, CT: Yale University Press, 2001.

Borromeo, Elisabetta. "The Ottomans and Hunting, according to Julien Bordier's Travelogue (1604–1612)." In *Animals and People in the Ottoman Empire*. Edited by Suraiya Faroqhi. Istanbul: Eren, 2010.

Borsch, Stuart J. *The Black Death in Egypt and England: A Comparative Study*. Austin: University of Texas Press, 2005.

———. "Environment and Population: The Collapse of Large Irrigation Systems Reconsidered." *Comparative Studies in Society and History* 46 (2004): 451–68.

Bostan, İdris. "An Ottoman Base in Eastern Mediterranean: Alexandria of Egypt in the 18th Century." In *Proceedings of the International Conference on Egypt during the Ottoman Era: 26–30 November 2007, Cairo, Egypt*. Edited by Research Centre for Islamic History, Art and Culture. Istanbul: IRCICA, 2010.

Bough, Jill. *Donkey*. London: Reaktion Books, 2011.

Bousquet, G. H. "Des animaux et de leur traitement selon le judaïsme, le christianisme et l'islam." *Studia Islamica* 9 (1958): 31–48.

Boyajian, James C. *Portuguese Trade in Asia under the Habsburgs, 1580–1640*. Baltimore: Johns Hopkins University Press, 1993.

Boyce, James. "Canine Revolution: The Social and Environmental Impact of the Introduction of the Dog to Tasmania." *Environmental History* 11 (2006): 102–29.

Boyd, James. "Horse Power: The Japanese Army, Mongolia and the Horse, 1927–43." *Japan Forum* 22 (2010): 23–42.

Brandes, Stanley. "The Meaning of American Pet Cemetery Gravestones." *Ethnology* 48 (2009): 99–118.

Brandt, Keri. "A Language of Their Own: An Interactionist Approach to Human-Horse Communication." *Society and Animals* 12 (2004): 299–316.

Braudel, Fernand. *The Wheels of Commerce*. Translated by Siân Reynolds. Vol. 2 of *Civilization and Capitalism, 15th–18th Century*. London: Collins, 1982.

Braverman, Harry. *Labor and Monopoly Capital: The Degradation of Work in the Twentieth Century*. New York: Monthly Review Press, 1974.

Braverman, Irus. *Zooland: The Institution of Captivity*. Stanford, CA: Stanford University Press, 2013.

Brewer, Douglas. "Hunting, Animal Husbandry and Diet in Ancient Egypt." In *A History of the Animal World in the Ancient Near East*. Edited by Billie Jean Collins. Leiden: Brill, 2002.

Brittlebank, Kate. "Sakti and Barakat: The Power of Tipu's Tiger, An Examination of the Tiger Emblem of Tipu Sultan of Mysore." *Modern Asian Studies* 29 (1995): 257–69.

———. *Tipu Sultan's Search for Legitimacy: Islam and Kingship in a Hindu Domain*. Delhi: Oxford University Press, 1997.

Brown, Karen, and Daniel Gilfoyle, eds. *Healing the Herds: Disease, Livestock Economies, and the Globalization of Veterinary Medicine*. Athens: Ohio University Press, 2010.

Brummett, Palmira. "Dogs, Women, Cholera, and Other Menaces in the Streets: Cartoon Satire in the Ottoman Revolutionary Press, 1908–11." *International Journal of Middle East Studies* 27 (1995): 433–60.

———. "What Sidi Ali Saw." *Portuguese Studies Review* 9 (2002): 232–53.

Buecker, Thomas R. "The Fort Robinson War Dog Reception and Training Center, 1942–46." *Military History of the West* 38 (2008): 115–40.

Buettinger, Craig. "Women and Antivivisection in Late Nineteenth-Century America." *Journal of Social History* 30 (1997): 857–72.

Bulliet, Richard W. "The Camel and the Watermill." *International Journal of Middle East Studies* 42 (2010): 666–68.

———. *The Camel and the Wheel*. Cambridge, MA: Harvard University Press, 1975.

———. *Cotton, Climate, and Camels in Early Islamic Iran: A Moment in World History*. New York: Columbia University Press, 2009.

———. "History and Animal Energy in the Arid Zone." In *Water on Sand: Environmental Histories of the Middle East and North Africa*. Edited by Alan Mikhail. New York: Oxford University Press, 2013.

———. *Hunters, Herders, and Hamburgers: The Past and Future of Human-Animal Relationships*. New York: Columbia University Press, 2005.

Buquet, Thierry. "*La belle captive*. La girafe dans les ménageries princières au Moyen Âge." In *La bête captive au Moyen Âge et à l'époque moderne*. Edited by Corinne Beck and Fabrice Guizard. Amiens: Encrage, 2012.

Burke, Edmund, III. "The Big Story: Human History, Energy Regimes, and the Environment." In *The Environment and World History*. Edited by Edmund Burke III and Kenneth Pomeranz. Berkeley: University of California Press, 2009.

Burleigh, Nina. *Mirage: Napoleon's Scientists and the Unveiling of Egypt*. New York: HarperCollins, 2007.

Butzer, Karl W. *Early Hydraulic Civilization in Egypt: A Study in Cultural Ecology*. Chicago: University of Chicago Press, 1976.

Caferoğlu, Ahmet. "Türk Onomastiğinde 'Köpek' Kültü." *Belleten* 25 (1961): 1–11.

Cahn, Théophile. *La vie et l'oeuvre d'Étienne Geoffroy Saint-Hilaire*. Paris: Presses universitaires de France, 1962.

Carrière, Charles, Marcel Courdurié, and Ferréol Rebuffat. *Marseille ville morte: la peste de 1720*. Marseille: M. Garçon, 1968.

Cartmill, Matt. *A View to a Death in the Morning: Hunting and Nature through History*. Cambridge, MA: Harvard University Press, 1993.

Casale, Giancarlo. "The Ottoman Administration of the Spice Trade in the Sixteenth-Century Red Sea and Persian Gulf." *Journal of the Economic and Social History of the Orient* 49 (2006): 170–98.

———. *The Ottoman Age of Exploration*. New York: Oxford University Press, 2010.

Chaiklin, Martha. "Exotic-Bird Collecting in Early-Modern Japan." In *JAPANimals: History and Culture in Japan's Animal Life*. Edited by Gregory M. Pflugfelder and Brett L. Walker. Ann Arbor: Center for Japanese Studies, University of Michigan, 2005.

Chakrabarty, Dipesh. "The Climate of History: Four Theses." *Critical Inquiry* 35 (2009): 197–222.

———. "The Muddle of Modernity." *American Historical Review* 116 (2011): 663–75.

Chakravarti, Ranabir. "Early Medieval Bengal and the Trade in Horses: A Note." *Journal of the Economic and Social History of the Orient* 42 (1999): 194–211.

———. "Horse Trade and Piracy at Tana (Thana, Maharashtra, India): Gleanings from Marco Polo." *Journal of the Economic and Social History of the Orient* 34 (1991): 159–82.

Chalcraft, John T. *The Striking Cabbies of Cairo and Other Stories: Crafts and Guilds in Egypt, 1863–1914*. Albany: State University of New York Press, 2004.

Chalfen, Richard. "Celebrating Life after Death: The Appearance of Snapshots in Japanese Pet Gravesites." *Visual Studies* 18 (2003): 144–56.

Chaudhuri, K. N. *Trade and Civilisation in the Indian Ocean: An Economic History from the Rise of Islam to 1750*. Cambridge: Cambridge University Press, 1985.

———. *The Trading World of Asia and the English East India Company, 1660–1760.* Cambridge: Cambridge University Press, 1978.

Cheang, Sarah. "Women, Pets, and Imperialism: The British Pekingese Dog and Nostalgia for Old China." *Journal of British Studies* 45 (2006): 359–87.

Christ, Georg. *Trading Conflicts: Venetian Merchants and Mamluk Officials in Late Medieval Alexandria.* Leiden: Brill, 2012.

Chur-Hansen, Anna. "Cremation Services upon the Death of a Companion Animal: Views of Service Providers and Service Users." *Society and Animals* 19 (2011): 248–60.

Clarence-Smith, William G. "Elephants, Horses, and the Coming of Islam to Northern Sumatra." *Indonesia and the Malay World* 32 (2004): 271–84.

Clark, Stephen R. L. *The Nature of the Beast: Are Animals Moral?* Oxford: Oxford University Press, 1982.

Cobb, B. Francis. "Report on Raw Silk." *The Journal of the Society of Arts and Institutions in Union, and Official Record of Annual International Exhibitions* 21 (1872–73): 727–30.

Cole, Juan. *Napoleon's Egypt: Invading the Middle East.* New York: Palgrave Macmillan, 2007.

Coleman, Jon T. *Vicious: Wolves and Men in America.* New Haven, CT: Yale University Press, 2004.

Colla, Elliott. *Conflicted Antiquities: Egyptology, Egyptomania, Egyptian Modernity.* Durham, NC: Duke University Press, 2007.

Coller, Ian. *Arab France: Islam and the Making of Modern Europe, 1798–1831.* Berkeley: University of California Press, 2011.

Collins, Robert O. *The Nile.* New Haven, CT: Yale University Press, 2002.

Comaroff, Jean and John L. "Goodly Beasts, Beastly Goods: Cattle and Commodities in a South African Context." *American Ethnologist* 17 (1990): 195–216.

Cook, Michael. "Ibn Qutayba and the Monkeys." *Studia Islamica* 89 (1999): 43–74.

Cooper, Frederick, and Ann Laura Stoler, eds. *Tensions of Empire: Colonial Cultures in a Bourgeois World.* Berkeley: University of California Press, 1997.

Cooper, Jilly. *Animals in War.* London: Heinemann, 1983.

Cooper, Ross Gordon, and Jaroslaw Olav Horbañczuk. "Anatomical and Physiological Characteristics of Ostrich (*Struthio camelus var. domesticus*) Meat Determine Its Nutritional Importance for Man." *Animal Science Journal* 73 (2002): 167–73.

Coppinger, Raymond, and Lorna Coppinger. *Dogs: A Startling New Understanding of Canine Origin, Behavior, and Evolution.* New York: Scribner, 2001.

Corvi, Steven J. "Men of Mercy: The Evolution of the Royal Army Veterinary Corps and the Soldier-Horse Bond During the Great War." *Journal of the Society for Army Historical Research* 76 (1998): 272–84.

Creager, Angela N. H., and William Chester Jordan, eds. *The Animal/Human Boundary: Historical Perspectives.* Rochester, NY: University of Rochester Press, 2002.

Crecelius, Daniel. "The Importance of Qusayr in the Late Eighteenth Century." *Journal of the American Research Center in Egypt* 24 (1987): 53–60.

———. *The Roots of Modern Egypt: A Study of the Regimes of ʿAli Bey al-Kabir and Muhammad Bey Abu al-Dhahab, 1760–1775.* Minneapolis: Bibliotheca Islamica, 1981.

Cronon, William. *Changes in the Land: Indians, Colonists, and the Ecology of New England.* Rev. ed. New York: Hill and Wang, 2003.

———. *Nature's Metropolis: Chicago and the Great West.* New York: W. W. Norton, 1991.

———. "The Trouble with Wilderness; or, Getting Back to the Wrong Nature." *Environmental History* 1 (1996): 7–28.

Crosby, Alfred W. *Ecological Imperialism: The Biological Expansion of Europe, 900–1900.* Cambridge: Cambridge University Press, 2004.

———. *Throwing Fire: Projectile Technology Through History.* Cambridge: Cambridge University Press, 2002.

Crutzen, Paul J. "Geology of Mankind." *Nature* 415 (2002): 23.

Cummins, John. *The Hound and the Hawk: The Art of Medieval Hunting.* London: Weidenfeld and Nicolson, 1988.

Cuneo, Pia F. "(Un)Stable Identities: Hippology and the Professionalization of Scholarship and Horsemanship in Early Modern Germany." In *Early Modern Zoology: The Construction of*

Animals in Science, Literature and the Visual Arts. Edited by Karl A. E. Enenkel and Paul J. Smith. 2 vols. Leiden: Brill, 2007.

Cuno, Kenneth M. "Commercial Relations between Town and Village in Eighteenth and Early Nineteenth-Century Egypt." *Annales Islamologiques* 24 (1988): 111–35.

———. *The Pasha's Peasants: Land, Society, and Economy in Lower Egypt, 1740–1858.* Cambridge: Cambridge University Press, 1992.

Curl, James Stevens. *Egyptomania: The Egyptian Revival, a Recurring Theme in the History of Taste.* Manchester: Manchester University Press, 1994.

Curth, Louise Hill. *The Care of Brute Beasts: A Social and Cultural Study of Veterinary Medicine in Early Modern England.* Leiden: Brill, 2010.

Cvetkova, Bistra. "La fauconnerie dans les sancaks de Nicopol et de Vidin aux XVe et XVIe siècles." *İstanbul Üniversitesi Edebiyat Fakültesi Tarih Dergisi* 32 (1979): 795–818.

Çelik, Zeynep. *Displaying the Orient: Architecture of Islam at Nineteenth-Century World's Fairs.* Berkeley: University of California Press, 1992.

Çoruhlu, Yaşar. "Türk Sanatı ve Sembolizminde Kedi: Selçuklu ve Osmanlı Sanatında Kedi." *Toplumsal Tarih* 124 (2004): 40–45.

Dale, Stephen F. *Indian Merchants and Eurasian Trade, 1600–1750.* Cambridge: Cambridge University Press, 1994.

———. *The Muslim Empires of the Ottomans, Safavids, and Mughals.* Cambridge: Cambridge University Press, 2010.

Dallapiccola, Anna Libera, ed., with Stephanie Zingel-Avé Lallemant. *Vijayanagara, City and Empire: New Currents of Research.* 2 vols. Stuttgart: Steiner Verlag Wiesbaden, 1985.

Daly, M. W., ed. *Modern Egypt, from 1517 to the End of the Twentieth Century.* Vol. 2 of *The Cambridge History of Egypt.* Cambridge: Cambridge University Press, 1998.

Dankoff, Robert. "Animal Traits in the Army Commander." *Journal of Turkish Studies* 1 (1977): 95–112.

Darling, Linda T. "'Do Justice, Do Justice, For That is Paradise': Middle Eastern Advice for Indian Muslim Rulers." *Comparative Studies of South Asia, Africa and the Middle East* 22 (2002): 3–19.

———. *A History of Social Justice and Political Power in the Middle East: The Circle of Justice from Mesopotamia to Globalization.* New York: Routledge, 2013.

———. "Medieval Egyptian Society and the Concept of the Circle of Justice." *Mamlūk Studies Review* 10 (2006): 1–17.

Darnton, Robert. *The Great Cat Massacre and Other Episodes in French Cultural History.* New York: Basic Books, 1999.

Darwin, Charles. *The Expression of the Emotions in Man and Animals.* Edited by Joe Cain and Sharon Messenger. London: Penguin, 2009.

———. *On the Origin of Species by Means of Natural Selection, or the Preservation of Favoured Races in the Struggle for Life.* London: John Murray, 1859.

Das, Asok Kumar. "The Elephant in Mughal Painting." In *Flora and Fauna in Mughal Art.* Edited by Som Prakash Verma. Mumbai: Marg Publications, 1999.

Das Gupta, Ashin. *Malabar in Asian Trade: 1740–1800.* Cambridge: Cambridge University Press, 1967.

———. *The World of the Indian Ocean Merchant, 1500–1800: Collected Essays of Ashin Das Gupta.* Compiled by Uma Das Gupta. New Delhi: Oxford University Press, 2001.

Davis, Diana K. "Brutes, Beasts, and Empire: Veterinary Medicine and Environmental Policy in French North Africa and British India." *Journal of Historical Geography* 34 (2008): 242–67.

———. "Potential Forests: Degradation Narratives, Science, and Environmental Policy in Protectorate Morocco, 1912–1956." *Environmental History* 10 (2005): 211–38.

———. "Prescribing Progress: French Veterinary Medicine in the Service of Empire." *Veterinary Heritage* 29 (2006): 1–7.

———. *Resurrecting the Granary of Rome: Environmental History and French Colonial Expansion in North Africa.* Athens: Ohio University Press, 2007.

Davis, Diana K., and Denys Frappier. "The Social Context of Working Equines in the Urban Middle East: The Example of Fez Medina." *Journal of North African Studies* 5 (2000): 51–68.

Davis, Natalie Zemon. *The Gift in Sixteenth-Century France*. Madison: University of Wisconsin Press, 2000.

Dean, Charles L. *Soldiers and Sled Dogs: A History of Military Dog Mushing*. Lincoln: University of Nebraska Press, 2005.

El Deeb, Sarah. "Egypt's Garbage Problem Continues to Grow." *Huffington Post*. September 1, 2012.

Demircan, Yasemin. "XV–XVII. Yüzyıllarda Ege Adaları'nda Çeşitli Yönleriyle Küçükbaş Hayvancılık." *Ankara Üniversitesi Osmanlı Tarihi Araştırma ve Uygulama Merkezi Dergisi* 24 (2008): 71–95.

Denevan, William M. "The Pristine Myth: The Landscape of the Americas in 1492." *Annals of the Association of American Geographers* 82 (1992): 369–85.

Derr, Mark. *A Dog's History of America: How Our Best Friend Explored, Conquered, and Settled a Continent*. New York: North Point Press, 2004.

———. *How the Dog Became the Dog: From Wolves to Our Best Friends*. New York: Overlook Press, 2011.

Derrida, Jacques. *The Animal That Therefore I Am*. Edited by Marie-Louise Mallet. Translated by David Wills. New York: Fordham University Press, 2008.

Derry, Margaret E. *Bred for Perfection: Shorthorn Cattle, Collies, and Arabian Horses since 1800*. Baltimore: Johns Hopkins University Press, 2003.

Dickens, Peter. "The Labor Process: How the Underdog is Kept Under." *Society and Animals* 11 (2003): 69–72.

Digard, Jean-Pierre. "Cat II. Persian Cat." *Encyclopaedia Iranica*. New York: Iranica Online, 2013.

———. "Chah des Chats, Chat de Chah? Sur les traces du chat persan." In *Hommes et terres d'Islam: Mélanges offerts à Xavier de Planhol*. Edited by Daniel Balland. 2 vols. Tehran: Institut de recherche en Iran, 2000.

Digby, Simon. *War-horse and Elephant in the Delhi Sultanate: A Study of Military Supplies*. Oxford: Orient Monographs, 1971.

DiMarco, Louis A. *War Horse: A History of the Military Horse and Rider*. Yardley: Westholme Publishing, 2008.

DiNardo, R. L., and Austin Bay. "Horse-Drawn Transport in the German Army." *Journal of Contemporary History* 23 (1988): 129–42.

Dittrich, Lothar, and Annelore Rieke-Müller. *Carl Hagenbeck (1844–1913): Tierhandel und Schaustellungen im Deutschen Kaiserreich*. Frankfurt: Peter Lang, 1998.

Divyabhanusinh. *The End of a Trail: The Cheetah in India*. New Delhi: Banyan Books, 1995.

Dixon, B. A. *Animals, Emotion, and Morality: Marking the Boundary*. Amherst, NY: Prometheus Books, 2008.

Dodwell, Henry. *The Founder of Modern Egypt: A Study of Muhammad ʿAli*. Cambridge: The University Press, 1931.

Dohner, Janet Vorwald. *Livestock Guardians: Using Dogs, Donkeys, and Llamas to Protect Your Herd*. North Adams, MA: Storey Publishing, 2007.

Dols, Michael W. *The Black Death in the Middle East*. Princeton, NJ: Princeton University Press, 1977.

———. "The General Mortality of the Black Death in the Mamluk Empire." In *The Islamic Middle East, 700–1900: Studies in Social and Economic History*. Edited by Abraham L. Udovitch. Princeton, NJ: Darwin Press, 1981.

Doumani, Beshara, ed. *Family History in the Middle East: Household, Property, and Gender*. Albany: State University of New York Press, 2003.

Downey, Fairfax Davis. *Dogs for Defense: American Dogs in the Second World War, 1941–45*. New York: Daniel P. McDonald, 1955.

Drews, Robert. *Early Riders: The Beginnings of Mounted Warfare in Asia and Europe*. New York: Routledge, 2004.

Dumas, M. *Eloge historique de Isidore Geoffroy Saint-Hilaire*. Paris: Typ. de F. Didot frères, fils et Cie, 1872.

Dursteler, Eric R. *Venetians in Constantinople: Nation, Identity, and Coexistence in the Early Modern Mediterranean*. Baltimore: Johns Hopkins University Press, 2006.

Edmonds, David, and John Eidinow. *Rousseau's Dog: Two Great Thinkers at War in the Age of Enlightenment*. New York: HarperCollins, 2006.
Eisenstadt, Shmuel N., and Wolfgang Schluchter. "Introduction: Paths to Early Modernities—A Comparative View." *Daedalus* 127 (1998): 1–18.
Eisenstein, Herbert. *Einführung in die arabische Zoographie: Das tierkundliche Wissen in der arabisch-islamischen Literatur*. Berlin: Dietrich Reimer, 1990.
Elzein, Intisar. "Ottoman Archaeology of the Middle Nile Valley in the Sudan." In *The Frontiers of the Ottoman World*. Edited by A. C. S. Peacock. Oxford: Oxford University Press, 2009.
Emiralioğlu, Pınar. "Cognizance of the Ottoman World: Visual and Textual Representations in the Sixteenth-Century Ottoman Empire (1514–1596)." Ph.D. diss., University of Chicago, 2006.
———. "Relocating the Center of the Universe: China and the Ottoman Imperial Project in the Sixteenth Century." *Journal of Ottoman Studies* 39 (2012): 161–87.
Emiroğlu, Kudret, and Ahmet Yüksel. *Yoldaşımız At*. Istanbul: Yapı Kredi Yayınları, 2009.
"Empress Eugenie's Turtle Dies in Cairo Zoo at 90." *New York Times*. February 12, 1949.
Enenkel, Karl A. E., and Paul J. Smith, eds. *Early Modern Zoology: The Construction of Animals in Science, Literature and the Visual Arts*. 2 vols. Leiden: Brill, 2007.
The Entomologist's Annual for 1871. London: John Van Voorst, 1871.
Ergene, Boğaç A. *Local Court, Provincial Society and Justice in the Ottoman Empire: Legal Practice and Dispute Resolution in Çankırı and Kastamonu (1652–1744)*. Leiden: Brill, 2003.
Erler, Mehmet Yavuz. *Osmanlı Devleti'nde Kuraklık ve Kıtlık Olayları, 1800–1880*. Istanbul: Libra Kitap, 2010.
Esmeir, Samera. *Juridical Humanity: A Colonial History*. Stanford, CA: Stanford University Press, 2012.
Evans, E. P. *The Criminal Prosecution and Capital Punishment of Animals*. New York: E. P. Dutton, 1906.
Evans, Katy M., and Vicki J. Adams. "Proportion of Litters of Purebred Dogs Born by Caesarean Section." *Journal of Small Animal Practice* 51 (2010): 113–18.
Fahmy, Khaled. *All the Pasha's Men: Mehmed Ali, His Army and the Making of Modern Egypt*. Cambridge: Cambridge University Press, 1997.
———. "For Cavafy, with Love and Squalor: Some Critical Notes on the History and Historiography of Modern Alexandria." In *Alexandria, Real and Imagined*. Edited by Anthony Hirst and Michael Silk. Aldershot, UK: Ashgate, 2004.
———. "The Era of Muhammad ʿAli Pasha, 1805–1848." In *Modern Egypt, from 1517 to the End of the Twentieth Century*. Vol. 2 of *The Cambridge History of Egypt*. Edited by M. W. Daly. Cambridge: Cambridge University Press, 1998.
———. "Medical Conditions in Egyptian Prisons in the Nineteenth Century." In *Marginal Voices in Literature and Society: Individual and Society in the Mediterranean Muslim World*. Edited by Robin Ostle. Strasbourg: European Science Foundation in collaboration with Maison méditerranéenne des sciences de l'homme d'Aix-en-Provence, 2000.
———. *Mehmed Ali: From Ottoman Governor to Ruler of Egypt*. Oxford: Oneworld, 2009.
———. "An Olfactory Tale of Two Cities: Cairo in the Nineteenth Century." In *Historians in Cairo: Essays in Honor of George Scanlon*. Edited by Jill Edwards. Cairo: American University in Cairo Press, 2002.
———. "The Police and the People in Nineteenth-Century Egypt." *Die Welt des Islams* 39 (1999): 340–77.
———. "al-Raʾīs Mursī wa Akhlāqiyāt al-Naẓāfa." *Akhbār al-Adab*. August 14, 2012.
———. "Towards a Social History of Modern Alexandria." In *Alexandria, Real and Imagined*. Edited by Anthony Hirst and Michael Silk. Aldershot, UK: Ashgate, 2004.
———. "Women, Medicine, and Power in Nineteenth-Century Egypt." In *Remaking Women: Feminism and Modernity in the Middle East*. Edited by Lila Abu-Lughod. Princeton, NJ: Princeton University Press, 1998.
Farhad, Massumeh. "An Artist's Impression: Muʿin Musavvir's *Tiger Attacking a Youth*." *Muqarnas* 9 (1992): 116–23.
Farooqi, Naimur Rahman. *Mughal-Ottoman Relations: A Study of Political and Diplomatic Relations between Mughal India and the Ottoman Empire, 1556–1748*. Delhi: Idarah-i Adabiyat-i Delli, 1989.

Faroqhi, Suraiya. "Agriculture and Rural Life in the Ottoman Empire (ca 1500–1878) (A Report on Scholarly Literature Published between 1970 and 1985)." *New Perspectives on Turkey* 1 (1987): 3–34.

———. "Camels, Wagons, and the Ottoman State in the Sixteenth and Seventeenth Centuries." *International Journal of Middle East Studies* 14 (1982): 523–39.

———. "Coffee and Spices: Official Ottoman Reactions to Egyptian Trade in the Later Sixteenth Century." *Wiener Zeitschrift für die Kunde des Morgenlandes* 76 (1986): 87–93.

———. "Exotic Animals at the Sultan's Court." In *Another Mirror for Princes: The Public Image of the Ottoman Sultans and Its Reception*. Piscataway: Gorgias Press; Istanbul: Isis Press, 2009.

———. "Fish and Fishermen in Ottoman Istanbul." In *Water on Sand: Environmental Histories of the Middle East and North Africa*. Edited by Alan Mikhail. New York: Oxford University Press, 2013.

———. "Horses Owned by Ottoman Officials and Notables: Means of Transportation but also Sources of Pride and Joy." In *Animals and People in the Ottoman Empire*. Edited by Suraiya Faroqhi. Istanbul: Eren, 2010.

———. "Ottoman Peasants and Rural Life: The Historiography of the Twentieth Century." *Archivum Ottomanicum* 18 (2000): 153–82.

———. "The Peasants of Saideli in the Late Sixteenth Century." *Archivum Ottomanicum* 8 (1983): 215–50.

———. "Red Sea Trade and Communications as Observed by Evliya Çelebi (1671–72)." *New Perspectives on Turkey* 5–6 (1991): 87–105.

———. "Rural Society in Anatolia and the Balkans During the Sixteenth Century, I." *Turcica* 9 (1977): 161–95.

———. "Rural Society in Anatolia and the Balkans During the Sixteenth Century, II." *Turcica* 11 (1979): 103–53.

———. "Trade Controls, Provisioning Policies, and Donations: The Egypt-Hijaz Connection during the Second Half of the Sixteenth Century." In *Süleymân the Second and His Time*. Edited by Halil İnalcık and Cemal Kafadar. Istanbul: Isis Press, 1993.

Fattah, Hala. *The Politics of Regional Trade in Iraq, Arabia, and the Gulf, 1745–1900*. Albany: State University of New York Press, 1997.

Al-Fayez, Ghenaim, Abdelwahid Awadalla, Donald I. Templer, and Hiroko Arikawa. "Companion Animal Attitude and its Family Pattern in Kuwait." *Society and Animals* 11 (2003): 17–28.

Ferguson, Heather Lynn. "The Circle of Justice as Genre, Practice, and Objectification: A Discursive Re-Mapping of the Early Modern Ottoman Empire." Ph.D. diss., University of California, Berkeley, 2009.

Findlen, Paula. "Inventing Nature: Commerce, Art, and Science in the Early Modern Cabinet of Curiosities." In *Merchants and Marvels: Commerce, Science, and Art in Early Modern Europe*. Edited by Pamela H. Smith and Paula Findlen. New York: Routledge, 2002.

Fleischer, Cornell H. *Bureaucrat and Intellectual in the Ottoman Empire: The Historian Mustafa Âli (1541–1600)*. Princeton, NJ: Princeton University Press, 1986.

———. "Royal Authority, Dynastic Cyclism, and 'Ibn Khaldûnism' in Sixteenth-Century Ottoman Letters." *Journal of Asian and African Studies* 18 (1983): 198–220.

Floor, Willem. "A Note on Persian Cats." *Iranian Studies* 36 (2003): 27–42.

———. *The Persian Gulf: A Political and Economic History of Five Port Cities, 1500–1730*. Washington, DC: Mage Publishers, 2006.

———. "The Talar-i Tavila or Hall of Stables, a Forgotten Safavid Palace." *Muqarnas* 19 (2002): 19–63.

Flynn, Clifton P. "Animal Abuse in Childhood and Later Support for Interpersonal Violence in Families." *Society and Animals* 7 (1999): 161–72.

Foltz, Richard C. *Animals in Islamic Tradition and Muslim Cultures*. Oxford: Oneworld, 2006.

———. "Zoroastrian Attitudes toward Animals." *Society and Animals* 18 (2010): 367–78.

Forrest, S. K. Mendoza. *Witches, Whores, and Sorcerers: The Concept of Evil in Early Iran*. Austin: University of Texas Press, 2011.

Foucault, Michel. *Discipline and Punish: The Birth of the Prison*. Translated by Alan Sheridan. New York: Pantheon Books, 1978.

———. *History of Madness*. Edited by Jean Khalfa. Translated by Jonathan Murphy and Jean Khalfa. London: Routledge, 2006.

———. *An Introduction*. Vol. 1 of *The History of Sexuality*. Translated by Robert Hurley. New York: Vintage Books, 1978.

———. *Madness and Civilization: A History of Insanity in the Age of Reason*. Translated by Richard Howard. New York: Vintage Books, 1988.

———. *The Order of Things: An Archaeology of the Human Sciences*. New York: Vintage Books, 1994.

———. "What is Enlightenment?" In *Ethics: Subjectivity and Truth*. Vol. 1 of *The Essential Works of Michel Foucault: 1954–1984*. Edited by Paul Rabinow. New York: New Press, 1997.

Franklin, Adrian. *Animals and Modern Cultures: A Sociology of Human-Animal Relations in Modernity*. London: Sage, 1999.

Franklin, Julian H. *Animal Rights and Moral Philosophy*. New York: Columbia University Press, 2004.

Frembgen, Jürgen Wasim. "The Scorpion in Muslim Folklore." *Asian Folklore Studies* 63 (2004): 95–123.

Fritz, John M. "Vijayanagara: Authority and Meaning of a South Indian Imperial Capital." *American Anthropologist* 88 (1986): 44–55.

Fudge, Bruce. "Dog." *Encyclopaedia of the Qur'ān*. Leiden: Brill Online, 2013.

Fudge, Erica. *Brutal Reasoning: Animals, Rationality, and Humanity in Early Modern England*. Ithaca, NY: Cornell University Press, 2006.

———. *Perceiving Animals: Humans and Beasts in Early Modern English Culture*. New York: St. Martin's Press, 1999.

———. *Pets*. Stocksfield: Acumen, 2008.

Fudge, Erica, ed. *Renaissance Beasts: Of Animals, Humans, and Other Wonderful Creatures*. Urbana: University of Illinois Press, 2004.

Gallerani, Paola. *The Menagerie of Pieter Boel: Animal Painter in the Age of Louis XIV*. Translated by Catherine Bolton. Milan: Officina Libraria, 2011.

Genç, Mehmet. "Contrôle et taxation du commerce du café dans l'Empire ottoman fin XVIIe–première moitié du XVIIIe siècle." In *Le commerce du café avant l'ère des plantations coloniales: espaces, réseaux, sociétés (XVe–XIXe siècle)*. Edited by Michel Tuchscherer. Cairo: Institut français d'archéologie orientale, 2001.

Geoffroy Saint-Hilaire, Isidore. *Vie, travaux et doctrine scientifique d'Étienne Geoffroy Saint-Hilaire*. Paris: P. Bertrand, 1847.

Gibson, Mary. "Global Perspective on the Birth of the Prison." *American Historical Review* 116 (2011): 1040–63.

Gladitz, Charles. *Horse Breeding in the Medieval World*. Dublin: Four Courts Press, 1997.

Glaholt, Hayley Rose. "Vivisection as War: The 'Moral Diseases' of Animal Experimentation and Slavery in British Victorian Quaker Pacifist Ethics." *Society and Animals* 20 (2012): 154–72.

Goitein, S. D., and Mordechai Akiva Friedman. *India Traders of the Middle Ages: Documents from the Cairo Geniza, "India Book."* Leiden: Brill, 2008.

Goldstone, Jack A. "The Problem of the 'Early Modern' World." *Journal of the Economic and Social History of the Orient* 41 (1998): 249–84.

Goldziher, Ignaz. "Islamisme et Parsisme." *Revue de L'Histoire des Religions* 43 (1901): 1–29.

Gommans, Jos. "The Horse Trade in Eighteenth-Century South Asia." *Journal of the Economic and Social History of the Orient* 37 (1994): 228–250.

———. "Warhorse and Post-Nomadic Empire in Asia, c. 1000–1800." *Journal of Global History* 2 (2007): 1–21.

Goodavage, Maria. *Soldier Dogs: The Untold Story of America's Canine Heroes*. New York: Dutton, 2012.

Gorgas, Michael. "Animal Trade Between India and Western Eurasia in the Sixteenth Century—The Role of the Fuggers in Animal Trading." In *Indo-Portuguese Trade and the Fuggers of Germany, Sixteenth Century*. Edited by K. S. Mathew. New Delhi: Manohar, 1997.

Gran, Peter. *Islamic Roots of Capitalism: Egypt, 1760–1840*. Austin: University of Texas Press, 1979.

———. "Tahtawi in Paris." *al-Ahram Weekly*. January 10–16, 2002.

Grandin, Temple, ed. *Improving Animal Welfare: A Practical Approach*. Cambridge: CAB International, 2010.

Grandin, Temple. *Thinking in Pictures: And Other Reports from My Life with Autism*. New York: Doubleday, 1995.

Grandin, Temple, with Mark Deesing. *Humane Livestock Handling: Understanding Livestock Behavior and Building Facilities for Healthier Animals*. North Adams, MA: Storey Publishing, 2008.

Grandin, Temple, and Catherine Johnson. *Animals in Translation: Using the Mysteries of Autism to Decode Animal Behavior*. New York: Scribner, 2005.

Grassby, Richard. "The Decline of Falconry in Early Modern England." *Past and Present* 157 (1997): 37–62.

Greene, Ann Norton. *Horses at Work: Harnessing Power in Industrial America*. Cambridge, MA: Harvard University Press, 2008.

Greene, Molly. *Catholic Pirates and Greek Merchants: A Maritime History of the Mediterranean*. Princeton, NJ: Princeton University Press, 2010.

———. "The Ottoman Experience." *Daedalus* 134 (2005): 88–99.

Greenwood, Antony. "Istanbul's Meat Provisioning: A Study of the *Celepkeşan* System." Ph.D. diss., University of Chicago, 1988.

Greenwood, Ned H. *The Sinai: A Physical Geography*. Austin: University of Texas Press, 1997.

Grier, Katherine C. *Pets in America: A History*. Chapel Hill: University of North Carolina Press, 2006.

Gruen, Lori. *Ethics and Animals: An Introduction*. Cambridge: Cambridge University Press, 2011.

Gschwend, Annemarie Jordan. *The Story of Süleyman: Celebrity Elephants and Other Exotica in Renaissance Portugal*. Philadelphia: Pachyderm, 2010.

Guerrini, Anita. *Experimenting with Humans and Animals: From Galen to Animal Rights*. Baltimore: Johns Hopkins University Press, 2003.

Gupta, Maya. "Functional Links Between Intimate Partner Violence and Animal Abuse: Personality Features and Representations of Aggression." *Society and Animals* 16 (2008): 223–42.

Habib, Irfan, ed. *Confronting Colonialism: Resistance and Modernization under Haidar Ali and Tipu Sultan*. London: Anthem, 2002.

Hahn, Daniel. *The Tower Menagerie: The Amazing 600-Year History of the Royal Collection of Wild and Ferocious Beasts Kept at the Tower of London*. New York: Jeremy P. Tarcher/Penguin, 2004.

Hamadeh, Shirine. *The City's Pleasures: Istanbul in the Eighteenth Century*. Seattle: University of Washington Press, 2008.

Hämäläinen, Pekka. *The Comanche Empire*. New Haven, CT: Yale University Press, 2008.

Hamilton, Alastair. *The Copts and the West, 1439–1822: The European Discovery of the Egyptian Church*. Oxford: Oxford University Press, 2006.

Hanna, Nelly. *Artisan Entrepreneurs in Cairo and Early Modern Capitalism (1600–1800)*. Syracuse, NY: Syracuse University Press, 2011.

———. *Making Big Money in 1600: The Life and Times of Isma'il Abu Taqiyya, Egyptian Merchant*. Syracuse, NY: Syracuse University Press, 1998.

———. *In Praise of Books: A Cultural History of Cairo's Middle Class, Sixteenth to the Eighteenth Century*. Syracuse, NY: Syracuse University Press, 2003.

Hansen, Valerie. *The Open Empire: A History of China to 1600*. New York: W. W. Norton, 2000.

Haque, Nadeem, and Al-Hafiz Basheer Ahmad Masri. "The Principles of Animal Advocacy in Islam: Four Integrated Ecognitions." *Society and Animals* 19 (2011): 279–90.

Haraway, Donna J. *The Companion Species Manifesto: Dogs, People, and Significant Otherness*. Chicago: Prickly Paradigm Press, 2003.

———. "Teddy Bear Patriarchy: Taxidermy in the Garden of Eden, New York City, 1908–1936." *Social Text* 11 (1984): 19–64.

———. *When Species Meet*. Minneapolis: University of Minnesota Press, 2008.

Harris, Gardiner. "A Bin Laden Hunter on Four Legs." *New York Times*. May 4, 2011.

Harris, Marvin. "The Cultural Ecology of India's Sacred Cattle." *Current Anthropology* 7 (1966): 51–66.

Harrison, Brian. "Animals and the State in Nineteenth-Century England." *English Historical Review* 88 (1973): 786–820.

Hatch, Charles E., Jr. "Mulberry Trees and Silkworms: Sericulture in Early Virginia." *Virginia Magazine of History and Biography* 65 (1957): 3–61.

Hathaway, Jane. *Beshir Agha: Chief Eunuch of the Ottoman Imperial Harem*. Oxford: Oneworld, 2005.

———. "Eunuch Households in Istanbul, Medina, and Cairo during the Ottoman Era." *Turcica* 41 (2009): 291–303.
———. *The Politics of Households in Ottoman Egypt: The Rise of the Qazdağlıs*. Cambridge: Cambridge University Press, 1997.
———. "The Role of the Kızlar Ağası in 17th–18th Century Ottoman Egypt." *Studia Islamica* 75 (1992): 141–58.
———. *A Tale of Two Factions: Myth, Memory, and Identity in Ottoman Egypt and Yemen*. Albany: State University of New York Press, 2003.
———. "The Wealth and Influence of an Exiled Ottoman Eunuch in Egypt: The *Waqf* Inventory of ʿAbbās Agha." *Journal of the Economic and Social History of the Orient* 37 (1994): 293–317.
Hearne, Vicki. *Adam's Task: Calling Animals by Name*. New York: Knopf, 1986.
Hehenberger, Susanne. "Dehumanised Sinners and their Instruments of Sin: Men and Animals in Early Modern Bestiality Cases, Austria 1500–1800." In *Early Modern Zoology: The Construction of Animals in Science, Literature and the Visual Arts*. Edited by Karl A. E. Enenkel and Paul J. Smith. 2 vols. Leiden: Brill, 2007.
Henninger-Voss, Mary J., ed. *Animals in Human Histories: The Mirror of Nature and Culture*. Rochester, NY: University of Rochester Press, 2002.
Heywood, Colin. "The Ottoman *Menzilhane* and *Ulak* System in Rumeli in the Eighteenth Century." In *Türkiye'nin Sosyal ve Ekonomik Tarihi, 1071–1920: Birinci Uluslararası Türkiye'nin Sosyal ve Ekonomik Tarihi Kongresi Tebliğleri*. Edited by Osman Okyar and Halil İnalcık. Ankara: Meteksan Şirketi, 1980.
———. "Some Turkish Archival Sources for the History of the Menzilhane Network in Rumeli during the Eighteenth Century (Notes and Documents on the Ottoman Ulak, I)." *Boğaziçi Üniversitesi Dergisi* 4–5 (1976–77): 39–54.
———. "The Via Egnatia in the Ottoman Period: The Menzilḫānes of the Ṣol Ḳol in the Late 17th/Early 18th Century." In *The Via Egnatia under Ottoman Rule (1380–1699): Halcyon Days in Crete II*. Edited by Elizabeth Zachariadou. Rethymnon: Crete University Press, 1996.
Heyworth-Dunne, James. *An Introduction to the History of Education in Modern Egypt*. London: Luzac, 1939.
———. "Printing and Translations under Muḥammad ʿAlī of Egypt: The Foundation of Modern Arabic." *Journal of the Royal Asiatic Society* 72 (1940): 325–49.
Hinschaw-Fischli, Robert Elena. "Some Italian and German Representations of India in the Early Renaissance—with a Predilection for Elephants." In *Maritime Malabar and the Europeans, 1500–1962*. Edited by K. S. Mathew. Gurgaon: Hope India Publications, 2003.
Hinz, Walther. *Islamische Masse und Gewichte umgerechnet ins metrische System*. Leiden: Brill, 1955.
Hoage, R. J., and William A. Deiss, eds. *New Worlds, New Animals: From Menagerie to Zoological Park in the Nineteenth Century*. Baltimore: Johns Hopkins University Press, 1996.
Hoage, R. J., Anne Roskell, and Jane Mansour. "Menageries and Zoos to 1900." In *New Worlds, New Animals: From Menagerie to Zoological Park in the Nineteenth Century*. Edited by R. J. Hoage and William A. Deiss. Baltimore: Johns Hopkins University Press, 1996.
Hodgson, Marshall G. S. *The Gunpowder Empires and Modern Times*. Vol. 3 of *The Venture of Islam: Conscience and History in a World Civilization*. Chicago: University of Chicago Press, 1974.
Holden, Clare, and Ruth Mace. "Phylogenetic Analysis of the Evolution of Lactose Digestion in Adults." *Human Biology* 81 (2009): 597–619.
Holt, P. M. *Egypt and the Fertile Crescent, 1516–1922: A Political History*. London: Longmans Green, 1966.
Holt, P. M., ed. *Political and Social Change in Modern Egypt: Historical Studies from the Ottoman Conquest to the United Arab Republic*. London: Oxford University Press, 1968.
Horowitz, Alexandra. *Inside of a Dog: What Dogs See, Smell, and Know*. New York: Scribner, 2009.
Horowitz, Roger. *Putting Meat on the American Table: Taste, Technology, Transformation*. Baltimore: Johns Hopkins University Press, 2006.
Houlihan, Patrick F. "Animals in Egyptian Art and Hieroglyphs." In *A History of the Animal World in the Ancient Near East*. Edited by Billie Jean Collins. Leiden: Brill, 2002.
Hourani, Albert. *Arabic Thought in the Liberal Age, 1798–1939*. Cambridge: Cambridge University Press, 1983.

Howell, Philip. "A Place for the Animal Dead: Pets, Pet Cemeteries and Animal Ethics in Late Victorian Britain." *Ethics, Place and Environment* 5 (2002): 5–22.

Hribal, Jason C. "Animals, Agency, and Class: Writing the History of Animals from Below." *Human Ecology Review* 14 (2007): 101–12.

Hubbell, Harry M. "Ptolemy's Zoo." *The Classical Journal* 31 (1935): 68–76.

Humes, Edward. *Garbology: Our Dirty Love Affair with Trash*. New York: Avery, 2012.

Humphreys, R. Stephen. "Egypt in the World System of the later Middle Ages." In *Islamic Egypt, 640–1517*. Vol. 1 of *The Cambridge History of Egypt*. Edited by Carl F. Petry. Cambridge: Cambridge University Press, 1998.

Hunter, F. Robert. *Egypt under the Khedives, 1805–1879: From Household Government to Modern Bureaucracy*. Pittsburgh: University of Pittsburgh Press, 1984.

Husain, Iqbal. "The Diplomatic Vision of Tipu Sultan: Briefs for Embassies to Turkey and France, 1785–86." In *State and Diplomacy under Tipu Sultan: Documents and Essays*. Edited by Irfan Habib. New Delhi: Tulika, 2001.

Ibrāhīm, Nāṣir Aḥmad. *al-Azamāt al-Ijtimāʿiyya fī Miṣr fī al-Qarn al-Sābiʿ ʿAshar*. Cairo: Dār al-Āfāq al-ʿArabiyya, 1998.

al-ʿĪd al-Miʾawī li-Ḥadāʾiq al-Ḥayawān, 1891–1991. Cairo: al-Idāra al-Markaziyya li-Ḥadāʾiq al-Ḥayawān wa al-Ḥifāẓ ʿalā al-Ḥayāh al-Barriyya, 1991.

Imber, Colin. *The Ottoman Empire, 1300–1650: The Structure of Power*. New York: Palgrave Macmillan, 2002.

Isenberg, Andrew C. *The Destruction of the Bison: An Environmental History, 1750–1920*. Cambridge: Cambridge University Press, 2000.

Ismāʿil, Muhammad Husām al-Din. "Le café dans la ville de Rosette à l'époque ottomane XVIe–XVIIe siècle." In *Le commerce du café avant l'ère des plantations coloniales: espaces, réseaux, sociétés (XVe–XIXe siècle)*. Edited by Michel Tuchscherer. Cairo: Institut français d'archéologie orientale, 2001.

Işın, Ekrem. "Street Dogs of Istanbul: The Four-Legged Municipality." In *Everyday Life in Istanbul: Social Historical Essays on People, Culture and Spatial Relations*. Translated by Virginia Taylor Saçlıoğlu. Istanbul: Yapı Kredi Yayınları, 2001.

İhsanoğlu, Ekmeleddin. *Mısır'da Türkler ve Kültürel Mirasları: Mehmed Ali Paşa'dan Günümüze Basılı Türk Kültürü Bibliyografyası ve Bir Değerlendirme*. Istanbul: İslam Tarih, Sanat, ve Kültür Araştırma Merkezi, 2006.

İnalcık, Halil. "'Arab' Camel Drivers in Western Anatolia in the Fifteenth Century." *Revue d'Histoire Maghrebine* 10 (1983): 256–70.

———. "Capital Formation in the Ottoman Empire." *Journal of Economic History* 29 (1969): 97–140.

———. "Doghandjı." *Encyclopaedia of Islam*. 2nd ed. Leiden: Brill Online, 2013.

———. "The Emergence of Big Farms, *Çiftliks*: State, Landlords, and Tenants." In *Landholding and Commercial Agriculture in the Middle East*. Edited by Çağlar Keyder and Faruk Tabak. Albany: State University of New York Press, 1991.

———. *The Ottoman Empire: The Classical Age, 1300–1600*. Translated by Norman Itzkowitz and Colin Imber. New York: Praeger, 1973.

———. "The Ottoman State: Economy and Society, 1300–1600." In *An Economic and Social History of the Ottoman Empire*. Edited by Halil İnalcık with Donald Quataert. 2 vols. Cambridge: Cambridge University Press, 1994.

İslamoğlu-İnan, Huri. *State and Peasant in the Ottoman Empire: Agrarian Power Relations and Regional Economic Development in Ottoman Anatolia during the Sixteenth Century*. Leiden: Brill, 1994.

Jackson, Kate. *Mean and Lowly Things: Snakes, Science, and Survival in the Congo*. Cambridge, MA: Harvard University Press, 2008.

Jacobs, Nancy J. "The Great Bophuthatswana Donkey Massacre: Discourse on the Ass and the Politics of Class and Grass." *American Historical Review* 106 (2001): 485–507.

Jagchid, S., and C. R. Bawden. "Some Notes on the Horse Policy of the Yüan Dynasty." *Central Asiatic Journal* 10 (1965): 246–68.

James, Lawrence. *Raj: The Making and Unmaking of British India*. New York: St. Martin's Press, 1997.

Jasanoff, Maya. *Edge of Empire: Lives, Culture, and Conquest in the East, 1750–1850.* New York: Vintage, 2006.

Jennison, George. *Animals for Show and Pleasure in Ancient Rome.* Manchester: Manchester University Press, 1937.

Jila, Namu. "Myths and Traditional Beliefs about the Wolf and the Crow in Central Asia: Examples from the Turkic Wu-Sun and the Mongols." *Asian Folklore Studies* 65 (2006): 161–77.

Johnson, Christopher H., David Warren Sabean, Simon Teuscher, and Francesca Trivellato, eds. *Transregional and Transnational Families in Europe and Beyond: Experiences since the Middle Ages.* New York: Berghahn Books, 2011.

Jones, Susan D. *Death in a Small Package: A Short History of Anthrax.* Baltimore: Johns Hopkins University Press, 2010.

———. *Valuing Animals: Veterinarians and Their Patients in Modern America.* Baltimore: Johns Hopkins University Press, 2002.

Jordan, William Chester. *The Great Famine: Northern Europe in the Early Fourteenth Century.* Princeton, NJ: Princeton University Press, 1996.

Joshi, Maheshwar P., and C. W. Brown. "Some Dynamics of Indo-Tibetan Trade through Uttarākhaṇḍa (Kumaon-Garhwal), India." *Journal of the Economic and Social History of the Orient* 30 (1987): 303–17.

Kabadayı, M. Erdem. "The Introduction of Merino Sheep Breeding in the Ottoman Empire: Successes and Failures." In *Animals and People in the Ottoman Empire.* Edited by Suraiya Faroqhi. Istanbul: Eren, 2010.

Kafadar, Cemal. "The Myth of the Golden Age: Ottoman Historical Consciousness in the Post-Süleymanic Era." In *Süleymân the Second and His Time.* Edited by Halil İnalcık and Cemal Kafadar. Istanbul: Isis Press, 1993.

Kant, Immanuel. "An Answer to the Question: What is Enlightenment?" In *Perpetual Peace and Other Essays on Politics, History, and Morals.* Translated by Ted Humphrey. Indianapolis: Hackett Publishing, 1983.

Kasaba, Reşat. *The Ottoman Empire and the World Economy: The Nineteenth Century.* Albany: State University of New York Press, 1988.

Kauz, Ralph. "Horse Exports from the Persian Gulf until the Arrival of the Portuguese." In *Pferde in Asien: Geschichte, Handel Und Kultur.* Edited by Bert G. Fragner, Ralph Kauz, Roderich Ptak, and Angela Schottenhammer. Vienna: Österreichischen Akademie der Wissenschaften, 2009.

Kean, Hilda. "Imagining Rabbits and Squirrels in the English Countryside." *Society and Animals* 9 (2001): 163–75.

Kelekna, Pita. *The Horse in Human History.* Cambridge: Cambridge University Press, 2009.

Kerslake, Celia J. "The Correspondence between Selīm I and Ḳānṣūh al-Ġawrī." *Prilozi za Orijentalnu Filologiju* 30 (1980): 219–34.

Kete, Kathleen. *The Beast in the Boudoir: Petkeeping in Nineteenth-Century Paris.* Berkeley: University of California Press, 1994.

Khazeni, Arash. "Through an Ocean of Sand: Pastoralism and the Equestrian Culture of the Eurasian Steppe." In *Water on Sand: Environmental Histories of the Middle East and North Africa.* Edited by Alan Mikhail. New York: Oxford University Press, 2013.

Khoury, Dina Rizk. "The Introduction of Commercial Agriculture in the Province of Mosul and its Effects on the Peasantry, 1750–1850." In *Landholding and Commercial Agriculture in the Middle East.* Edited by Çağlar Keyder and Faruk Tabak. Albany: State University of New York Press, 1991.

Kirillina, Svetlana. "Representing the Animal World of the Ottoman Empire: The Accounts of Russian Orthodox Pilgrims (Sixteenth–Eighteenth Centuries)." In *Animals and People in the Ottoman Empire.* Edited by Suraiya Faroqhi. Istanbul: Eren, 2010.

Kirksey, S. Eben, and Stefan Helmreich. "The Emergence of Multispecies Ethnography." *Cultural Anthropology* 25 (2010): 545–76.

Koch, Ebba. *Mughal Art and Imperial Ideology: Collected Essays.* New Delhi: Oxford University Press, 2001.

Kortepeter, C. Max "The Life and Times of General Lew Wallace, Minister Extraordinary to the Ottoman Court, 1881–1885." In *Cultural Horizons: A Festschrift in Honor of Talat S. Halman.*

Edited by Jayne L. Warner. 2 vols. Syracuse, NY: Syracuse University Press; Istanbul: Yapı Kredi Yayınları, 2001.
Kosek, Jake. "Ecologies of Empire: On the New Uses of the Honeybee." *Cultural Anthropology* 25 (2010): 650–78.
Koslofsky, Craig. *Evening's Empire: A History of the Night in Early Modern Europe.* Cambridge: Cambridge University Press, 2011.
Krstić, Tijana. *Contested Conversions to Islam: Narratives of Religious Change in the Early Modern Ottoman Empire.* Stanford, CA: Stanford University Press, 2011.
Kruk, Remke. "Traditional Islamic Views of Apes and Monkeys." In *Ape, Man, Apeman: Changing Views since 1600.* Edited by Raymond Corbey and Bert Theunissen. Leiden: Leiden University, 1995.
Kuhnke, LaVerne. *Lives at Risk: Public Health in Nineteenth-Century Egypt.* Berkeley: University of California Press, 1990.
Kuru, Selim Sirri. "A Sixteenth-Century Ottoman Scholar, Deli Birader, and His Dāfiʿü'l-ġumūm ve Rāfiʿü'l-humūm." Ph.D. diss., Harvard University, 2000.
Kurz, O. *European Clocks and Watches in the Near East.* Leiden: Brill, 1975.
Kutluoğlu, Muhammad H. *The Egyptian Question (1831–1841): The Expansionist Policy of Mehmed Ali Paşa in Syria and Asia Minor and the Reaction of the Sublime Porte.* Istanbul: Eren, 1998.
LaCapra, Dominick. *History and Its Limits: Human, Animal, Violence.* Ithaca, NY: Cornell University Press, 2009.
Lambin, Eric. *An Ecology of Happiness.* Translated by Teresa Lavender Fagan. Chicago: University of Chicago Press, 2012.
Lane, Edward William. *An Arabic-English Lexicon.* 8 vols. Beirut: Librairie du Liban, 1968.
Lawson, Fred H. *The Social Origins of Egyptian Expansionism during the Muhammad ʿAli Period.* New York: Columbia University Press, 1992.
Le Guyader, Hervé. *Étienne Geoffroy Saint-Hilaire, 1772–1844: A Visionary Naturalist.* Translated by Marjorie Grene. Chicago: University of Chicago Press, 2004.
Lemish, Michael G. *War Dogs: Canines in Combat.* Washington, DC: Brassey's, 1996.
Lévi-Strauss, Claude. *Totemism.* Translated by Rodney Needham. Boston: Beacon Press, 1963.
Lewis, Bernard. "The Fatimids and the Route to India." *Revue de la Faculté des Sciences Économiques de l'Université d'Istanbul* 11 (1949–50): 50–54.
———. "The Mughals and the Ottomans." In *From Babel to Dragomans: Interpreting the Middle East.* New York: Oxford University Press, 2004.
Lilly, Robert J., and Michael B. Puckett. "Social Control and Dogs: A Sociohistorical Analysis." *Crime and Delinquency* 43 (1997): 123–47.
Limido, Luisa. *L'art des jardins sous le Second Empire: Jean-Pierre Barillet-Deschamps, 1824–1873.* Seyssel: Champ Vallon, 2002.
Linant de Bellefonds, M. A. *Mémoires sur les principaux travaux d'utilité publiqué éxécutés en Egypte depuis la plus haute antiquité jusqu'à nos jours: accompagné d'un atlas renfermant neuf planches grand in-folio imprimées en couleur.* Paris: Arthus Bertrand, 1872–73.
Lisān al-ʿArab. 4 vols. Beirut: Dār Lisān al-ʿArab, 1970.
Lockwood, Randall, and Frank R. Ascione, eds. *Cruelty to Animals and Interpersonal Violence: Readings in Research and Application.* West Lafayette, IN: Purdue University Press, 1998.
The London and China Telegraph. October 4, 1869.
Loureiro, Rudi Manuel. "Portuguese Involvement in Sixteenth Century Horse Trade through the Arabian Sea." In *Pferde in Asien: Geschichte, Handel Und Kultur.* Edited by Bert G. Fragner, Ralph Kauz, Roderich Ptak, and Angela Schottenhammer. Vienna: Österreichischen Akademie der Wissenschaften, 2009.
Lowry, Heath W., and İsmail E. Erünsal. *Remembering One's Roots: Mehmed Ali Paşa of Egypt's Links to the Macedonian Town of Kavala: Architectural Monuments, Inscriptions, and Documents.* Istanbul: Bahçeşehir University Press, 2011.
Lurker, Manfred. *The Gods and Symbols of Ancient Egypt: An Illustrated Dictionary.* London: Thames and Hudson, 1980.
Lynch, Dennis A. "Rhetorics of Proximity: Empathy in Temple Grandin and Cornel West." *Rhetoric Society Quarterly* 28 (1998): 5–23.

MacLean, Gerald. "The Sultan's Beasts: Encountering Ottoman Fauna." In *Looking East: English Writing and the Ottoman Empire before 1800*. New York: Palgrave Macmillan, 2007.
Maghen, Ze'ev. "Dead Tradition: Joseph Schacht and the Origins of 'Popular Practice.'" *Islamic Law and Society* 10 (2003): 276–347.
Maḥārīq, Yāsir ʿAbd al-Minʿam. *al-Minūfiyya fī al-Qarn al-Thāmin ʿAshar*. Cairo: al-Hayʾa al-Miṣriyya al-ʿĀmma lil-Kitāb, 2000.
Māhir, Suʿād. "Majrī Miyāh Famm al-Khalīj." *Egyptian Historical Review* 7 (1958): 134–57.
Makhlūf, Ḥasanayn Muḥammad. *al-Rifq bi-al-Ḥayawān fī Dīn al-Islām*. Cairo: Kashīda lil-Nashr wa al-Tawzīʿ, 2011.
Malamud, Randy. *Reading Zoos: Representations of Animals and Captivity*. New York: New York University Press, 1998.
Malekandathil, Pius. "Maritime Malabar and a Mercantile State: Polity and State Formation in Malabar under the Portuguese, 1498–1663." In *Maritime Malabar and the Europeans, 1500–1962*. Edited by K. S. Mathew. Gurgaon: Hope India Publications, 2003.
Manlius, Nicolas. "The Ostrich in Egypt: Past and Present." *Journal of Biogeography* 28 (2001): 945–53.
Manning, Joseph G. "A Ptolemaic Agreement Concerning a Donkey with an Unusual Warranty Clause. The Strange Case of *P. dem. Princ.* 1 (inv. 7524)." *Enchoria* 28 (2002/2003): 46–61.
Mansfield, Peter. *The British in Egypt*. London: Weidenfeld and Nicolson, 1971.
Manṣūr, Saʿīd Ḥ. *The World-View of al-Jāḥiẓ in* Kitāb al-Ḥayawān. Alexandria: Dar al-Maareff, 1977.
Marei, Nasr. *The Arabian Horse of Egypt*. Cairo: American University in Cairo Press, 2010.
Marks, Robert B. *The Origins of the Modern World: A Global and Ecological Narrative*. Lanham, MD: Rowman and Littlefield, 2002.
Marshall, J. E. *The Egyptian Enigma, 1890–1928*. London: John Murray, 1928.
Marsot, Afaf Lutfi al-Sayyid. "The Cartoon in Egypt." *Comparative Studies in Society and History* 13 (1971): 2–15.
———. *Egypt in the Reign of Muhammad Ali*. Cambridge: Cambridge University Press, 1984.
Martin, Edward C., Jr. *Dr. Johnson's Apple Orchard: The Story of America's First Pet Cemetery*. Hartsdale, NY: Hartsdale Canine Cemetery, 1997.
Martin, Jennifer Adams. "When Sharks (Don't) Attack: Wild Animal Agency in Historical Narratives." *Environmental History* 16 (2011): 451–55.
Martin, John. "Sheep and Enclosure in Sixteenth-Century Northamptonshire." *Agricultural History Review* 36 (1988): 39–54.
Marvin, Garry. *Wolf*. London: Reaktion Books, 2012.
Masri, Basheer Ahmad. *Animal Welfare in Islam*. Markfield: Islamic Foundation, 2007.
———. *Animals in Islam*. Petersfield: Athene Trust, 1989.
Masri, Al-Hafiz B. A. "Animal Experimentation: The Muslim Viewpoint." In *Animal Sacrifices: Religious Perspectives on the Use of Animals in Science*. Edited by Tom Regan. Philadelphia: Temple University Press, 1986.
Masson, Jeffrey Moussaieff. *Dogs Never Lie about Love: Reflections on the Emotional World of Dogs*. London: J. Cape, 1997.
———. *The Pig who Sang to the Moon: The Emotional World of Farm Animals*. New York: Ballantine Books, 2003.
Masson, Jeffrey Moussaieff, and Susan McCarthy. *When Elephants Weep: The Emotional Lives of Animals*. New York: Delacorte Press, 1995.
Mauss, Marcel. *The Gift: The Form and Reason for Exchange in Archaic Societies*. Translated by W. D. Halls. New York: W. W. Norton, 1990.
Maxwell-Stuart, P. G. "'Wild, Filthie, Execrabill, Detestabill, and Unnatural Sin': Bestiality in Early Modern Scotland." In *Sodomy in Early Modern Europe*. Edited by Tom Betteridge. Manchester: Manchester University Press, 2002.
Mayer, Andreas. "The Physiological Circus: Knowing, Representing, and Training Horses in Motion in Nineteenth-Century France." *Representations* 111 (2010): 88–120.
Mayers, Marilyn Anne. "A Century of Psychiatry: The Egyptian Mental Hospitals." Ph.D. diss., Princeton University, 1984.

Mayeur-Jaouen, Catherine. "Badawi and His Camel: An Animal as the Attribute of a Muslim Saint in Mamluk and Ottoman Egypt." Translated by Suraiya Faroqhi. In *Animals and People in the Ottoman Empire.* Edited by Suraiya Faroqhi. Istanbul: Eren, 2010.

McDonald, M. V. "Animal-Books as a Genre in Arabic Literature." *Bulletin (British Society for Middle Eastern Studies)* 15 (1988): 3–10.

McGowan, Bruce. "The Age of the Ayans, 1699–1812." In *An Economic and Social History of the Ottoman Empire.* Edited by Halil İnalcık with Donald Quataert. 2 vols. Cambridge: Cambridge University Press, 1994.

McHugh, Susan. *Dog.* London: Reaktion Books, 2004.

McNab, Brian K. *Extreme Measures: The Ecological Energetics of Birds and Mammals.* Chicago: University of Chicago Press, 2012.

McNeill, J. R. *The Mountains of the Mediterranean World: An Environmental History.* Cambridge: Cambridge University Press, 1992.

———. *Something New under the Sun: An Environmental History of the Twentieth-Century World.* New York: W. W. Norton, 2000.

McNeur, Catherine. "The 'Swinish Multitude': Controversies over Hogs in Antebellum New York City." *Journal of Urban History* 37 (2011): 639–60.

McShane, Clay. *Down the Asphalt Path: The Automobile and the American City.* New York: Columbia University Press, 1994.

McShane, Clay, and Joel A. Tarr. *The Horse in the City: Living Machines in the Nineteenth Century.* Baltimore: Johns Hopkins University Press, 2007.

Meacham, Sarah Hand. "Pets, Status, and Slavery in the Late-Eighteenth-Century Chesapeake." *Journal of Southern History* 77 (2011): 521–54.

Melosi, Martin V. *Garbage in the Cities: Refuse, Reform, and the Environment, 1880–1980.* College Station: Texas A&M University Press, 1981.

Meloy, John L. *Imperial Power and Maritime Trade: Mecca and Cairo in the Later Middle Ages.* Chicago: Middle East Documentation Center, 2010.

Melville, Elinor G. K. *A Plague of Sheep: Environmental Consequences of the Conquest of Mexico.* Cambridge: Cambridge University Press, 1997.

Melvin-Koushki, Matthew. "The Delicate Art of Aggression: Uzun Hasan's *Fathnama* to Qaytbay of 1469." *Iranian Studies* 44 (2011): 193–214.

Menache, Sophia. "Dogs: God's Worst Enemies?" *Society and Animals* 5 (1997): 23–44.

Merchant, Carolyn. *The Death of Nature: Women, Ecology, and the Scientific Revolution.* San Francisco: Harper & Row, 1980.

Meriwether, Margaret L. *The Kin Who Count: Family and Society in Ottoman Aleppo, 1770–1840.* Austin: University of Texas Press, 1999.

Midant-Reynes, Béatrix. *The Prehistory of Egypt: From the First Egyptians to the First Pharaohs.* Translated by Ian Shaw. Oxford: Blackwell, 2000.

Mikhail, Alan. "Anatolian Timber and Egyptian Grain: Things that Made the Ottoman Empire." In *Early Modern Things: Objects and Their Histories, 1500–1800.* Edited by Paula Findlen. London: Routledge, 2013.

———. "Animals as Property in Early Modern Ottoman Egypt." *Journal of the Economic and Social History of the Orient* 53 (2010): 621–52.

———. "The Heart's Desire: Gender, Urban Space and the Ottoman Coffee House." In *Ottoman Tulips, Ottoman Coffee: Leisure and Lifestyle in the Eighteenth Century.* Edited by Dana Sajdi. London: I. B. Tauris, 2007.

———. "An Irrigated Empire: The View from Ottoman Fayyum." *International Journal of Middle East Studies* 42 (2010): 569–90.

———. *Nature and Empire in Ottoman Egypt: An Environmental History.* Cambridge: Cambridge University Press, 2011.

———. "The Nature of Plague in Late Eighteenth-Century Egypt." *Bulletin of the History of Medicine* 82 (2008): 249–75.

Mikhail, Alan, and Christine M. Philliou. "The Ottoman Empire and the Imperial Turn." *Comparative Studies in Society and History* 54 (2012): 721–45.

Mikhail, E. M., N. S. Mansour, and H. N. Awadalla. "Identification of *Trichinella* Isolates from Naturally Infected Stray Dogs in Egypt." *Journal of Parasitology* 80 (1994): 151–54.

Miklósi, Ádám. *Dog Behaviour, Evolution, and Cognition*. New York: Oxford University Press, 2008.
Miller, Ian Jared. "The Nature of the Beast: The Ueno Zoological Gardens and Imperial Modernity in Japan, 1882–1945." Ph.D. diss., Columbia University, 2004.
Milton, Shaun. "Western Veterinary Medicine in Colonial Africa: A Survey, 1902–1963." *Argos* 18 (1998): 313–22.
Mitchell, Timothy. "Can the Mosquito Speak?" In *Rule of Experts: Egypt, Techno-Politics, Modernity*. Berkeley: University of California Press, 2002.
———. *Carbon Democracy: Political Power in the Age of Oil*. London: Verso, 2011.
———. *Colonising Egypt*. Berkeley: University of California Press, 1991.
Moazami, Mahnaz. "The Dog in Zoroastrian Religion: *Vidēvdād* Chapter XIII." *Indo-Iranian Journal* 49 (2006): 127–49.
———. "Evil Animals in the Zoroastrian Religion." *History of Religions* 44 (2005): 300–17.
Morgan, Philip D. "Slaves and Livestock in Eighteenth-Century Jamaica: Vineyard Pen, 1750–1751." *William and Mary Quarterly* 52 (1995): 47–76.
Mosier, Jennifer L. "The Big Attraction: The Circus Elephant and American Culture." *Journal of American Culture* 22 (1999): 7–18.
Mosley, Stephen. *The Environment in World History*. New York: Routledge, 2010.
el-Mouelhy, Ibrahim. *Organisation et fonctionnement des institutions ottomanes en Egypte (1517–1917): étude documentaire, d'après les sources archivistiques égyptiennes*. Ankara?: Imprimerie de la Société turque d'histoire, 1989.
"Mr. Roosevelt at Cairo." *Manchester Guardian*. March 26, 1910.
Mughayth, Kammāl Ḥāmid. *Miṣr fī al-ʿAṣr al-ʿUthmānī 1517–1798: al-Mujtamaʿ... wa al-Taʿlīm*. Cairo: Markaz al-Dirāsāt wa al-Maʿlūmāt al-Qānūniyya li-Ḥuqūq al-Insān, 1997.
Muḥammad, ʿIrāqī Yūsuf. *al-Wujūd al-ʿUthmānī al-Mamlūkī fī Miṣr fī al-Qarn al-Thāmin ʿAshar wa Awāʾil al-Qarn al-Tāsiʿ ʿAshar*. Cairo: Dār al-Maʿārif, 1985.
———. *al-Wujūd al-ʿUthmānī fī Miṣr fī al-Qarnayn al-Sādis ʿAshar wa al-Sābiʿ ʿAshar (Dirāsa Wathāʾiqiyya)*. Cairo: Markaz Kliyūbātrā lil-Kumbiyūtar, 1996.
Muhanna, Elias I. "The Sultan's New Clothes: Ottoman-Mamluk Gift Exchange in the Fifteenth Century." *Muqarnas* 27 (2010): 189–207.
Mukerji, Chandra. *Impossible Engineering: Technology and Territoriality on the Canal du Midi*. Princeton, NJ: Princeton University Press, 2009.
Murphey, Rhoads. *Ottoman Warfare, 1500–1700*. New Brunswick, NJ: Rutgers University Press, 1999.
Muslu, Emire Cihan. "Ottoman-Mamluk Relations: Diplomacy and Perceptions." Ph.D. diss., Harvard University, 2007.
Nagel, Thomas. "What Is It Like to Be a Bat?" *Philosophical Review* 83 (1974): 435–50.
El-Nahal, Galal H. *The Judicial Administration of Ottoman Egypt in the Seventeenth Century*. Minneapolis: Bibliotheca Islamica, 1979.
Nance, Susan. *Entertaining Elephants: Animal Agency and the Business of the American Circus*. Baltimore: Johns Hopkins University Press, 2013.
Nash, Linda. *Inescapable Ecologies: A History of Environment, Disease, and Knowledge*. Berkeley: University of California Press, 2006.
Naskali, Emine Gürsoy, and Hilal Oytun Altun, eds. *Av ve Avcılık Kitabı*. Istanbul: Kitabevi Yayınları, 2008.
Natterson-Horowitz, Barbara, and Kathryn Bowers. *Zoobiquity: The Astonishing Connection Between Human and Animal Health*. New York: Vintage Books, 2013.
Necipoğlu, Gülrü. "The Suburban Landscape of Sixteenth-Century Istanbul as a Mirror of Classical Ottoman Garden Culture." In *Gardens in the Time of the Great Muslim Empires: Theory and Design*. Edited by Attilio Petruccioli. Leiden: Brill, 1997.
Nelson, Amy. "The Legacy of Laika: Celebrity, Sacrifice, and the Soviet Space Dog." In *Beastly Natures: Animals, Humans, and the Study of History*. Edited by Dorothee Brantz. Charlottesville: University of Virginia Press, 2010.
Netz, Reviel. *Barbed Wire: An Ecology of Modernity*. Middletown, CT: Wesleyan University Press, 2004.
Newfield, Timothy P. "A Cattle Panzootic in Early Fourteenth-Century Europe." *Agricultural History Review* 57 (2009): 155–90.

Norton, Marcy. "Going to the Birds: Animals as Things and Beings in Early Modernity." In *Early Modern Things: Objects and their Histories, 1500–1800*. Edited by Paula Findlen. London: Routledge, 2013.

Nowak, Ronald M. *Walker's Mammals of the World*. 6th ed. 2 vols. Baltimore: Johns Hopkins University Press, 1999.

Oggins, Robin S. *The Kings and Their Hawks: Falconry in Medieval England*. New Haven, CT: Yale University Press, 2004.

Oman, Luke, Alan Robock, Georgiy L. Stenchikov, and Thorvaldur Thordarson. "High-Latitude Eruptions Cast Shadow over the African Monsoon and the Flow of the Nile." *Geophysical Research Letters* 33 (2006): L18711.

Omar, Omar Abdel-Aziz. "Anglo-Egyptian Relations and the Construction of the Alexandria-Cairo-Suez Railway (1833–1858)." D.Phil. thesis, University of London, 1966.

Orlean, Susan. *Rin Tin Tin: The Life and the Legend*. New York: Simon and Schuster, 2011.

Osborne, Michael A. *Nature, the Exotic, and the Science of French Colonialism*. Bloomington: Indiana University Press, 1994.

———. "Zoos in the Family: The Geoffroy Saint-Hilaire Clan and the Three Zoos of Paris." In *New Worlds, New Animals: From Menagerie to Zoological Park in the Nineteenth Century*. Edited by R. J. Hoage and William A. Deiss. Baltimore: Johns Hopkins University Press, 1996.

Owen, Roger. *Cotton and the Egyptian Economy, 1820–1914: A Study in Trade and Development*. Oxford: Clarendon Press, 1969.

———. *Lord Cromer: Victorian Imperialist, Edwardian Proconsul*. New York: Oxford University Press, 2004.

Özbaran, Salih. *Ottoman Expansion towards the Indian Ocean in the 16th Century*. Istanbul: Bilgi University Press, 2009.

———. "Ottoman Naval Power in the Indian Ocean in the 16th Century." In *The Kapudan Pasha, His Office and His Domain: Halcyon Days in Crete IV*. Edited by Elizabeth Zachariadou. Rethymnon: Crete University Press, 2002.

———. *The Ottoman Response to European Expansion: Studies on Ottoman-Portuguese Relations in the Indian Ocean and Ottoman Administration in the Arab Lands during the Sixteenth Century*. Istanbul: Isis Press, 1994.

———. "A Turkish Report on the Red Sea and the Portuguese in the Indian Ocean (1525)." *Arabian Studies* 4 (1978): 81–88.

Pachirat, Timothy. *Every Twelve Seconds: Industrialized Slaughter and the Politics of Sight*. New Haven, CT: Yale University Press, 2011.

Pagliero, Roberto, ed. *Aegyptica Animalia: Il bestiario del Nilo*. Torino: Museo di Antropologia ed Etnografia dell'Università di Torino, 2000.

Palmer, Clare. "Madness and Animality in Michel Foucault's *Madness and Civilization*." In *Animal Philosophy: Essential Readings in Continental Thought*. Edited by Matthew Calarco and Peter Atterton. London: Continuum, 2004.

———. "'Taming the Wild Profusion of Existing Things'? A Study of Foucault, Power, and Human/Animal Relationships." *Environmental Ethics* 23 (2001): 339–58.

Palsetia, Jesse S. "Mad Dogs and Parsis: The Bombay Dog Riots of 1832." *Journal of the Royal Asiatic Society* 11 (2001): 13–30.

Panzac, Daniel. "Alexandrie: évolution d'une ville cosmopolite au XIXe siècle." In *Population et santé dans l'Empire ottoman (XVIIIe–XXe siècles)*. Istanbul: Isis, 1996.

———. "International and Domestic Maritime Trade in the Ottoman Empire during the 18th Century." *International Journal of Middle East Studies* 24 (1992): 189–206.

———. *La Peste dans l'Empire Ottoman, 1700–1850*. Louvain: Association pour le développement des études turques, 1985.

———. *Quarantaines et lazarets: l'Europe et la peste d'Orient (XVIIe–XXe siècles)*. Aix-en-Provence: Édisud, 1986.

Paraskevas, Philippe. *Breeding the Arabian Horse*. Vol. 1 of *The Egyptian Alternative*. Exeter: Obelisque Publications, 2010.

———. *In Search of the Identity of the Egyptian Arabian Bloodlines*. Vol. 2 of *The Egyptian Alternative*. Exeter: Obelisque Publications, 2012.

Paton, David. *Animals of Ancient Egypt*. Princeton, NJ: Princeton University Press, 1925.

Pearson, Susan J. *The Rights of the Defenseless: Protecting Animals and Children in Gilded Age America*. Chicago: University of Chicago Press, 2011.

Pedani, Maria Pia, ed., based on the materials compiled by Alessio Bombaci. *Inventory of the* Lettere e Scritture Turchesche *in the Venetian State Archives*. Leiden: Brill, 2010.

Pedani Fabris, Maria Pia, with Alessio Bombaci. *I "Documenti Turchi" dell'Archivio di Stato di Venezia*. Rome: Ministero per i Beni Culturali e Ambientali, Ufficio Centrale per i Beni Archivistici, 1994.

Peirce, Leslie. *The Imperial Harem: Women and Sovereignty in the Ottoman Empire*. Oxford: Oxford University Press, 1993.

———. *Morality Tales: Law and Gender in the Ottoman Court of Aintab*. Berkeley: University of California Press, 2003.

Perdue, Peter C. *Exhausting the Earth: State and Peasant in Hunan, 1500–1850*. Cambridge, MA: Harvard University Press, 1987.

Perlo, Katherine Wills. *Kinship and Killing: The Animal in World Religions*. New York: Columbia University Press, 2009.

———. "Marxism and the Underdog." *Society and Animals* 10 (2002): 303–18.

Peters, Rudolph. "Controlled Suffering: Mortality and Living Conditions in 19th-Century Egyptian Prisons." *International Journal of Middle East Studies* 36 (2004): 387–407.

———. "Egypt and the Age of the Triumphant Prison: Legal Punishment in Nineteenth Century Egypt." *Annales Islamologiques* 36 (2002): 253–85.

———. "'For His Correction and as a Deterrent Example for Others': Meḥmed 'Alī's First Criminal Legislation (1829–1830)." *Islamic Law and Society* 6 (1999): 164–92.

———. "Prisons and Marginalisation in Nineteenth-century Egypt." In *Outside In: On the Margins of the Modern Middle East*. Edited by Eugene Rogan. London: I. B. Tauris, 2002.

Petry, Carl F. *Protectors or Praetorians?: The Last Mamlūk Sultans and Egypt's Waning as a Great Power*. Albany: State University of New York Press, 1994.

———. *Twilight of Majesty: The Reigns of the Mamlūk Sultans al-Ashrāf Qāytbāy and Qanṣūh al-Ghawrī in Egypt*. Seattle: University of Washington Press, 1993.

Pflugfelder, Gregory M., and Brett L. Walker, eds. *JAPANimals: History and Culture in Japan's Animal Life*. Ann Arbor: Center for Japanese Studies, University of Michigan, 2005.

Philliou, Christine M. *Biography of an Empire: Governing Ottomans in an Age of Revolution*. Berkeley: University of California Press, 2011.

Phillips, Jacke. "Ostrich Eggshells." In *Ancient Egyptian Materials and Technology*. Edited by Paul T. Nicholson and Ian Shaw. Cambridge: Cambridge University Press, 2000.

Pincus, Steve. *1688: The First Modern Revolution*. New Haven, CT: Yale University Press, 2009.

Pinguet, Catherine. *Les chiens d'Istanbul: des rapports entre l'homme et l'animal de l'Antiquité à nos jours*. Saint-Pourçain-sur-Soule: Bleu autour, 2008.

———. "Istanbul's Street Dogs at the End of the Ottoman Empire: Protection or Extermination." In *Animals and People in the Ottoman Empire*. Edited by Suraiya Faroqhi. Istanbul: Eren, 2010.

Piterberg, Gabriel. *An Ottoman Tragedy: History and Historiography at Play*. Berkeley: University of California Press, 2003.

Pluhar, Evelyn B. *Beyond Prejudice: The Moral Significance of Human and Nonhuman Animals*. Durham, NC: Duke University Press, 1995.

Plumb, Christopher. "'Strange and Wonderful': Encountering the Elephant in Britain, 1675–1830." *Journal for Eighteenth-Century Studies* 33 (2010): 525–43.

Podobnik, Bruce. "Toward a Sustainable Energy Regime: A Long-Wave Interpretation of Global Energy Shifts." *Technological Forecasting and Social Change* 62 (1999): 155–72.

Poliquin, Rachel. *The Breathless Zoo: Taxidermy and the Cultures of Longing*. University Park: Pennsylvania State University Press, 2012.

Pollan, Michael. *The Botany of Desire: A Plant's-Eye View of the World*. New York: Random House, 2001.

Pomeranz, Kenneth. *The Great Divergence: China, Europe, and the Making of the Modern World Economy*. Princeton, NJ: Princeton University Press, 2000.

Popper, William. *The Cairo Nilometer: Studies in Ibn Taghrî Birdî's Chronicles of Egypt, I*. Berkeley: University of California Press, 1951.

Powell, Eve M. Troutt. *A Different Shade of Colonialism: Egypt, Great Britain, and the Mastery of the Sudan*. Berkeley: University of California Press, 2003.
Prager, Ellen. *Sex, Drugs, and Sea Slime: The Oceans' Oddest Creatures and Why They Matter*. Chicago: University of Chicago Press, 2011.
Preece, Rod. "Thoughts out of Season on the History of Animal Ethics." *Society and Animals* 15 (2007): 365–78.
Ptak, Roderich. "Pferde auf See: Ein vergessener Aspekt des maritimen chinesischen Handels im frühen 15. Jahrhundert." *Journal of the Economic and Social History of the Orient* 34 (1991): 199–233.
Purcell, Natalie. "Cruel Intimacies and Risky Relationships: Accounting for Suffering in Industrial Livestock Production." *Society and Animals* 19 (2011): 59–81.
Putney, William W. *Always Faithful: A Memoir of the Marine Dogs of WWII*. New York: Free Press, 2001.
Pycior, Helena. "The Public and Private Lives of 'First Dogs': Warren G. Harding's Laddie Boy and Franklin D. Roosevelt's Fala." In *Beastly Natures: Animals, Humans, and the Study of History*. Edited by Dorothee Brantz. Charlottesville: University of Virginia Press, 2010.
Pyne, Stephen J. *Vestal Fire: An Environmental History, Told through Fire, of Europe and Europe's Encounter with the World*. Seattle: University of Washington Press, 1997.
———. *World Fire: The Culture of Fire on Earth*. Seattle: University of Washington Press, 1997.
Quammen, David. *Spillover: Animal Infections and the Next Human Pandemic*. New York: W. W. Norton, 2012.
Quinn, Stephen Christopher. *Windows on Nature: The Great Habitat Dioramas of the American Museum of Natural History*. New York: Harry N. Abrams, 2006.
Raḍwān, Abū al-Futūḥ. *Tārīkh Maṭbaʿat Būlāq wa Lamḥa fī Tārīkh al-Ṭibāʿa fī Buldān al-Sharq al-Awsaṭ*. Cairo: al-Maṭbaʿa al-Amīriyya, 1953.
al-Rāfʿī, ʿAbd al-Raḥman. *ʿAṣr Muḥammad ʿAlī*. Cairo: Dār al-Maʿārif, 1989.
Ramaswamy, Sumathi. "Conceit of the Globe in Mughal Visual Practice." *Comparative Studies in Society and History* 49 (2007): 751–82.
Ramzī, Muḥammad. *al-Qāmūs al-Jughrāfī lil-Bilād al-Miṣriyya min ʿAhd Qudamāʾ al-Miṣriyyīn ilā Sanat 1945*. 6 vols. in 2 pts. Cairo: al-Hayʾa al-Miṣriyya al-ʿĀmma lil-Kitāb, 1994.
Rao, Velcheru Narayana, and Sanjay Subrahmanyam. "Notes on Political Thought in Medieval and Early Modern South India." *Modern Asian Studies* 43 (2009): 175–210.
al-Rawashdeh, Odeh F., Falah K. al-Ani, Labib A. Sharrif, Khaled M. al-Qudah, Yasin al-Hami, and Nicolas Frank. "A Survey of Camel (*Camelus dromedarius*) Diseases in Jordan." *Journal of Zoo and Wildlife Medicine* 31 (2000): 335–38.
Ray, Aniruddha, ed. *Tipu Sultan and His Age: A Collection of Seminar Papers*. Kolkata: Asiatic Society, 2001.
Raymond, André. *Artisans et commerçants au Caire au XVIIIe siècle*. 2 vols. Damascus: Institut français de Damas, 1973–74.
———. *Cairo: City of History*. Translated by Willard Wood. Cairo: American University in Cairo Press, 2001.
———. "A Divided Sea: The Cairo Coffee Trade in the Red Sea Area during the Seventeenth and Eighteenth Centuries." In *Modernity and Culture: From the Mediterranean to the Indian Ocean*. Edited by Leila Tarazi Fawaz and C. A. Bayly. New York: Columbia University Press, 2002.
———. *Égyptiens et Français au Caire (1798–1801)*. Cairo: Institut français d'archéologie orientale, 2004.
———. "Une famille de grands négociants en café au Caire dans la première moitié du XVIIIe siècle: les Sharāybī." In *Le commerce du café avant l'ère des plantations coloniales: espaces, réseaux, sociétés (XVe–XIXe siècle)*. Edited by Michel Tuchscherer. Cairo: Institut français d'archéologie orientale, 2001.
———. "La population du Caire et de l'Égypte à l'époque ottomane et sous Muḥammad ʿAlī." In *Mémorial Ömer Lûtfi Barkan*. Paris: Librairie d'Amérique et d'Orient Adrien Maisonneuve, 1980.
Rees, Paul A. *An Introduction to Zoo Biology and Management*. Chichester: Wiley-Blackwell, 2011.
Reichenbach, Herman. "A Tale of Two Zoos: The Hamburg Zoological Garden and Carl Hagenbeck's Tierpark." In *New Worlds, New Animals: From Menagerie to Zoological Park in the*

Nineteenth Century. Edited by R. J. Hoage and William A. Deiss. Baltimore: Johns Hopkins University Press, 1996.

Reid, Anthony. "Sixteenth-Century Turkish Influence in Western Indonesia." *Journal of South East Asian History* 10 (1969): 395–414.

Reimer, Michael J. "Ottoman Alexandria: The Paradox of Decline and the Reconfiguration of Power in Eighteenth-Century Arab Provinces." *Journal of the Economic and Social History of the Orient* 37 (1994): 107–46.

Reindl-Kiel, Hedda. "The Chickens of Paradise: Official Meals in the Seventeenth-Century Ottoman Palace." In *The Illuminated Table, the Prosperous House: Food and Shelter in Ottoman Material Culture.* Edited by Suraiya Faroqhi and Christoph Neumann. Istanbul: Orient-Institut, 2003.

———. "Dogs, Elephants, Lions, a Ram and a Rhino on Diplomatic Mission: Animals as Gifts to the Ottoman Court." In *Animals and People in the Ottoman Empire.* Edited by Suraiya Faroqhi. Istanbul: Eren, 2010.

———. "No Horses for the Enemy: Ottoman Trade Regulations and Horse Gifting." In *Pferde in Asien: Geschichte, Handel Und Kultur.* Edited by Bert G. Fragner, Ralph Kauz, Roderich Ptak, and Angela Schottenhammer. Vienna: Österreichischen Akademie der Wissenschaften, 2009.

Reisner, George A. "The Dog which was Honored by the King of Upper and Lower Egypt." *Bulletin of the Museum of Fine Arts* 34 (1936): 96–99.

Reports from Her Majesty's Consuls on the Manufactures, Commerce, &c., of their Consular Districts. Vol. 22. London: Harrison and Sons, 1872.

Richards, Alan R. "Primitive Accumulation in Egypt, 1798–1882." In *The Ottoman Empire and the World-Economy.* Edited by Huri İslamoğlu-İnan. Cambridge: Cambridge University Press, 1987.

Richards, John F. "Early Modern India and World History." *Journal of World History* 8 (1997): 197–209.

———. "Toward a Global System of Property Rights in Land." In *The Environment and World History.* Edited by Edmund Burke III and Kenneth Pomeranz. Berkeley: University of California Press, 2009.

———. *The Unending Frontier: An Environmental History of the Early Modern World.* Berkeley: University of California Press, 2003.

Ritvo, Harriet. *The Animal Estate: The English and Other Creatures in the Victorian Age.* Cambridge, MA: Harvard University Press, 1987.

———. "Animal Planet." *Environmental History* 9 (2004): 204–20.

———. "Going Forth and Multiplying: Animal Acclimatization and Invasion." *Environmental History* 17 (2012): 404–14.

———. *The Platypus and the Mermaid and Other Figments of the Classifying Imagination.* Cambridge, MA: Harvard University Press, 1997.

———. "Pride and Pedigree: The Evolution of the Victorian Dog Fancy." *Victorian Studies* 29 (1986): 227–53.

Rivlin, Helen Anne B. *The Agricultural Policy of Muḥammad ʿAlī in Egypt.* Cambridge, MA: Harvard University Press, 1961.

———. "The Railway Question in the Ottoman-Egyptian Crisis of 1850–1852." *Middle East Journal* 15 (1961): 365–88.

Robbins, Louise E. *Elephant Slaves and Pampered Parrots: Exotic Animals in Eighteenth-Century Paris.* Baltimore: Johns Hopkins University Press, 2002.

Rogan, Eugene. "Madness and Marginality: The Advent of the Psychiatric Asylum in Egypt and Lebanon." In *Outside In: On the Margins of the Modern Middle East.* Edited by Eugene Rogan. London: I. B. Tauris, 2002.

Rolt, L. T. C. *George and Robert Stephenson: The Railway Revolution.* London: Longmans, 1960.

Root, Nina J. "Victorian England's Hippomania." *Natural History* 102 (1993): 34–39.

Rosenthal, Jean-Laurent, and R. Bin Wong. *Before and Beyond Divergence: The Politics of Economic Change in China and Europe.* Cambridge, MA: Harvard University Press, 2011.

Rossabi, Morris. "The Tea and Horse Trade with Inner Asia During the Ming." *Journal of Asian History* 4 (1970): 136–68.

Rossi, Ettore. "A Turkish Map of the Nile River, about 1685." *Imago Mundi* 6 (1949): 73–75.

Rothfels, Nigel, ed. *Representing Animals*. Bloomington: Indiana University Press, 2002.

Rothfels, Nigel. *Savages and Beasts: The Birth of the Modern Zoo*. Baltimore: Johns Hopkins University Press, 2002.

Rothman, E. Natalie. *Brokering Empire: Trans-Imperial Subjects between Venice and Istanbul*. Ithaca, NY: Cornell University Press, 2012.

el-Rouayheb, Khaled. "Was There a Revival of Logical Studies in Eighteenth-Century Egypt?" *Die Welt des Islams* 45 (2005): 1–19.

Rudacille, Deborah. *The Scalpel and the Butterfly: The War between Animal Research and Animal Protection*. New York: Farrar, Straus, and Giroux, 2000.

Russell, Edmund. *Evolutionary History: Uniting History and Biology to Understand Life on Earth*. New York: Cambridge University Press, 2011.

Russell, Leslie A. "Decoding Equine Emotions." *Society and Animals* 11 (2003): 265–66.

Russell, Mona L. *Creating the New Egyptian Woman: Consumerism, Education, and National Identity, 1863–1922*. New York: Palgrave Macmillan, 2004.

Rustom, Asad J. *The Royal Archives of Egypt and the Origins of the Egyptian Expedition to Syria, 1831–1841*. Beirut: American Press, 1936.

al-Ṣabbāgh, ʿAbd al-Laṭīf Muḥammad. "al-Ṭibb al-Bayṭarī fī Miṣr (1828–1849): Dirāsa Wathāʾiqiyya." *Miṣr al-Ḥadītha* 8 (2009): 17–55.

Sabev, Orlin. *İbrahim Müteferrika ya da İlk Osmanlı Matbaa Serüveni, 1726–1746: Yeniden Değerlendirme*. Istanbul: Yeditepe, 2006.

Sahlins, Peter. "The Royal Menageries of Louis XIV and the Civilizing Process Revisited." *French Historical Studies* 35 (2012): 237–67.

Salisbury, Joyce E. *The Beast Within: Animals in the Middle Ages*. New York: Routledge, 1994.

Sanders, Jeffrey C. "Animal Trouble and Urban Anxiety: Human-Animal Interaction in Post-Earth Day Seattle." *Environmental History* 16 (2011): 226–61.

Sapontzis, S. F. *Morals, Reason, and Animals*. Philadelphia: Temple University Press, 1987.

al-Ṣarrāf, Maḥmūd Ḥasan ʿAbd al-ʿAzīz. *Maʿrakat Chāldārān, 920 H/1514 M: Ūlā Ṣafaḥāt al-Ṣirāʿ al-ʿUthmānī al-Fārisī: al-Asbāb wa al-Natāʾij*. Cairo: Maktabat al-Nahḍa al-Miṣriyya, 1991.

Saurer, Karl, and Elena Hinshaw-Fischli. "They Called Him Suleyman: The Adventurous Journey of an Indian Elephant from the Forests of Kerala to the Capital of Vienna in the Middle of the Sixteenth Century." In *Maritime Malabar and the Europeans, 1500–1962*. Edited by K. S. Mathew. Gurgaon: Hope India Publications, 2003.

Sax, Boria. *Animals in the Third Reich: Pets, Scapegoats, and the Holocaust*. New York: Continuum, 2000.

al-Sayyid, Mirfat Aḥmad. "Idārat al-Shurṭa fī Miṣr fī al-ʿAṣr al-ʿUthmānī." *Annales Islamologiques* 40 (2006): 51–70.

Sbeinati, Mohamed Reda, Ryad Darawcheh, and Mikhail Mouty. "The Historical Earthquakes of Syria: An Analysis of Large and Moderate Earthquakes from 1365 B.C. to 1900 A.D." *Annals of Geophysics* 48 (2005): 347–435.

Schacht, Joseph. *The Origins of Muḥammadan Jurisprudence*. Oxford: Clarendon Press, 1950.

Schafer, Edward H. "The Conservation of Nature under the T'ang Dynasty." *Journal of the Economic and Social History of the Orient* 5 (1962): 279–308.

Schick, İrvin Cemil. "Evliya Çelebi'den Köpeklere Dair." *Toplumsal Tarih* 202 (2010): 34–44.

———. "İstanbul'da 1910'da Gerçekleşen Büyük Köpek İtlâfı: Bir Mekan Üzerinde Çekişme Vakası." *Toplumsal Tarih* 200 (2010): 22–33.

Schiebinger, Londa. *Nature's Body: Gender in the Making of Modern Science*. New Brunswick, NJ: Rutgers University Press, 2004.

———. "Why Mammals are Called Mammals: Gender Politics in Eighteenth-Century Natural History." *American Historical Review* 98 (1993): 382–411.

Schimmel, Annemarie. *The Empire of the Great Mughals: History, Art and Culture*. Translated by Corinne Attwood. Edited by Burzine K. Waghmar. London: Reaktion Books, 2004.

———. *Islam and the Wonders of Creation: The Animal Kingdom*. London: al-Furqān Islamic Heritage Foundation, 2003.

———. *Die Orientalische Katze*. Köln: Diederichs, 1983.

Schivelbusch, Wolfgang. *Disenchanted Night: The Industrialization of Light in the Nineteenth Century*. Translated by Angela Davies. Berkeley: University of California Press, 1988.

Schmitt, Jean-Claude. *The Holy Greyhound: Guinefort, Healer of Children since the Thirteenth Century*. Translated by Martin Thom. Cambridge: Cambridge University Press, 1983.
Schönig, Hanne. "Reflections on the Use of Animal Drugs in Yemen." *Quaderni di Studi Arabi* 20/21 (2002–03): 157–84.
Schwartz, Marion. *A History of Dogs in the Early Americas*. New Haven, CT: Yale University Press, 1977.
Scott, James C. *Seeing Like a State: How Certain Schemes to Improve the Human Condition Have Failed*. New Haven, CT: Yale University Press, 1998.
Seguin, Marilyn W. *Dogs of War: And Stories of Other Beasts of Battle in the Civil War*. Boston: Branden, 1998.
Serpell, James, ed. *The Domestic Dog: Its Evolution, Behaviour, and Interactions with People*. Cambridge: Cambridge University Press, 1995.
Serpell, James. "From Paragon to Pariah: Some Reflections on Human Attitudes to Dogs." In *The Domestic Dog: Its Evolution, Behaviour, and Interactions with People*. Edited by James Serpell. Cambridge: Cambridge University Press, 1995.
Sertoğlu, Midhat. *Osmanlı Tarih Lûgatı*. Istanbul: Enderun Kitabevi, 1986.
El Shakry, Omnia. "Youth as Peril and Promise: The Emergence of Adolescent Psychology in Postwar Egypt." *International Journal of Middle East Studies* 43 (2011): 591–610.
Sharkey, Heather J. *Living with Colonialism: Nationalism and Culture in the Anglo-Egyptian Sudan*. Berkeley: University of California Press, 2003.
Shaw, Ian. "Egypt and the Outside World." In *The Oxford History of Ancient Egypt*. Edited by Ian Shaw. Oxford: Oxford University Press, 2000.
Shaw, Stanford J. *Between Old and New: The Ottoman Empire under Sultan Selim III, 1789–1807*. Cambridge, MA: Harvard University Press, 1971.
———. "Cairo's Archives and the History of Ottoman Egypt." *Middle East Institute Report on Current Research* (1956): 59–72.
———. *The Financial and Administrative Organization and Development of Ottoman Egypt, 1517–1798*. Princeton, NJ: Princeton University Press, 1962.
———. "Landholding and Land-Tax Revenues in Ottoman Egypt." In *Political and Social Change in Modern Egypt: Historical Studies from the Ottoman Conquest to the United Arab Republic*. Edited by P. M. Holt. London: Oxford University Press, 1968.
———. "The Ottoman Archives as a Source for Egyptian History." *Journal of the American Oriental Society* 83 (1963): 447–52.
Shehada, Housni Alkhateeb. *Mamluks and Animals: Veterinary Medicine in Medieval Islam*. Leiden: Brill, 2013.
Shukin, Nicole. *Animal Capital: Rendering Life in Biopolitical Times*. Minneapolis: University of Minnesota Press, 2009.
Shukrī, Muḥammad Fu'ād, 'Abd al-Maqṣūd al-'Inānī, and Sayyid Muḥammad Khalīl. *Binā' Dawlat Miṣr Muḥammad 'Alī*. 2 vols. Cairo: Dār al-Kutub wa al-Wathā'iq al-Qawmiyya, 2009.
Shukry, M. F. *The Khedive Ismail and Slavrey in the Sudan (1863–1879)*. Cairo: Librairie La Renaissance d'Égypte, 1937.
Silvera, Alain. "The First Egyptian Student Mission to France under Muhammad Ali." *Middle Eastern Studies* 16 (1980): 1–22.
Silverstein, Adam J. *Postal Systems in the Pre-Modern Islamic World*. Cambridge: Cambridge University Press, 2010.
Simpson, Marianna Shreve. *Persian Poetry, Painting & Patronage: Illustrations in a Sixteenth-Century Masterpiece*. New Haven, CT: Yale University Press, 1998.
Sinopoli, Carla M. "From the Lion Throne: Political and Social Dynamics of the Vijayanagara Empire." *Journal of the Economic and Social History of the Orient* 43 (2000): 364–98.
Sivasundaram, Sujit. "Trading Knowledge: The East India Company's Elephants in India and Britain." *The Historical Journal* 48 (2005): 27–63.
Skabelund, Aaron. "Breeding Racism: The Imperial Battlefields of the 'German' Shepherd." *Society and Animals* 16 (2008): 354–71.
———. "Can the Subaltern Bark? Imperialism, Civilization, and Canine Cultures in Nineteenth-Century Japan." In *JAPANimals: History and Culture in Japan's Animal Life*.

Edited by Gregory M. Pflugfelder and Brett L. Walker. Ann Arbor: Center for Japanese Studies, University of Michigan, 2005.

———. *Empire of Dogs: Canines, Japan, and the Making of the Modern Imperial World*. Ithaca, NY: Cornell University Press, 2011.

Slavin, Philip. "The Great Bovine Pestilence and its Economic and Environmental Consequences in England and Wales, 1318–50." *Economic History Review* 65 (2012): 1239–66.

Smil, Vaclav. *Energy in Nature and Society: General Energetics of Complex Systems*. Cambridge: Massachusetts Institute of Technology Press, 2008.

———. *Energy in World History*. Boulder, CO: Westview Press, 1994.

Smith, Jay M. *Monsters of the Gévaudan: The Making of a Beast*. Cambridge, MA: Harvard University Press, 2011.

Smith, Julie A., and Robert W. Mitchell, eds. *Experiencing Animal Minds: An Anthology of Animal-Human Encounters*. New York: Columbia University Press, 2012.

Smith, Paul J. "On Toucans and Hornbills: Readings in Early Modern Ornithology from Belon to Buffon." In *Early Modern Zoology: The Construction of Animals in Science, Literature and the Visual Arts*. Edited by Karl A. E. Enenkel and Paul J. Smith. 2 vols. Leiden: Brill, 2007.

Sonbol, Amira el-Azhary. *The Creation of a Medical Profession in Egypt, 1800–1922*. Syracuse, NY: Syracuse University Press, 1991.

Sood, Gagan D. S. "'Correspondence is Equal to Half a Meeting': The Composition and Comprehension of Letters in Eighteenth-Century Islamic Eurasia." *Journal of the Economic and Social History of the Orient* 50 (2007): 172–214.

Starn, Randolph. "The Early Modern Muddle." *Journal of Early Modern History* 6 (2002): 296–307.

Steeves, H. Peter, ed. *Animal Others: On Ethics, Ontology, and Animal Life*. Albany: State University of New York Press, 1999.

Steffen, Will, Paul J. Crutzen, and John R. McNeill. "The Anthropocene: Are Humans Now Overwhelming the Great Forces of Nature?" *Ambio* 36 (2007): 614–21.

Stein, Sarah Abrevaya. *Plumes: Ostrich Feathers, Jews, and a Lost World of Global Commerce*. New Haven, CT: Yale University Press, 2008.

Steinberg, Ted. "Down to Earth: Nature, Agency, and Power in History." *American Historical Review* 107 (2002): 798–820.

Sterchx, Roel. *The Animal and the Daemon in Early China*. Albany: State University of New York Press, 2002.

Stiner, Mary C., and Gillian Feeley-Harnik. "Energy and Ecosystems." In *Deep History: The Architecture of Past and Present*. Edited by Andrew Shryock and Daniel Lord Smail. Berkeley: University of California Press, 2011.

Stow, Kenneth. *Jewish Dogs: An Image and Its Interpreters: Continuity in the Catholic-Jewish Encounter*. Stanford, CA: Stanford University Press, 2006.

Subrahmanyam, Sanjay. "Connected Histories: Notes towards a Reconfiguration of Early Modern Eurasia." *Modern Asian Studies* 31 (1997): 735–62.

———. *Explorations in Connected History: Mughals and Franks*. Delhi: Oxford University Press, 2005.

———. *Explorations in Connected History: From the Tagus to the Ganges*. Delhi: Oxford University Press, 2005.

———. "Hearing Voices: Vignettes of Early Modernity in South Asia, 1400–1750." *Daedalus* 127 (1998): 75–104.

———, ed. *Merchants, Markets and the State in Early Modern India*. Delhi: Oxford University Press, 1990.

———. *The Political Economy of Commerce: Southern India, 1500–1650*. Cambridge: Cambridge University Press, 1990.

———. "Reflections on State-Making and History-Making in South India, 1500–1800." *Journal of the Economic and Social History of the Orient* 41 (1998): 382–416.

———. "On World Historians in the Sixteenth Century." *Representations* 91 (2005): 26–57.

Sulaymān, ʿAbd al-Ḥamīd. "al-Sukhra fī Miṣr fī al-Qarnayn al-Sābiʿ ʿAshar wa al-Thāmin ʿAshar, Dirāsa fī al-Asbāb wa al-Natāʾij." In *al-Rafḍ wa al-Iḥtijāj fī al-Mujtamaʿ al-Miṣrī fī al-ʿAṣr al-ʿUthmānī*. Edited by Nāṣir Ibrāhīm and Raʾūf ʿAbbās. Cairo: Markaz al-Buḥūth wa al-Dirāsāt al-Ijtimāʿiyya, 2004.

Swabe, Joanna. *Animals, Disease, and Human Society: Human-Animal Relations and the Rise of Veterinary Medicine*. London: Routledge, 1999.
Szpakowska, Kasia. *Daily Life in Ancient Egypt: Recreating Lahun*. Malden: Blackwell, 2008.
Tabak, Faruk. *The Waning of the Mediterranean, 1550–1870: A Geohistorical Approach*. Baltimore: Johns Hopkins University Press, 2008.
Tague, Ingrid H. "Dead Pets: Satire and Sentiment in British Elegies and Epitaphs for Animals." *Eighteenth-Century Studies* 41 (2008): 289–306.
Tanman, M. Baha, ed. *Nil Kıyısından Boğaziçi'ne: Kavalalı Mehmed Ali Paşa Hanedanı'nın İstanbul'daki İzleri*. Istanbul: İstanbul Araştırmaları Enstitüsü, 2011.
Taylor, Joseph E., III. *Making Salmon: An Environmental History of the Northwest Fisheries Crisis*. Seattle: University of Washington Press, 1999.
Teeter, Emily. "Animals in Egyptian Literature." In *A History of the Animal World in the Ancient Near East*. Edited by Billie Jean Collins. Leiden: Brill, 2002.
———. "Animals in Egyptian Religion." In *A History of the Animal World in the Ancient Near East*. Edited by Billie Jean Collins. Leiden: Brill, 2002.
Tester, Keith. *Animals and Society: The Humanity of Animal Rights*. London: Routledge, 1991.
Tezcan, Baki. *The Second Ottoman Empire: Political and Social Transformation in the Early Modern World*. Cambridge: Cambridge University Press, 2010.
Thomas, Angela P. *Egyptian Gods and Myths*. Aylesbury: Shire Publications, 1986.
Thomas, Keith. *Man and the Natural World: Changing Attitudes in England, 1500–1800*. London: Allen Lane, 1983.
The Times. November 29, 1895.
Timur, Taner. "XIX. Yüzyılda İstanbul'un Köpekleri." *Tarih ve Toplum* 20: 117 (September 1993): 10–14.
Tlili, Sarra. *Animals in the Qur'an*. Cambridge: Cambridge University Press, 2012.
Toledano, Ehud R. "Late Ottoman Concepts of Slavery (1830s–1880s)." *Poetics Today* 14 (1993): 477–506.
———. "Mehmet Ali Paşa or Muhammad Ali Basha? An Historiographic Appraisal in the Wake of a Recent Book." *Middle Eastern Studies* 21 (1985): 141–59.
———. *Slavery and Abolition in the Ottoman Middle East*. Seattle: University of Washington Press, 1998.
———. *State and Society in Mid-Nineteenth-Century Egypt*. Cambridge: Cambridge University Press, 1990.
Tooley, Angela M. J. "Coffin of a Dog from Beni Hasan." *Journal of Egyptian Archaeology* 74 (1988): 207–11.
Totman, Conrad. *The Green Archipelago: Forestry in Preindustrial Japan*. Berkeley: University of California Press, 1989.
Toynbee, Arnold J. "The Roman Revolution from the Flora's Point of View." In *Rome and Her Neighbours after Hannibal's Exit*. Vol. 2 of *Hannibal's Legacy: The Hannibalic War's Effects on Roman Life*. London: Oxford University Press, 1965.
Trautmann, Thomas R. "Elephants and the Mauryas." In *India, History and Thought: Essays in Honour of A. L. Basham*. Edited by S. N. Mukherjee. Calcutta: Subarnarekha, 1982.
Trivellato, Francesca. *The Familiarity of Strangers: The Sephardic Diaspora, Livorno, and Cross-Cultural Trade in the Early Modern Period*. New Haven, CT: Yale University Press, 2009.
Tropp, Jacob. "Dogs, Poison, and the Meaning of Colonial Intervention in the Transkei, South Africa." *Journal of African History* 43 (2002): 451–72.
Trut, Lyudmila N. "Early Canid Domestication: The Farm-Fox Experiment." *American Scientist* 87 (1999): 160–69.
Tsugitaka, Sato. *State and Rural Society in Medieval Islam: Sultans, Muqta'ʿs and Fallahun*. Leiden: Brill, 1997.
Tuan, Yi-Fu. *Dominance and Affection: The Making of Pets*. New Haven, CT: Yale University Press, 1984.
Tuchscherer, Michel. "Commerce et production du café en mer Rouge au XVIe siècle." In *Le commerce du café avant l'ère des plantations coloniales: espaces, réseaux, sociétés (XVe–XIXe siècle)*. Edited by Michel Tuchscherer. Cairo: Institut français d'archéologie orientale, 2001.
———. "La flotte impériale de Suez de 1694 à 1719." *Turcica* 29 (1997): 47–69.

———. "Some Reflections on the Place of the Camel in the Economy and Society of Ottoman Egypt." Translated by Suraiya Faroqhi. In *Animals and People in the Ottoman Empire*. Edited by Suraiya Faroqhi. Istanbul: Eren, 2010.

Tucker, Judith E. *Women in Nineteenth-Century Egypt*. Cambridge: Cambridge University Press, 1985.

Tucker, Richard P. *Insatiable Appetite: The United States and the Ecological Degradation of the Tropical World*. Berkeley: University of California Press, 2000.

Tucker, Richard P., and J. F. Richards, eds. *Global Deforestation and the Nineteenth-Century World Economy*. Durham, NC: Duke University Press, 1983.

Tucker, William F. "Natural Disasters and the Peasantry in Mamlūk Egypt." *Journal of the Economic and Social History of the Orient* 24 (1981): 215–24.

Turner, E. S. "Animals and Humanitarianism." In *Animals and Man in Historical Perspective*. Edited by Joseph and Barrie Klaits. New York: Harper and Row, 1974.

Turner, Howard R. *Science in Medieval Islam: An Illustrated Introduction*. Austin: University of Texas Press, 1995.

Turner, James. *Reckoning with the Beast: Animals, Pain, and Humanity in the Victorian Mind*. Baltimore: Johns Hopkins University Press, 1980.

Uexhüll, Jacob von. *A Foray into the Worlds of Animals and Humans; with* A Theory of Meaning. Translated by Joseph D. O'Neil. Minneapolis: University of Minnesota Press, 2010.

Ullmann, Manfred. *Die Natur- und Geheimwissenschaften im Islam*. Leiden: Brill, 1972.

Um, Nancy. *The Merchant Houses of Mocha: Trade and Architecture in an Indian Ocean Port*. Seattle: University of Washington Press, 2009.

ʿUmrān, Jīhān Aḥmad. "Dirāsa Diblūmātiyya li-Wathāʾiq Wafāʾ al-Nīl bi-Sijilāt al-Dīwān al-ʿĀlī maʿa Nashr Namādhij minhā." *Waqāʾiʿ Tārīkhiyya: Dauriyya ʿIlmiyya Muḥakkama* (2004): 347–81.

Uther, Hans-Jörg. "The Fox in World Literature: Reflections on a 'Fictional Animal.'" *Asian Folklore Studies* 65 (2006): 133–60.

ʿUthmān, Nāṣir. "Maḥkamat Rashīd ka-Maṣdar li-Dirāsat Tijārat al-Nasīj fī Madīnat al-Iskandariyya fī al-ʿAṣr al-ʿUthmānī." *al-Rūznāma: al-Ḥauliyya al-Miṣriyya lil-Wathāʾiq* 3 (2005): 355–85.

Van Sittert, Lance, and Sandra Swart, eds. *Canis Africanis: A Dog History of Southern Africa*. Leiden: Brill, 2008.

Varner, John Grier, and Jeannette Johnson Varner. *Dogs of the Conquest*. Norman: University of Oklahoma Press, 1983.

Veblen, Thorstein. *The Theory of the Leisure Class*. Boston: Houghton Mifflin, 1973.

Veinstein, Gilles. "On the *Çiftlik* Debate." In *Landholding and Commercial Agriculture in the Middle East*. Edited by Çağlar Keyder and Faruk Tabak. Albany: State University of New York Press, 1991.

———. "Falconry in the Ottoman Empire of the Mid-Sixteenth Century." In *Animals and People in the Ottoman Empire*. Edited by Suraiya Faroqhi. Istanbul: Eren, 2010.

———. "La fauconnerie dans l'Empire ottoman au milieu du XVIe siècle: une institution en péril." In *Hommes et terres d'Islam: Mélanges offerts à Xavier de Planhol*. Edited by Daniel Balland. 2 vols. Tehran: Institut de recherche en Iran, 2000.

Veltre, Thomas. "Menageries, Metaphors, and Meanings." In *New Worlds, New Animals: From Menagerie to Zoological Park in the Nineteenth Century*. Edited by R. J. Hoage and William A. Deiss. Baltimore: Johns Hopkins University Press, 1996.

Verdery, Richard N. "The Publications of the Būlāq Press under Muḥammad ʿAlī of Egypt." *Journal of the American Oriental Society* 91 (1971): 129–32.

Verghese, Anila, and Anna L. Dallapiccola, eds. *South India under Vijayanagara: Art and Archaeology*. New York: Oxford University Press, 2011.

Vialles, Noëlie. *Animal to Edible*. Translated by J. A. Underwood. Cambridge: Cambridge University Press, 1994.

Vilà, Carles, et al. "Multiple and Ancient Origins of the Domestic Dog." *Science* 276 (1997): 1687–89.

Viré, François. "La chasse au guépard d'après sources arabes et les oeuvres d'art musulman par Ahmad Abd ar-Raziq." *Arabica* 21 (1974): 85–88.

———. "À propos des chiens de chasse *salûqî* et *zagârî*." *Revue des études islamiques* 41 (1973): 331–40.
———. "Essai du détermination des oiseaux-de-vol mentionnés dans les principaux manuscrits arabes médiévaux sur la fauconnerie." *Arabica* 24 (1977): 138–49.
———. "Kalb." *Encyclopaedia of Islam*. 2nd ed. Leiden: Brill Online, 2013.
de Vries, Jan. *The Dutch Rural Economy in the Golden Age, 1500–1700*. New Haven, CT: Yale University Press, 1974.
Vroom, Joanita. "'Mr. Turkey Goes to Turkey,' Or: How an Eighteenth-Century Dutch Diplomat Lunched at Topkapı Palace." In *Starting with Food: Culinary Approaches to Ottoman History*. Edited by Amy Singer. Princeton, NJ: Markus Wiener Publishers, 2011.
Walker, Brett L. *The Lost Wolves of Japan*. Seattle: University of Washington Press, 2005.
Walker, Elaine. *Horse*. London: Reaktion Books, 2008.
Walker, Suzanne J. "Making and Breaking the Stag: The Construction of the Animal in the Early Modern Hunting Treatise." In *Early Modern Zoology: The Construction of Animals in Science, Literature and the Visual Arts*. Edited by Karl A. E. Enenkel and Paul J. Smith. 2 vols. Leiden: Brill, 2007.
Walker-Meikle, Kathleen. *Medieval Pets*. Woodbridge, UK: Boydell Press, 2012.
Wallen, Martin. *Fox*. London: Reaktion Books, 2006.
Waller, Richard. "'Clean' and 'Dirty': Cattle Disease and Control Policy in Colonial Kenya, 1900–40." *Journal of African History* 45 (2004): 45–80.
Walton, John K. "Mad Dogs and Englishmen: The Conflict over Rabies in Late Victorian England." *Journal of Social History* 13 (1979): 219–39.
Walz, Terence. *Trade between Egypt and Bilād as-Sūdān, 1700–1820*. Cairo: Institut français d'archéologie orientale du Caire, 1978.
Walz, Terence, and Kenneth M. Cuno, eds. *Race and Slavery in the Middle East: Histories of Trans-Saharan Africans in Nineteenth-Century Egypt, Sudan, and the Ottoman Mediterranean*. Cairo: American University in Cairo Press, 2010.
Warde, Paul. *Ecology, Economy and State Formation in Early Modern Germany*. Cambridge: Cambridge University Press, 2006.
Weber, Max. "Politics as a Vocation." In *From Max Weber: Essays in Sociology*. Translated and edited by H. H. Gerth and C. Wright Mills. New York: Oxford University Press, 1958.
White, Richard. *Railroaded: The Transcontinentals and the Making of Modern America*. New York: W. W. Norton, 2011.
White, Sam. *The Climate of Rebellion in the Early Modern Ottoman Empire*. Cambridge: Cambridge University Press, 2011.
———. "Rethinking Disease in Ottoman History." *International Journal of Middle East Studies* 42 (2010): 549–67.
Williams, A. R. "Animal Mummies." *National Geographic*. November 2009.
Williams, Howard. "Ashes to Asses: An Archaeological Perspective on Death and Donkeys." *Journal of Material Culture* 16 (2011): 219–39.
Wilson, R. T. "The Incidence and Control of Livestock Diseases in Darfur, Anglo-Egyptian Sudan, during the Period of the Condominium, 1916–56." *International Journal of African Historical Studies* 12 (1979): 62–82.
Wing, John T. "Keeping Spain Afloat: State Forestry and Imperial Defense in the Sixteenth Century." *Environmental History* 17 (2012): 116–45.
———. "Roots of Empire: State Formation and the Politics of Timber Access in Early Modern Spain, 1556–1759." Ph.D. diss., University of Minnesota, 2009.
Winter, Michael. *Egyptian Society under Ottoman Rule, 1517–1798*. London: Routledge, 1992.
Wittrock, Björn. "Early Modernities: Varieties and Transitions." *Daedalus* 127 (1998): 19–40.
Wolfe, Cary. *What is Posthumanism?* Minneapolis: University of Minnesota Press, 2010.
Wolfe, Nathan. *The Viral Storm: The Dawn of a New Pandemic Age*. New York: Times Books, 2011.
Wolloch, Nathaniel. *Subjugated Animals: Animals and Anthropocentrism in Early Modern European Culture*. Amherst, NY: Humanity Books, 2006.
Wood, Amy Louise. "'Killing the Elephant': Murderous Beasts and the Thrill of Retribution, 1885–1930." *Journal of the Gilded Age and Progressive Era* 11 (2012): 405–44.

Wrigley, E. A. *Continuity, Chance, and Change: The Character of the Industrial Revolution in England.* Cambridge: Cambridge University Press, 1988.

Yang, Bin. "Horses, Silver, and Cowries: Yunnan in Global Perspective." *Journal of World History* 15 (2004): 281–322.

Yasuhiro, Yokkaichi. "Horses in the East-West Trade between China and Iran under Mongol Rule." In *Pferde in Asien: Geschichte, Handel Und Kultur.* Edited by Bert G. Fragner, Ralph Kauz, Roderich Ptak, and Angela Schottenhammer. Vienna: Österreichischen Akademie der Wissenschaften, 2009.

Yaycıoğlu, Ali. "The Provincial Challenge: Regionalism, Crisis, and Integration in the Late Ottoman Empire (1792–1812)." Ph.D. diss., Harvard University, 2008.

———. "Provincial Power-Holders and the Empire in the Late Ottoman World: Conflict or Partnership?" In *The Ottoman World.* Edited by Christine Woodhead. New York: Routledge, 2012.

Zachariadou, Elizabeth, ed. *Natural Disasters in the Ottoman Empire.* Rethymnon: Crete University Press, 1999.

al-Zaybaq, Muḥammad. *The Animal: Its Particulars and Its Rights within Islam.* Translated by Ṣaḥeeḥ International. Jeddah: Abul-Qasim, 2001.

INDEX